# MATHEMATICS FOR LARGE SCALE COMPUTING

# MATHEMATICS FOR LARGE SCALE COMPUTING

Edited by

## J. C. DÍAZ

*Center for Parallel and Scientific Computing*
*The University of Tulsa*
*Tulsa, Oklahoma*

## CRC Press
Taylor & Francis Group
Boca Raton London New York

CRC Press is an imprint of the
Taylor & Francis Group, an **informa** business

CRC Press
Taylor & Francis Group
6000 Broken Sound Parkway NW, Suite 300
Boca Raton, FL 33487-2742

First issued in hardback 2019

© 1989 by Taylor & Francis Group, LLC
CRC Press is an imprint of Taylor & Francis Group, an Informa business

No claim to original U.S. Government works

ISBN-13: 978-0-8247-8122-4 (pbk)
ISBN-13: 978-1-138-41328-3 (hbk)

Library of Congress Cataloging-in-Publication Data

Mathematics for large scale computing / edited by J.C. Díaz.
    p. cm.
   Includes index.
   ISBN 0-8247-8122-8
   1. Electronic data processing--Mathematics. I. Díaz, J. C.
QA76.9.M35M388  1989
004' .01'51–dc20                                    89-34808
                                                    CIP

Visit the Taylor & Francis Web site at
http://www.taylorandfrancis.com

and the CRC Press Web site at
http://www.crcpress.com

# Preface

During recent years a great deal of interest has been devoted to large scale computing applications. This has occurred in great part because of the introduction of advanced high performance computer architectures. These new computational engines have brought about a renewed emphasis in the development of scientific computing techniques that exploit the specialized features of these advanced architectures, such as parallelism and vectorization.

Several meetings have focused on various aspects of this expanding discipline. At the 830th meeting of the American Mathematical Society, held in Denton, Texas, October 31 to November 1, 1986, Professor Mary F. Wheeler of Rice University gave an invited address and a special session on Mathematics for Large Scale Computing was also held. Together the invited address and this special session covered a broad spectrum of mathematical issues related to large scale computing, including physical and engineering applications, algorithm development and analysis, performance evaluation of algorithms in advanced architectures, and special properties of advanced architectures. The special session was aimed at presenting large scale computing issues and ideas from a broad spectrum to stimulate mathematical discussion through comparisons and synergism.

Given the success and interest generated by the invited address and the special session, and in coordination with the publisher, it was decided to produce this volume, covering a broad range of topics related to large scale computing. The speakers at the AMS special session and other contibutors were invited to submit a manuscript. The chapters selected for publication went through a rigorous review process by expert referees.

The book contains survey articles as well as chapters on specific research applications, development and analysis of numerical algorithms, and performance evaluation of algorithms on advanced architectures. The effect of specialized architectural features on the performance of large scale computation is also considered by several authors. Several areas of applications are represented, including the numerical solution of partial differential equations, iterative techniques for large structured problems, the numerical solution of boundary value problems for ordinary differential equations, numerical optimization, and numerical quadrature. Mathematical issues in computer architecture are also presented, including the description of gray codes for generalized hypercubes. The results presented in this volume give, in our opinion, a representative picture of today's state of the art in several aspects of large scale computing.

We wish to thank the authors for their diligence in producing the quality manuscripts that are included here and for their patience as this project progressed through its own slow pace. We also wish to thank K. E. Jordan for overseeing the review of the manuscript jointly authored by the editor of this volume. We wish to thank the referees for their blind reviews. They are the unsung heroes of this effort and shall remain anonymous.

J. C. Díaz

# Contents

# Contributors

**L. S. Barasch**

*Parallel Processing Institute*
*University of Oklahoma EECS*
*202 W. Boyd, Suite 219*
*Norman, Oklahoma 73019*


**C. N. Dawson**

*Department of Mathematics*
*University of Chicago*
*5734 S. University Avenue*
*Chicago, Illinois 60637*


**S. K. Dhall**

*Parallel Processing Institute*
*University of Oklahoma EECS*
*202 W. Boyd, Suite 219*
*Norman, Oklahoma 73019*


**J. C. Díaz**

*Center for Parallel and Scientific Computing*
*Computer and Mathematical Sciences*
*The University of Tulsa*
*600 S. College Ave.*
*Tulsa, Oklahoma 74109-3189*


**T. F. Dupont**

*Department of Computer Sciences*
*University of Chicago*
*5734 S. University Avenue*
*Chicago, Illinois 60637*

**R. E. Ewing**                    *Departments of Mathematics,*
                                   *Petroleum, and Chemical Engineering*
                                   *University of Wyoming*
                                   *Laramie, WY 82070*


**G. Fairweather**                 *Center for Computational Sciences*
                                   *University of Kentucky*
                                   *Lexington, KY 40506*


**A. Genz**                        *Computer Science Department*
                                   *Washington State University*
                                   *Pullman, WA 99164-1210*


**S. Gómes**                       *IIMAS-UNAM,*
                                   *Ap. Postal 20-726*
                                   *01000 México DF,*
                                   *México*


**A. Griewank**                    *Mathematics and Computer Science Division*
                                   *Argonne National Laboratory*
                                   *9700 South Cass Avenue*
                                   *Argonne, Ill 60439*


**L. J. Hayes**                    *Texas Institute for Computational Mechanics*
                                   *Department of Aerospace Engineering*
                                   *and Engineering Mechanics*
                                   *University of Texas at Austin*
                                   *Austin, TX 78712-1085*


**K. E. Jordan**                   *Computer Technology and Services Division*
                                   *EXXON Research and Enginering Company*
                                   *Clinton Township*
                                   *Rte. 22 East*
                                   *Annandale, NJ  08801*

**H. B. Keller**

*Department of Applied Mathematics*
*California Institute of Technology*
*Pasadena, California 91125*


**S. V. Krishnamachari**

*Texas Institute for Computational Mechanics*
*Department of Aerospace Engineering*
*and Engineering Mechanics*
*University of Texas at Austin*
*Austin, TX 78712-1085*


**S. Lakshmivarahan**

*Parallel Processing Institute*
*University of Oklahoma EECS*
*202 W. Boyd, Suite 219*
*Norman, Oklahoma 73019*


**G. K. Leaf**

*Argonne National Laboratory*
*Mathematics and Computer Science Division*
*9700 South Cass Avenue*
*Argonne, Ill 60439*


**G. Li**

*Department of Mathematics*
*University of Wyoming*
*Laramie, WY 82070*


**D. Meade**

*Instituto Tecnologico*
*y de Estudios Superiores de Monterrey*
*Sucursal de Correos "J"*
*Monterrey, N. L.*
*64849 México*


**M. Minkoff**

*Argonne National Laboratory*
*Mathematics and Computer Science Division*
*9700 South Cass Avenue*
*Argonne, Ill 60439*

**J. L. Morales**
*Instituto de Investigaciones Eléctricas*
*Departamento de Simulación*
*Interior Internado Palmira*
*Cuernavaca, Morelos*
*México.*


**P. Nelson**
*Department of Applied Mathematics*
*California Institute of Technology*
*Pasadena, California 91125*
*and*
*Department of Nuclear Engineering*
*Texas A&M University*
*College Station, TX 77843*


**J. S. Scroggs**
*ICASE, Mail Stop 132C*
*NASA Langley Research Center*
*Hampton, VA 23665*


**L. Sheng**
*Department of Mathematics*
*Dedman College*
*Southern Methodist University*
*Dallas, Texas 75275*


**D. C. Sorensen**
*Argonne National Laboratory*
*Mathematics and Computer Science Division*
*9700 South Cass Avenue*
*Argonne, Ill 60439*


**M. F. Wheeler**
*Department of Mathematical Sciences*
*Rice University*
*Houston, TX 75252-1892*
*and*
*Department of Mathematics*
*University of Houston-University Park*
*Houston, TX 77004*

**Y. Yuan**    *Department of Mathematics*
*Shandong, Jinan*
*Shandong, China*
*and*
*Department of Mathematics*
*University of Wyoming*
*Laramie, WY 80720*

# MATHEMATICS FOR
# LARGE SCALE
# COMPUTING

# On the Gauss-Broyden Method for Nonlinear Least Squares

ANDREAS GRIEWANK Mathematics and Computer Science Division, Argonne National Laboratory, Argonne, Illinois 60439

LAIHUA SHENG Department of Mathematics, Dedman College, Southern Methodist University, Dallas, Texas 75275

## 1 GENERAL ASSUMPTIONS AND TERMINOLOGY

Nonlinear least squares problems arise in many applications, in particular data fitting and parameter estimation [3]. Mathematically we have the optimization problem

$$Min \ \gamma(x) \ \equiv \ \tfrac{1}{2}\|g(x)\| \quad for \ g : \mathbf{R}^n \to \mathbf{R}^m.$$

Throughout we will assume that the vector function $g$ has a Lipschitz-continuous Jacobian $G(x) \equiv g'(x) \in \mathbf{R}^{m \times n}$ with full column rank $n \leq m$ at all points $x$ in some bounded level set

$$\mathcal{L} \ \equiv \ \{x \in \mathbf{R}^n : \gamma(x) \leq \bar{\gamma}\}$$

with a non-empty interior $\mathcal{L}^0$. Finally we will assume that the restriction of $g$ to $\mathcal{L}$ is injective, i.e.

$$x, z \in \mathcal{L} \ , \ g(x) = g(z) \quad \Rightarrow \quad x = z. \tag{1.1}$$

While these regularity assumptions appear quite strong, they do not exclude the possibility that $\gamma(x)$ attains several local minima and other stationary points in $\mathcal{L}$. Throughout let $x_*$ denote a local minimizer in the interior of $\mathcal{L}$. Like all other stationary points of $\gamma$, it must solve the vector equation

$$\nabla \gamma(x) \equiv G(x)^T g(x) = 0. \tag{1.2}$$

---

This work was supported by the Applied Mathematical Sciences subprogram of the Office of Energy Research, U.S. Department of Energy, under contracts W-31-109-Eng-38.

The most serious objection to the use of approximating Jacobians in nonlinear least squares problems is that apparently one cannot decide whether a current iterate $x_k$ is a stationary point of $\gamma$ without knowing the exact Jacobian $G_k \equiv G(x_k)$ or at least its range $R_k \equiv R(x_k)$. This difficulty will be of concern throughout the design and analysis of the method.

With $B_k \in \mathbf{R}^{m \times n}$ an approximation to $G_k$ one can expect that for small displacement vectors $s \in \mathbf{R}^n$

$$g(x_k + s) \sim g_k + B_k s$$

where naturally $g_k \equiv g(x_k)$. Provided $B_k$ has full rank, the right hand side has a linear least squares solution $s_k$, which can be used as search direction for the nonlinear problem. This approach yields a Gauss-Newton-like algorithm of the following form.

(0) Pick $x_0 \in \mathcal{L}^0$, $B_0 \in \mathbf{R}^{m \times n}$ and set $k = 0$.

(1) Compute a nonzero search direction $s_k$ s.t. $B_k^T B_k s_k = -B_k^T g_k$.

(2) Find a step multiplier $\alpha_k \in \mathbf{R}$ such that
$\gamma_{k+1} \ll \gamma_k$ for $x_{k+1} \equiv x_k + \alpha_k s_k$.

(3) Obtain $B_{k+1} \sim G_{k+1}$ by evaluation and/or updating.

(4) Unless $g_{k+1} \sim 0$ set $k \leftarrow k + 1$ and go to (1).

Even though step (1) is formulated in terms of the normal equation, the search direction $s_k$ should be computed via the QR-factorization $B_k = Q_k R_k$. A potential advantage of low rank update methods is that the factors $Q_{k+1}$ and $R_{k+1}$ can be obtained directly from $Q_k$ and $R_k$ at the cost of $O(mn)$ arithmetic operations [8]. In contrast the refactorization of a completely new version $B_{k+1}$, e.g. $B_{k+1} = G_{k+1}$ in the Gauss-Newton method, requires $n^2(m - n/3)$ arithmetic operations [22]. In the unlikely event that any of the approximating Jacobians $B_k$ is rank deficient, the step $s_k$ may be selected as any nonzero nullvector.

In step (2) we have not yet specified in what sense the next residual norm should be *significantly* smaller than the current one, i.e. $\gamma_{k+1} \ll \gamma_k$. This line search problem will be discussed in Sections 3 and 4. Ideally one would hope that once $x_k \sim x_*$ and $B_k \sim G_* \equiv G(x_*)$ the full *quasi-Gauss-Newton* steps $s_k$ with $\alpha_k = 1$ should be successful, i.e. achieve convergence to $x_*$. As a consequence of a result by Ostrowski (Theorem 10.1.3 in [20]), even the Gauss-Newton choice $B_k = G_k$ ensures this desirable property only when the problem has a *small residual* in that

$$\lambda_* \equiv \lambda_*[I - (G_*^T G_*)^{-1} \nabla^2 \gamma(x_*)] < 1. \tag{1.3}$$

When the spectral norm $\lambda_*$ is greater than 1 the iteration is almost certain to diverge, unless infinitely many of the step multipliers $\alpha_k$ are chosen smaller than 1. While we will introduce a line-search to cope with bad search directions $s_k$ resulting from poor approximations $B_k$, our main target are small residual problems. More specifically we address the situation where the Gauss-Newton method converges theoretically quite rapidly but is costly to implement because the Jacobians $G_k$ are hard to evaluate and/or factorize.

## 2 MOTIVATION AND THEORY OF GAUSS-BROYDEN METHOD

Extending the philosophy of quasi-Newton methods to the overdetermined case we will replace $G_k$ by approximations $B_k$ that can be evaluated and factorized more economically. Naturally this modification will lead to an increase in the number of steps required in order to achieve a certain solution accuracy. However, the savings at each iteration may result in a more economical calculation overall. For example in the square case $m = n$, one often finds that the computation of a Broyden step is $\mathcal{O}(n)$ times cheaper than that of a Newton step, but that the number of steps needed to reach an acceptable iterate grows only by a factor of 2 or 3. Mathematically one can show that the convergence is no longer quadratic (as in the case of Newton's method) but still *Q-superlinear* in that

$$\lim_{k \to \infty} \|x_{k+1} - x_*\| / \|x_k - x_*\| = 0. \tag{2.1}$$

When $m \geq n$ the Gauss-Broyden method to be proposed exhibits local and Q-superlinear convergence, but only when the residual is zero, so that $\lambda_* = 0$. This is not really surprising since on nonzero residual problems the Gauss-Newton method itself converges only linearly with the R-factor

$$\lim_{k \to \infty} [\|x_k - x_*\| / \|x_0 - x_*\|]^{1/k} = \lambda_* . \tag{2.2}$$

Here we have assumed that the starting point $x_0$ is *general*, i.e. does not belong to the lower dimensional set of initial points from which the convergence is theoretically superlinear. Obviously the best we can expect from a quasi-Gauss-Newton method is that it will also converge linearly with an R-factor between $\lambda_*$ and 1.

While the Broyden formula [5] is often thought of as a matrix update, it is really a method for approximating the gradients of the otherwise unrelated component functions of $g$. Therefore it readily generalizes to rectangular systems and may be written as

$$B_{k+1} \equiv B_k + (y_k - B_k s_k) s_k^T / s_k^T s_k. \tag{2.3}$$

where $y_k \in \mathbf{R}^m$ is defined in the *secant condition*

$$B_{k+1} s_k = y_k \equiv (g_{k+1} - g_k)/\alpha_k \sim G_k s_k \tag{2.4}$$

with the convention

$$y_k \equiv G_k s_k \quad if \quad \alpha_k = 0.$$

Obviously there are many other $m \times n$ matrices that satisfy this system of $m$ linear equations. Theoretically it is convenient to characterize Broyden's choice as the one that minimizes the Frobenius norm $Tr(C_k^T C_k)$ of the correction matrix

$$C_k \equiv B_{k+1} - B_k \in \mathbf{R}^{m \times n}.$$

This *least change* property [4] allows us to prove the following result.

**THEOREM 1** If the steps are chosen such that

$$\sum_{k=0}^{\infty} \|x_{k+1} - x_k\|^2 < \infty$$

then

$$\lim_{k \to \infty} \frac{1}{k} \sum_{j=0}^{k-1} \|C_j\|^2 = 0 \tag{2.5}$$

which implies the existence of an index sequence $\mathcal{J} \subset \mathcal{N}$ satisfying

$$\lim_{\mathcal{J} \not\ni k \to \infty} \|y_k - B_k s_k\| / \|s_k\| = 0 \tag{2.6}$$

and

$$\lim_{k \to \infty} |\{j \in \mathcal{J} : j < k\}| / k = 0 \tag{2.7}$$

where $|\cdot|$ denotes the cardinality of the sets in question.

Proof: The first assertion (2.5) follows from an extension of the proof of Theorem 4 in [12] to the rectangular case. Since as a consequence of the secant condition on $B_{k+1}$

$$\|y_k - B_k s_k\| = \|(B_{k+1} - B_k)s_k\| \le \|C_k\| \|s_k\|$$

it follows that also

$$\lim_{k \to \infty} \frac{1}{k} \sum_{j=0}^{k-1} \|y_k - B_k s_k\|^2 / \|s_k\|^2 = 0.$$

Now (2.6) and (2.7) hold by Lemma 2 in [11]. ∎

As we will see in the next Section, the square summability condition on the steps $x_{k+1} - x_k = s_k \alpha_k$ can be enforced by a suitable line-search criterion. Then the theorem says that, except for a few indices $k \in \mathcal{J}$, the approximating Jacobians $B_k$ provide increasingly accurate directional derivative approximations $B_k s_k \sim y_k$. In the square case $n = m$ this relation ensures that the relative errors between most steps $s_k$ and the corresponding Newton corrections $-G_k^{-1} g_k$ tend to zero, which implies in fact R-superlinear convergence. Unfortunately in the rectangular case $m > n$ even the relation $\lim \|y_k - B_k s_k\| / \|s_k\| = 0$ does not guarantee the convergence of our Gauss-Broyden scheme because the steps $s_k$ may remain substantially different from the corresponding Gauss-Newton corrections. This can be illustrated and practically observed on the Watson function from the test set of Moré et al [18].

After a suitable rearrangement Watson's function takes for $m = 3$ and $n = 2$ the form

$$g(\xi, \zeta) = \begin{bmatrix} \xi \\ h_1(\xi, \zeta) \\ h_2(\xi, \zeta) \end{bmatrix}$$

where $\partial h_j(0,0) / \partial \xi = 0$, *for* $j = 1, 2$. Consequently the Jacobian $G(\xi, \zeta)$ has everywhere the sparsity pattern

$$G(\xi, \zeta) = \begin{bmatrix} 1 & 0 \\ \times & \times \\ \times & \times \end{bmatrix}$$

and at the standard starting point $x_0^T = (\xi_0, \zeta_0) = (0,0)$ we have in particular

$$B_0 \equiv G_0 = \begin{bmatrix} 1 & 0 \\ 0 & \times \\ 0 & \times \end{bmatrix}$$

By induction it can be easily seen that for this initialization all subsequent $B_k$ have exactly the same sparsity pattern as $B_0$ since the first component $\xi_k$ remains constant at its initial value $\xi_0 = 0$. In other words the linear model always indicates that the first component $\xi$ has no effect on the second and third equation and should therefore stay at the value $\xi = 0$, which solves the first equation exactly. Consequently the first component of all steps is zero and the Broyden update leaves the first column unchanged. The limit point $x_*$ of the iterates $x_k$ is a solution of the least squares problem subject to the constraint that $\xi = 0$. Moreover we have $B_k^T g(a_*) \sim 0$ for all large k.

The obvious difficulty here is that the range of the approximating Jacobian $B_k$ is incorrect, i.e. differs from that of the actual Jacobian $G_k$. As a generalization of a result by Burmeister [6] and Gay [10] it was shown by Luk et al [17] that on linear problems, the full step Gauss-Broyden method reaches the solution after at most 2n iterations, *provided* the range of $B_0$ and consequently all subsequent approximations $B_k$ is correct. Clearly this crucial assumption is not very realistic for nonlinear problems. On the other hand we found that for almost any $B_0$ on a given nondegenerate linear problem, the Gauss-Broyden method with or without line-search always appeared to converge to the solution, though not in a finite number of steps. This was true even when the columns of $B_0$ were made orthogonal to the initial residual $g_0$ by a suitable elementary reflector [22]. Thus the arbitrary starting point $x_0$ formed theoretically the least squares solution of the initial linear model. Fortunately round-off errors ensured that $g_0^T B_0$ and consequently $s_0$ were not exactly equal to zero, and after the first tiny step the convergence was quite satisfactory. On Watson's function this desirable effect did not occur because the sparsity structure prevented any round-off from modifying $\xi_0$ or the first column of $B_0$. The slightest perturbation of either initial quantity enables the method to approach the correct solution.

To formalize this observed instability let us consider the iteration function

$$\mathcal{GB} : \mathcal{L} \times \mathbf{R}^{m \times n} \to \mathbf{R}^m \times \mathbf{R}^{m \times n}$$

defined by

$$\mathcal{GB} \begin{pmatrix} x_k \\ B_k \end{pmatrix} - \begin{pmatrix} x_k \\ B_k \end{pmatrix} \equiv \begin{pmatrix} s_k \\ C_k \end{pmatrix} \equiv \begin{pmatrix} -B_k^\dagger g_k \\ r_k s_k^\dagger \end{pmatrix} \tag{2.8}$$

where the superscript † denotes pseudoinverses [15] and the m-vector $r_k$ is given by

$$r_k \equiv y_k - B_k s_k = g_{k+1} - (I - B_k B_k^\dagger) g_k.$$

(Note that $s^\dagger = s^T / s^T s$ for any nonzero vector $s$ .) Through the use of pseudoinverses we have extended the definition of the Gauss-Broyden iteration function to the whole of $\mathcal{L} \times \mathbf{R}^{m \times n}$, but $\mathcal{GB}$ is only differentiable where $rank(B_k) = n$ and $g_k^T B_k \neq 0$ so that $s_k \neq 0$. As one checks easily the fixed points of $\mathcal{GB}$ form the set

$$\mathcal{F} \equiv \{(x, B) \in \mathcal{L} \times \mathbf{R}^{m \times n} : g(x) \perp range(B)\}. \tag{2.9}$$

This set is rather large since there exists for every $x \in \mathcal{L}$ a variety of $B$ such that $(x, B) \in \mathcal{F}$. Naturally we are only intested in computing points in the subset

$$
\begin{aligned}
\mathcal{F}_* &\equiv \{(x, B) \in \mathcal{F} : B = G(x)\} \\
&= \{(x, G(x)) : x \in \mathcal{L}, \nabla\gamma(x) = 0\}
\end{aligned}
\tag{2.10}
$$

whose elements represent *genuine* solutions of the underlying least-squares problem. Fortunately the undesirable fixed points in $\mathcal{F} - \mathcal{F}_*$ are all highly unstable, because the iteration function $\mathcal{GB}$ is not even continuous let alone contracting in their vicinity. Due to our nondegeneracy assumption on $g$ the converse is also true, and we can prove the following result under the simplifying assumption $m > n$.

**PROPOSITION 1** The Gauss-Broyden iteration function $\mathcal{GB}$ is continuous at a fixed point $(x_*, B_*) \in \mathcal{F}$ if and only if $(x_*, B_*) \in \mathcal{F}_*$, so that $B_* = G(x_*)$ and $x_*$ is a stationary point of the sum of squares residual $\gamma(x)$.

Moreover there are pairs $(x_0, B_0)$ arbitrarily close to any given $(x_*, B_*) \in \mathcal{F} - \mathcal{F}_*$ such that the resulting iterates

$$
(x_1, B_1) = \mathcal{GB}(x_0, B_0) \quad and \quad (x_2, B_2) = \mathcal{GB}(x_1, B_1)
$$

satisfy

$$
\|B_1 - B_*\| > c_*/\|g_*\| \quad and \quad \|x_2 - x_*\| > c_*/(\|B_*\| + \|G_*\|)^2
\tag{2.11}
$$

where $c_* \equiv .5\|P_* g_*\|/\|G_*^\dagger\|$ with $P_*$ the orthogonal projection onto the range of $G_*$.

Proof: $\Rightarrow$ Suppose $(x_*, B_*) \notin \mathcal{F}_*$ so that $B_* \neq G_* \equiv G(x_*)$. Then there exists a vector $s \in \mathcal{R}^n$ such that $0 \neq B_* s \neq G_* s$. With the rank one matrix $E \equiv g_* s^T B_*^T B_*$ we may now consider the perturbations

$$
B_0 \equiv B_* - \beta E = (I - \beta g_* s^T B_*^T) B_*,
$$

which represent rotations of $B$ since $det(I - \beta g_* s^T B_*^T) = 1$. The Gauss-Broyden step $s_0$ defined at $(x_0, B_0)$ with $x_0 \equiv x_*$ satisfies the normal equations $B_0^T B_0 s_0 = -B_0^T g_*$, which reduces to

$$
\left[ B_*^T B_* + \beta^2 \|g_*\|^2 B_*^T B_* s s^T B_*^T B_* \right] s_0 = \beta \|g_*\|^2 B_*^T B_* s.
$$

because $g_*$ and the range of $E$ are orthogonal to the columns of $B_*$. This square system has the unique solution $s_0 = s\beta/(\|g_*\|^{-2} + \beta^2) = \beta s + \mathcal{O}(\beta^3)$, provided $B_*$ has full rank as we will assume for the moment. Thus we obtain using Taylor's theorem

$$
r_0 \equiv g(x_* + s_0) - g_* - B_0 s_0 = \beta(G_* - B_*)s + \mathcal{O}(\beta^2).
$$

Correspondingly the matrix update is given by

$$
C_0 = r_0 s_0^\dagger = (G_* - B_*)s s^\dagger + \mathcal{O}(\beta).
$$

Since the leading constant term is nonzero by definition of $s$ we must have

$$
\lim_{\beta \to 0} \mathcal{GB}(x_0, B_0) \neq (x_*, B_*) = \mathcal{GB}\left[\lim_{\beta \to 0}(x_0, B_0)\right],
$$

which establishes the discontinuity of $\mathcal{GB}$ at its fixed point $(x_*, B_*)$. When $B_*$ is rank-deficient the same argument goes through for any $E$ satisfying

$$Es = \|B_* s\|^2 g_* \, , \ g_*^T E = \|g_*\|^2 s^T B_*^T B_*$$

$$range(E) \perp range(B_*) \quad and \quad kern(B_*) \cap kern(E) = \{0\},$$

where the last condition ensures the uniqueness of the given least squares solution $s_0$. A suitable $E$ can be found by adding to the rank-one version above to a matrix $\Delta E$ of rank $n - rank(B_*)$ satisfying

$$range(\Delta E) \perp range(B_*, g_*) \quad and \quad kern(\Delta E) \cap kern(B_*) = \{0\}.$$

Such $\Delta E$ are easily found since $m > n$ by assumption.

$\Leftarrow$ Since $rank(G_*) = n$ we have $B^\dagger = (B^T B)^{-1} B^T$ for all $B$ sufficiently close to $G_*$. Therefore

$$s = -B^\dagger g(x) = \mathcal{O}(\|B - G_*\| + \|x - x_*\|)$$

is differentiable at $(x_*, G_*)$. Consequently we have for all such small steps $s$ again by Taylor's theorem

$$\begin{aligned}
r/\|s\| &= [g(x+s) - g(x) - G_* s + (G_* - B)s]/\|s\| \\
&= \mathcal{O}(\|s\| + \|x - x_*\| + \|G_* - B\|) \\
&= \mathcal{O}(\|B - G_*\| + \|x - x_*\|).
\end{aligned}$$

Since always $\|C\| = \|r\|/\|s\|$ we obtain finally

$$\|\mathcal{GB}(x, B) - (x_*, G_*)\| = \mathcal{O}(\|C\| + \|s\|) = \mathcal{O}(\|B - G_*\| + \|x - x_*\|)$$

which shows that $\mathcal{GB}$ is in fact Lipschitz-continuous at $(x_*, G_*)$.

To establish the *Moreover* part we use the particular choice $s \equiv G_*^\dagger g_*$ so that

$$G_* s = P_* g_* \, , \ g_*^T G_* s = \|P_* g_*\|^2 \, , \ and \ \|s\| \le \|G_*^\dagger\| \|P_* g_*\|.$$

Using the orthogonality relation $g_*^T B_* = 0$ one can derive from the Cauchy-Schwarz inequality that

$$\|(G_* - B_*)s\| \ge \|P_* g_*\|^2 / \|g_*\| \quad .$$

Hence the correction matrix $C_0$ defined above satisfies

$$\|C_0\| + \mathcal{O}(\beta) \ge \|P_* g_*\|^2 \|s^\dagger\| / \|g_*\| \ge 2c_* \quad ,$$

where we have used that $1/\|s^\dagger\| = \|s\|$. Since $B_1 - B_0 = C_0$ and $B_0 - B_* = \mathcal{O}(\beta)$ the first inequality in (2.11) must hold for all sufficiently small $\beta$. In order to establish the second relation we note that as an immediate consequence of the normal equation

$$\|x_2 - x_1\| = \|B_1^\dagger g_1\| \ge \|B_1^T g_1\| / \|B_1\|^2 \quad .$$

No we have by definition of $B_1$ and the triangle inequality

$$\|B_1\| \le \|B_0(I - s s^\dagger)\| + \|G_*\| + \mathcal{O}(\beta)\|s s^\dagger\| \le \|B_*\| + \|G_*\| + \mathcal{O}(\beta).$$

Also, because $x_1 - x_0 = s_\beta = \mathcal{O}(\beta)$ we derive with the Lipschitz-continuity of $g$ for our particular choice of $s$ that

$$\|g_1^T B_1\| + \mathcal{O}(\beta) = |g_*^T G_* s|/\|s\| \geq 2c_* \quad ,$$

which completes the proof. ∎

According to the proposition tiny perturbations of the approximating Jacobian at a nonstationary fixed point can lead to a small step in the variable vector that results in a significant matrix correction, which in turn effects a sizable change of the variables on the subsequent iteration. In other words, near phony solutions the Gauss-Newton method is extremely unstable, not only with respect to the Jacobian approximation (as one might expect for a secant method) but also in terms of the variable vector itself. Hence we may conclude that in the presence of unbiased round-off, the full-step Gauss-Broyden method can only converge to desirable points, i.e. minima and possibly saddle points or even maxima of the sum of squares $\gamma$.

This statement begs the question whether the iteration is likely to converge at all. Since the $\gamma_k$ need not be reduced the Gauss-Broyden iterates $x_k$ may diverge towards infinity or oscillate between several limit points. We have even less control over the corresponding $B_k$, especially if $g(x)$ is highly nonlinear. Through the introduction of a line-search we will be able to guarantee that under our assumptions the steps $x_{k+1} - x_k$ are square summable and the matrix updates $B_{k+1} - B_k$ tend to zero except for a few special indices $k \in \mathcal{J}$. While this eliminates the possibility of oscillations in the iterates $x_k$, they could still creep along an infinite path approaching a nonoptimal level of $\gamma$. However, this seems not very likely since it can be shown that Proposition 1 is still valid if the iteration function $\mathcal{GB}$ is modified by the introduction of a nonunitary step multiplier $\alpha_k$.

Rather than finishing this section with a negative result, we proceed to establish that on small residual problems the full-step Gauss-Broyden method is guaranteed to achieve at least some reduction of the initial solution distance. As a corollary we obtain for zero residual problems local and Q-superlinear convergence; a result, which though not very surprising, has apparently not been published in the literature. Our convergence analysis will be based on the following relations near a stationary point $x_* \in \mathcal{L}$ of $\gamma(x)$.

**LEMMA 1** Suppose $\mu$ is a Lipschitz constant of $G(x)$ in $\mathcal{L}$ and denote the smallest singular value of $G_* \equiv G(x_*)$ by $\sigma_* \equiv 1/\|G_*^\dagger\| > 0$ with $G_*^\dagger \equiv (G_*^T G_*)^{-1} G_*^T$ the pseudoinverse as before. Then we find for $x \in \mathcal{L}$ and $B \in \mathbf{R}^{m \times n}$ that the conditions

$$\|x - x_*\| \leq \tfrac{1}{9}\sigma_*/\mu \quad and \quad \|B - G_*\| \leq \tfrac{1}{9}\sigma_* \tag{2.12}$$

imply

$$\|B^\dagger\| \leq \tfrac{9}{8}/\sigma_* \quad , \quad \|B^\dagger - G_*^\dagger\| \leq \tfrac{8}{5}\|B - G_*\|/\sigma_*^2 \tag{2.13}$$

$$.8 \leq \|B^\dagger[g(x) - g_*]\|/\|x - x_*\| \leq 1.2, \tag{2.14}$$

$$\big| \|B^\dagger g(x)\| - \|x - x_*\| \big| \leq .2\|x - x_*\| + 1.6\|B - G_*\| \, \|g_*\|/\sigma_*^2. \tag{2.15}$$

Proof: The two bounds in (2.12) follow directly from the relations (8.19) and (8.28) in [15]. By the fundamental theorem of calculus we have

$$B^\dagger[g(x) - g_*] = B^\dagger \int_0^1 G(x\tau + (1-\tau)x_*)(x - x_*)d\tau$$

$$= \left\{ B^\dagger B + B^\dagger [G_* - B] + B^\dagger \int_0^1 [G(x\tau + (1-\tau)x)_* - G_*]d\tau \right\} (x - x*)$$

so that by definition of the pseudoinverse and the triangle inequality

$$\|B^\dagger[g(x) - g_*)] - (x - x_*)\| \le \|B^\dagger\| \, [\|G_* - B\| + .5\mu\|x - x_*\|] \, \|x - x_*\| \le \|x - x_*\|/5$$

where the last inequality follows from the restrictions (2.11) on $x$ and $B$. Applying the inverse triangle inequality we obtain (2.13). In order to derive the last relation (2.14) we simply note that due to $x_*$ being a stationary point

$$\|B^\dagger g_*\| = \|(B^\dagger - G_*^\dagger)g_*\| \le 1.6\|B - G_*\| \, \|g_*\|/\sigma_*^2$$

where we have used the bound on the difference of the pseudoinverses. ∎

After these preparations and using the abbreviations

$$\delta_k \equiv \|(B_k - G_*)s_k\|/\|s_k\| \le \Delta_k \equiv \|B_k - G_*\| \tag{2.16}$$

we can now establish the following 'convergence' result.

**THEOREM 2** Suppose $\mathcal{L}$ contains the ball with radius $.1\sigma_*/\mu$ about $x_*$. Then the initial conditions

$$\|x_0 - x_*\| < .01\sigma_*/\mu \quad and \quad B_0 = G(x_0) \tag{2.17}$$

imply that the iterates $x_k$ generated by the full-step Gauss-Broyden method satisfy

$$\frac{\|x_{k+1} - x_*\|}{\|x_k - x_*\|} \le \frac{6(\delta_k + 2\mu\|x_k - x_*\|)}{\sigma_*} \le \frac{1}{3} \tag{2.18}$$

until for the first time

$$\frac{\|x_{k+1} - x_*\|}{\|x_0 - x_*\|} \le \frac{9\|g_*\|\mu}{\sigma_*^2}. \tag{2.19}$$

Subsequent iterates may leave the region $\mathcal{L}$ or be nonunique due to a rank loss in $B_k$.

Proof: In order to prove the result by induction let us assume that (2.17) holds for $0 \le j < k$ so that

$$\rho_j \equiv \|x_j - x_*\| \le \|x_0 - x_*\|3^{-j} \quad for \quad j \le k.$$

Then it follows from a purely notational generalization of Lemma 8.2.1 in [9] that for all $j \le k$ with $\Delta_j$ as defined in (2.15)

$$\Delta_j \le \Delta_{j-1} + \mu\rho_0 3^{1-j} \le \Delta_0 + 1.5\mu\rho_0 \le 2.5\mu\rho_0$$

where we have used the geometric series formula and the fact that $\Delta_0 = \|G_0 - G_*\| \le \mu\rho_0$ by definition of $B_0$. Because of the upper bound on $\rho_0$ we have $\Delta_j \le \sigma_*/9$ so that Lemma 1 applies for all $j \le k$. In particular we derive from its last assertion (2.14) for the next step $s_k$

$$\|s_k\| \equiv \|B_k^\dagger g_k\| \le 1.2\rho_k + 1.6\mu 2.5\rho_0\|g_*\|/\sigma_*^2 \le (74/45)\rho_k \le (5/3)\rho_k$$

where the last inequality holds provided $\rho_k$ is still greater $9\rho_0\|g_*\|\mu/\sigma_*^2$. Then we note that by the triangle inequality $\rho_{k+1} \le \rho_k + \|s_k\| \le (8/3)\rho_k$, which implies in particular that $x_{k+1}$ belongs also to $\mathcal{L}$. Hence we obtain from the mean-value theorem

$$\|g_{k+1} - g_k - G_*s_k\| \le \|G_k - G_*\| \, \|s_k\| + .5\mu\|s_k\|^2 \le \mu(\rho_k + \|s_k\|/2)\|s_k\| \le 3\mu\rho_k^2.$$

Premultiplying the vector on the left by $B_k^\dagger$ one finds using $B_k^\dagger(g_k - B_ks_k) = 0$ that

$$\|B_k^\dagger g_{k+1}\| \le \|B_k^\dagger\| \, [3\mu\rho_k^2 + \delta_k\|s_k\|] \le \rho_k[2\delta_k + 3.5\rho_k\mu]/\sigma_*.$$

On the other hand we have also by (2.14)

$$\|B_k^\dagger g_{k+1}\| \ge .8\rho_{k+1} - 1.6\Delta_k\|g_*\|/\sigma_*^2 \ge .8\rho_{k+1} - 4\rho_0\mu\|g_*\|/\sigma_*^2 \ge \rho_{k+1}/3$$

where the last inequality holds provided $\rho_{k+1}$ is also still greater than $9\rho_0\|g_*\|\mu/\sigma_*^2$. Combining the upper and lower bound for $\|B_k^\dagger g_{k+1}\|$ we obtain finally

$$\rho_{k+1}/\rho_k \le [6\delta_k + 10.5\mu\rho_k]/\sigma_* \le [15 + 10.5]\mu\rho_0/\sigma_* \le 1/3,$$

where the last inequality holds because of our rather stringent bound on $\rho_0$. ∎

 

The spectral norm $\lambda_*$ defined in (1.3), which determines the asymptotic convergence rate of the Gauss-Newton method is bounded below by the ratio $\|g_*\|\mu/\sigma_*^2$ that appears on the right hand side of (2.18). As the first step of the Gauss-Broyden method with $B_0 = G_0$ is identical to that of the Gauss-Newton method, it may appear that it should already achieve the asserted reduction so that simply $\|x_1 - x_*\|/\|x_0 - x_*\| \le 9\|g_*\|\mu/\sigma_*^2$. This would make the above result essentially trivial. However, since the three quantities $\sigma_*, \mu, \|g_*\|$ are completely independent of each other and the bound on $\|x_0 - x_*\|$ does not depend on $\|g_*\|$, the first step cannot achieve the asserted reduction if the residual is rather small. Because the upper bound on the smallest $\|x_k - x_*\|$ is solely based on the bounded deterioration of the Broyden update, we have essentially a result about the 'Gauss-chord' method where simply $B_k = G_0$ at all iterations. Naturally one expects that the updating does in fact improve the performance, which can certainly be observed in numerical experiments. Also, in the zero residual case benefits in the form of superlinear convergence can be established theoretically.

**COROLLARY 1** If the assumptions of Theorem 2 hold with $g_* = 0$ then the Gauss-Broyden iterates satisfy for some constant $c$ and all $k$

$$\frac{\|x_k - x_*\|}{\|x_0 - x_*\|} \le \left(\frac{c}{k}\right)^{k/2} \quad and \quad \lim_{k\to\infty} \frac{\|x_{k+1} - x_*\|}{\|x_k - x_*\|} = 0$$

unless the iteration terminates because an $x_k$ happens to coincide with the solution $x_*$.

 

Proof: Since the last inequality (2.18) of Theorem 2 can never be satisfied, the $x_k$ must converge at least linearly to $x_*$. Consequently the steps $s_k$ are summable and it follows as in the proof of Theorem 4 in [12] that the update matrices $C_k$ are square summable. Since always $\|r_k\| = \|C_k\|$ and

$$|\delta_k - \|r_k\|| \le \|y_k - G_*s_k\|/\|s_k\| \le \mu\rho_k$$

the $\delta_k$ are also square summable. Hence the central inequality (2.17) of Theorem 2 implies the first assertion of the corollary by the inequality of the means. The second assertion is a consequence of the same estimate since obviously $lim_{k\to\infty}\delta_k = 0$. ∎

According to Corollary 1 the Gauss-Broyden method has in the zero residual case exactly the same local convergence properties as its specialization, the classical Broyden method, in the usual square case $m = n$. As was done in [13] similar superlinear convergence results can be established under considerably weaker differentiability assumptions on $g$.

## 3 A DERIVATIVE-FREE LINE-SEARCH CRITERION

Given a search direction $s_k \neq 0$ at the current iterate $x_k$, we would like to find a step multiplier $\alpha_k \in \mathbf{R}$ that minimizes the residual

$$\gamma_k(\alpha) \equiv \gamma(x_k + \alpha s_k).$$

Since the iterative solution of this one-dimensional minimization problem could be quite costly, one prefers to accept values for $\alpha_k$ that satisfy considerably weaker line-search conditions. Because most standard step-size rules were designed for unconstrained optimization, they typically involve the directional derivative

$$\left.\frac{d}{d\alpha}\gamma_k(\alpha)\right|_0 = \nabla\gamma(x_k)^T s_k.$$

Moreover, a search direction $s_k$ is usually only accepted if it can be ensured explicitly or implicitly that the angle between $s_k$ and the direction of steepest descent $-\nabla\gamma(x_k)$ is sufficiently acute.

Convergence of a subsequence to a stationary point of $\gamma$ can be established if the line search is *efficient* as defined in [23], and the search directions satisfy the Zoutendijk [25] condition

$$\sum_{k=0}^{\infty}\left(\frac{\nabla\gamma(x_k)^T s_k}{\|\nabla\gamma(x_k)\|\,\|s_k\|}\right)^2 = \infty. \tag{3.1}$$

Here it is usually assumed that $\nabla\gamma(x_k)^T s_k \leq 0$ for all $k$, which cannot be guaranteed in our case, as the gradient takes the form

$$\nabla\gamma(x)^T = g(x)^T G(x)$$

and is therefore, like the Jacobian $G(x) = \nabla g(x)$, unknown. Hence, we cannot implement customary line-search criteria, e.g. the popular Goldstein test [24], because they explicitly involve the directional derivative $\nabla\gamma(x_k)^T s_k = g_k^T G_k s_k$. Despite this apparent lack of information it will be shown that the line-search criterion proposed in [12] for square problems is efficient in that for some positive constant $c$ and all $k$

$$\gamma_{k+1} - \gamma_k \leq -c(g_k^T G_k s_k/\|G_k s_k\|)^2.$$

After developing the line-search criterion in the current context we will give a line-search algorithm for computing a suitable step multiplier in the subsequent Section 4.

Following a suggestion of Lindström and Wedin [16], as well as Al-Baali and Fletcher, [1] we interpolate the vector function

$$g_k(\alpha) \equiv g(x_k + \alpha s_k) : \mathbf{R} \to \mathbf{R}^n$$

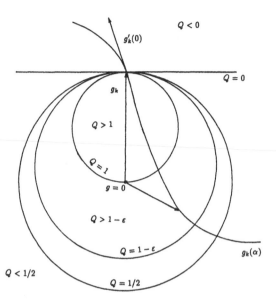

Figure 1: Circular contours of multiplier ratio $Q = Q_k(g)$ for fixed $g_k$.

rather than just the scalar $\gamma_k(\alpha)$. Based on the values of $g_k(\alpha)$ for $\alpha = 0$ and some $\alpha_c \neq 0$ we can form the linear approximation

$$g_k(\alpha) \approx g_k + [g_k(\alpha_c) - g_k]\alpha/\alpha_c$$

where $g_k = g(x_k) = g_k(0)$ as before. It is easily seen that the Euclidean norm of the right-hand side has the unique minimizer $q_k(\alpha_c) \cdot \alpha_c$ where

$$q_k(\alpha) \equiv \frac{g_k^T(g_k - g_k(\alpha))}{\|g_k - g_k(\alpha)\|^2} : \mathbf{R} - \{0\} \quad \rightarrow \quad \mathbf{R}. \tag{3.2}$$

This ratio between the supposedly optimal multiplier and the current guess plays a central role in our analysis. To obtain some geometrical understanding of $q_k(\alpha)$ we may write it in the form

$$q_k(\alpha) = Q(g_k(\alpha)) \quad ,$$

where for fixed base vector $g_k$

$$Q_k(g) \equiv g_k^T(g_k - g)/\|g_k - g\| : \mathbf{R}^n - \{g\} \quad \rightarrow \quad \mathbf{R}.$$

As depicted in Figure 1 the vectors $g$ yielding a certain value $Q$ of $Q_k$ form spheres with radius $1/\,|\,2Q\,|$ about the center $(1 - \frac{1}{2}/Q)g_k$.

Now our first and main requirement on the step multipliers $\alpha_k$ is that for some constant $\varepsilon \in (0, \frac{1}{2})$

$$q_k(\alpha) \geq 1 - \varepsilon \quad or \quad \alpha_k = 0. \tag{3.3}$$

Since the reduction of $\|g\|$ can be expressed as

$$\|g_k\|^2 - \|g_k(\alpha_k)\|^2 = [2q_k(\alpha_k) - 1] \cdot \|g_k - g_k(\alpha_k)\|^2 \geq 0 \tag{3.4}$$

it follows for $x_{k+1} = x_k + \alpha_k s_k$ that

$$\sigma^2 \|\alpha_k s_k\|^2 \leq \|g_k - g_{k+1}\|^2 \leq (\gamma_k - \gamma_{k+1})/(\tfrac{1}{2} - \varepsilon) \tag{3.5}$$

where

$$\sigma \equiv \inf\{\|g(x) - g(z)\|/\|x - z\| : x, z \in \mathcal{L}, x \neq z\}. \tag{3.6}$$

As a consequence of our assumption that $g$ is injective and its Jacobian has no singularities on $\mathcal{L}$, the infimum $\sigma$ is in fact a positive minimum. Hence one finds by summing over k that

$$\sum_{k=0}^{\infty} \|x_{k+1} - x_k\|^2 \leq \gamma_0 \sigma^{-2}/(\tfrac{1}{2} - \varepsilon). \tag{3.7}$$

Thus we see that our regularity assumptions on $g$ and the requirement $q_k(\alpha_k) \geq 1 - \varepsilon$ imply the main hypothesis of Theorem 1, namely the square summability of the steps. As was noted in Section 1, the resulting fact that $B_k s_k \sim y_k$ for almost all k does not guarantee that any one of the steps $s_k$ comes close to the Gauss-Newton correction $-G_k^\dagger g_k$. This follows only when $B_k$ and $G_k$ have the same range, which is trivially the case if $m = n$ and $det(B_k) \neq 0$. Otherwise, not even the highly desirable property $(B_k - G_k) \to 0$ implies that all later steps $s_k = -B_k^\dagger g_k$ will be similar to the Gauss-Newton corrections $-G_k^\dagger g_k$. However, in combination with our derivative-free line-search the condition $(B_k - G_k) \to 0$ does imply convergence of a subsequence to a stationary point.

In any case we want to ensure that good use is made of all search directions. To this end we have to prevent $\alpha_k$ from being chosen too small, since the first condition ensures only that $\alpha_k$ is not too large and has the right sign. Since $q_k(\alpha_k) = 1$ is the ideal value in the sense of the linear model, one may require that

$$q_k(\alpha_k) < 1/(1 - \epsilon) < 2 \quad or \quad g_k^T G_k s_k = 0. \tag{3.8}$$

The second alternative implicitly restricts the use of the trivial choice $\alpha_k = 0$ to the exceptional case where the search direction $s_k$ is exactly perpendicular to the gradient $\nabla \gamma_k$. By varying the parameter $\varepsilon \in (0, \tfrac{1}{2})$ we can make the line-search conditions more or less *accurate*. However, it should be noted that the limiting choice $\varepsilon = 0$ does not even theoretically ensure that the line-search is *exact* in the sense that $\alpha_k$ actually minimizes $\gamma_k(\alpha)$. This is only the case for linear $g$. Otherwise the multipliers $\alpha_k \in q_k^{-1}(1)$ are typically somewhat larger than the minimizers of $\gamma_k(\alpha)$. It is well known that in curved valleys, stepping beyond the minimum is sometimes beneficial overall. However, the main advantage of the line-search aim $q_k(\alpha_k) \sim 1$ is that it does not involve directional derivatives. These would have to be estimated by differences and an interpolatory algorithm for minimizing $\gamma_k(\alpha)$ would in fact involve a second level of differencing.

Unfortunately, as in case of the Goldstein test, the iterative calculation of an $\alpha_k$ that satisfies the lower and upper bound on $q_k(\alpha_k)$ simultaneously can be quite lengthy. Therefore we may introduce an Armijo type relaxation [2] by accepting $\alpha_k$ already as sufficiently large if for some $\hat{\alpha}_k$

$$0 < |\frac{\hat{\alpha}_k}{\alpha_k}| \leq \tilde{\tau}^{-2} \quad and \quad q_k(\hat{\alpha}_k) \in \begin{cases} (-\infty, 2] & \text{if } \hat{\alpha}_k \alpha_k > 0 \\ [-2, +2] & \text{if } \hat{\alpha}_k \alpha_k < 0 \end{cases} \tag{3.9}$$

Here $\tilde{\tau}$ is a positive constant less than 1. Often a suitable $\hat{\alpha}_k$ is found as an earlier iterate in the computation of $\alpha_k$, or we may of course use $\hat{\alpha}_k = \alpha_k$ if $q_k(\alpha_k) \leq 2$. Now we can show that the line search conditions developed above are *efficient* in the sense of Warth and Werner [23].

**PROPOSITION 2** There exists a positive constant $c$ such that for any point $x_k \in \mathcal{L}$ and any nonzero search direction $s_k$ the iterate $x_{k+1} = x_k + \alpha_k s_k$ satisfies

$$\gamma_{k+1} - \gamma_k \le -c \left( g_k^T G_k s_k / \|G_k s_k\| \right)^2 \tag{3.10}$$

provided $\alpha_k$ is chosen such that (3.3) and (3.9) hold except when $g_k^T G_k s_k = 0$ and $\alpha_k = 0$.

Proof: Because of our regularity and continuity assumptions there exists a neighborhood

$$\mathcal{L}_\rho \equiv \{x \in \mathbf{R}^n : \rho \ge \|x - z\|, z \in \mathcal{L}\}$$

such that for positive constants $\sigma_1, \sigma_2$ and all distinct pairs $x, z \in \mathcal{L}_\rho$

$$\sigma_1 \le \|g(x) - g(z)\| / \|x - z\| \le \sigma_2$$

as well as

$$\|g(z) - g(x) - G(x)(z - x)\| \le .5\mu \|z - x\|^2 \tag{3.11}$$

where $\mu$ is a Lipschitz constant of $g$ on the convex hull of $\mathcal{L}_\rho$, which is also bounded. Since by (3.4)

$$\gamma_{k+1} - \gamma_k \le -(\tfrac{1}{2} - \varepsilon)\|g_k - g_{k+1}\|^2 \le 0 \tag{3.12}$$

the iterate $x_{k+1}$ also belongs to $\mathcal{L}$, and the assertion must be true if we can show that for some positive constant $\hat{c}$

$$\|g_k - g_{k+1}\|/\hat{c} \ge |g_k^T G_k s_k| / \|G_k s_k\|. \tag{3.13}$$

Because of the Cauchy-Schwarz inequality the right hand side is bounded above by $\sqrt{(2\bar{\gamma})} \ge \|g_k\|$, where $\bar{\gamma}$ is the upper bound that defines the level set $\mathcal{L}$. Consequently difficulties can only arise when $\|g_k - g_{k+1}\|$ is sufficiently small, say less than $\rho\sigma_1\tilde{\tau}^2$. Then we find

$$\|\hat{\alpha}_k s_k\| \le \|x_{k+1} - x_k\|\tilde{\tau}^2 \le \|g_k - g_{k+1}\|/\sigma_1\tilde{\tau}^2 \le \rho$$

so that

$$\hat{x}_{k+1} \equiv x_k + \hat{\alpha}_k s_k \in \mathcal{L}_\rho.$$

Hence we derive for $\hat{g}_{k+1} \equiv g(\hat{x}_{k+1})$ the upper bound

$$
\begin{aligned}
\mu^{-2}\|\hat{g}_{k+1} - g_k\|^2/|\hat{\alpha}_k| \quad & \le |\hat{\alpha}_k|\|s_k\|^2 \le \|\alpha_k\|\|s_k\|\|G_k s_k\|/(\tilde{\tau}^2 \sigma_1) \\
& \le \|g_k - g_{k+1}\|\|G_k s_k\|/(\tilde{\tau}\sigma_1)^2
\end{aligned}
\tag{3.14}
$$

which will be used later. By Taylor's theorem

$$\|g_k - \hat{g}_{k+1} + \hat{\alpha}_k G_k s_k\| \le \mu(\hat{\alpha}_k \|s_k\|)^2 \le \mu\|g_k - \hat{g}_{k+1}\|^2/\sigma_1^2. \tag{3.15}$$

Premultiplying by $g_k^T$ and then dividing by $\|g_k - \hat{g}_{k+1}\|^2$ we obtain the crucial estimate

$$\left| q_k(\hat{\alpha}_k) + \hat{\alpha}_k g_k^T G_k s_k / \|g_k - \hat{g}_{k+1}\|^2 \right| \le \|g_k\|\mu/\sigma_1^2 \le \hat{\mu} - 2 \tag{3.16}$$

where $\hat{\mu} \equiv 2 + \sqrt{(2\bar{\gamma})}\mu/\sigma_1^2$ . Now, if $|q_k(\hat{\alpha}_k)| \le 2$ it follows directly that

$$\frac{|g_k^T G_k s_k|}{\|G_k s_k\|} \le \frac{\hat{\mu}\|g_k - \hat{g}_{k+1}\|^2}{|\hat{\alpha}_k|\|G_k s_k\|} \le \frac{\hat{\mu}\|g_k - g_{k+1}\|^2 \mu^2}{(\sigma\tilde{\tau})^2} \tag{3.17}$$

where the last inequality holds by (3.14). For the other possible case, where $\hat{\alpha}_k \alpha_k > 0$ and $q_k(\hat{\alpha}_k) < -2$ we derive the same inequality as follows.

If $\alpha_k$ has the opposite sign of $g_k^T G_k s_k$, then (3.16) implies

$$0 > \hat{\alpha}_k g_k^T G_k s_k / \|g_k - \hat{g}_{k+1}\|^2 \geq -\hat{\mu} + 2 - q_k(\hat{\alpha}_k) > -\hat{\mu}.$$

Consequently the two inequalities in (3.17) are still valid. Finally, when $\alpha_k g_k^T G_k s_k > 0$ we find as in (3.16) but this time for $\alpha_k$ instead of $\hat{\alpha}_k$ and with $q_k(\alpha_k) > 0$ that

$$0 < \alpha_k g_k^T G_k s_k / \|g_k - g_{k+1}\|^2 \leq \hat{\mu} - 2 - q_k(\alpha_k) \leq \hat{\mu}.$$

This yields directly

$$\frac{|g_k^T G_k s_k|}{\|G_k s_k\|} \leq \hat{\mu} \|g_k - g_{k+1}\| \frac{\|g_k - g_{k+1}\|}{|\alpha_k| \|G_k s_k\|} \leq \frac{\hat{\mu} \|g_k - g_{k+1}\| \mu}{\sigma}.$$

Hence, we conclude that (3.13) holds in all possible cases with

$$\hat{c} \equiv \sigma_1 \tilde{\tau}^2 / \max\{\mu \hat{\mu}, \sqrt{(2\hat{\gamma})}/\rho\}.$$

Thus the proposition is true for $c \leq (\frac{1}{2} - \varepsilon)\hat{c}^2$. ∎

As an immediate consequence of Proposition 2 we see that

$$\sum_{k=0}^{\infty} (\|P_k g_k\| cos\theta_k)^2 \leq \gamma_0/c < \infty \tag{3.18}$$

where $P_k$ denotes the orthogonal projection onto the range $R_k$ of $G_k$, and the angle $\theta_k \in (-\pi/2, \pi/2)$ is defined by

$$cos\theta_k \equiv -g_k^T G_k s_k / (\|P_k g_k\| \|G_k s_k\|). \tag{3.19}$$

In case of a Gauss-Newton step $-G_k^\dagger g_k$ the projected gradient $P_k g_k$ coincides with the expected change in the residual vector, namely $-G_k s_k \sim g_k - g_{k+1}$. Hence all $\theta_k$ are exactly zero in case of the Gauss-Newton method. Then (3.18) implies that the projected gradients $P_k g_k$ are square-summable and must in particular go to zero. More generally, we find that a subsequence of the $x_k$ must converge to a stationary point of $\|g(x)\|$ whenever the cosines are not square summable, i.e.

$$\sum_{k=0}^{\infty} cos^2\theta_k = \infty \quad \Rightarrow \quad \liminf_k \|P_k g_k\| = 0. \tag{3.20}$$

As a nontrivial application of this relation we can prove under the assumptions of Proposition 2 the following convergence result.

**COROLLARY 2** If the approximating Jacobians $B_k$ satisfy $\lim_{k \to \infty}(B_k - G_k) = 0$, then a subsequence of the quasi-Gauss-Newton iterates $x_k$ converges to a stationary point of $\|g(x)\|$.

Proof: Because everything happens in the compact domain $\mathcal{L}$ it follows from (2.12) in Lemma 1 that

$$\lim_{k \to \infty} \|G_k s_k - P_k g_k\| = \lim_{k \to \infty} \|G_k(B_k^\dagger - G_k^\dagger)g_k\| = 0.$$

Now, if the assertion were false the norms $\|P_k g_k\|$ would be bounded away from zero, and consequently the angles $\theta_k$ between the vectors $P_k g_k$ and $G_k s_k$ would converge to zero. Thus the Zoutendijk condition would hold, which contradicts our assumption that $\inf \|P_k g_k\| > 0$. Hence the assertion is indeed true. ∎

Clearly the main assumption of Corollary 2 could be replaced by the considerably weaker condition $\liminf(B_k - G_k) = 0$. Unfortunately it appears currently that neither property can be guaranteed for secant approximations $B_k$, unless special steps are taken to make the sequence of updating directions uniformly linearly independent in the sense of Powell [21]. Moreover, it should be noted that even the property $\lim(B_k - G_k) = 0$ does not imply $(G_k^\dagger - B_k^\dagger)g_k/\|P_k g_k\| \to 0$, so that the relative error between the quasi-Gauss-Newton steps and the Gauss-Newton steps need not tend to zero. Thus we have to expect a *rough ride* in the sense of never ending line-search activity, even when the Broyden formula or any other secant update does a good job of approximating the exact Jacobian.

## 4 A DERIVATIVE-FREE LINE-SEARCH PROCEDURE

So far we have only claimed that our line-search conditions are consistent, and it has not been demonstrated how a suitable step multiplier can be computed. The following line-search procedure generates for fixed $x_k \in \mathcal{L}$ and $s_k \in \mathbf{R}^n - \{0\}$ a sequence of trial values $\left\langle \alpha_k^{(i)} \right\rangle_{i=0}^\infty$ with $\alpha_k^{(0)} = 1$ such that the limit

$$\alpha_k^{(*)} \equiv \lim_{i \to \infty} \alpha_k^{(i)}$$

is finite and satisfies

$$q_k(\alpha_k^{(*)}) = 1 \quad or \quad \alpha_k^{(*)} = 0 = g_k^T G_k s_k. \tag{4.1}$$

Except in the second case, where we have to set $\alpha_k \equiv \alpha_k^{(*)} = 0$, this implies obviously that for some first iterate $\alpha_k^{(i)}$ the ratio $q_k(\alpha_k^{(i)})$ lies in the interval $[1 - \varepsilon, (1 - \varepsilon)^{-1}]$ so that $\alpha_k \equiv \alpha_k^{(i)}$ satisfies our line-search criterion. Quite likely, an earlier iterate does already satisfy the Armijo condition (3.9) and is therefore an acceptable step multiplier.

Under our general assumptions the functions $q_k(\alpha)$ is differentiable at all $\alpha \neq 0$ and has the limiting properties

$$\lim_{\alpha \to 0} \alpha q_k(\alpha) = g_k^T G_k s_k / \|G_k s_k\|^2 \tag{4.2}$$

and

$$\lim_{|\alpha| \to \infty} \sup q_k(\alpha) \leq .5 \tag{4.3}$$

as well as for any $\underline{\alpha} > 0$

$$\sup_{\alpha > \underline{\alpha}} |q_k(\alpha)| < \infty \quad . \tag{4.4}$$

The first relation (4.2) means that $q_k(\alpha)$ has a simple pole at $\alpha = 0$ whenever $g_k^T G_k s_k \neq 0$. The second (4.3) relation follows from (3.4), and the third (4.4) is a consequence of the Cauchy-Schwarz inequality $|q_k(\alpha)| \leq \|g_k\|/\|g_k - g_k(\alpha)\|$ and the injectivity of $g$ on the bounded level set $\mathcal{L}$. The same inequality yields the implication

$$\lim_{\|x\| \to \infty} \|g(x)\| = \infty \quad \Rightarrow \quad \lim_{|\alpha| \to \infty} q_k(\alpha) = 0. \tag{4.5}$$

Even when this coercivity condition is not met, we will set

$$q_k(-\infty) \equiv 0 \equiv q_k(+\infty). \tag{4.6}$$

While $q_k(\alpha)$ is quite volatile near the origin, the function

$$h_k(\alpha) \equiv \begin{cases} \alpha[1 - q_k(\alpha)] & if \alpha \neq 0 \\ g_k^T G_k s_k / \|G_k s_k\|^2 & if \alpha = 0 \end{cases}. \tag{4.7}$$

is Lipschitz-continuous at $\alpha = 0$ and differentiable everywhere else. Moreover, because of (4.3) we have

$$\lim_{\alpha \to -\infty} h_k(\alpha) = -\infty \quad and \quad \lim_{\alpha \to +\infty} h_k(\alpha) = +\infty \tag{4.8}$$

so that there exists at least one root $\alpha_k^{(*)}$ satisfying (4.1).

Consequently one may use any safeguarded scheme to solve the scalar equation $h_k(\alpha) = 0$ starting from a sufficiently large interval containing the origin. When the underlying vector function $g$ is twice Lipschitz continuously differentiable, then $h_k$ has even at the origin a Lipschitz continuous first derivative, which is in general also nonzero. Hence we see that (even in the special case $g_k^T G_k s_k = 0$) superlinear convergence can theoretically be achieved by the general purpose implementations of the secant method [7]. However, because $h_k(\alpha)$ is unknown at $\alpha = 0$ and its numerical values are severely effected by cancellation errors when $\alpha$ is small, rapid convergence to the trivial step size $\alpha_k^{(*)} = 0$ will usually not occur and is not even desirable. Also, rather than just using the values of the scalar function $h_k(\alpha)$ one may prefer to interpolate the underlying vector function $g_k(\alpha)$ to compute new trial values for $\alpha_k$. To this end we develop the following bracketing scheme based on regula falsi.

Throughout the calculation we maintain lower and upper bounds $\underline{\alpha}_k^{(i)} \neq 0$ and $\overline{\alpha}_k^{(i)} \neq 0$ such that

$$-\infty \leq \underline{\alpha}_k^{(i)} < \overline{\alpha}_k^{(i)} \leq +\infty \tag{4.9}$$

and

$$h_k(\underline{\alpha}_k^{(i)}) < 0 < h_k(\overline{\alpha}_k^{(i)}). \tag{4.10}$$

Because of (4.3) the first condition is met initially by $(\underline{\alpha}_k^{(0)}, \overline{\alpha}_k^{(0)}) \equiv (-\infty, +\infty)$. Having evaluated $q_k$ at the i-th trial value

$$\alpha_k^{(i)} \in (\underline{\alpha}_k^{(i)}, \overline{\alpha}_k^{(i)}) - \{0\} \tag{4.11}$$

we set $(\underline{\alpha}_k^{(i+1)}, \overline{\alpha}_k^{(i+1)})$ either to $(\underline{\alpha}_k^{(i)}, \alpha_k^{(i)})$, or $(\alpha_k^{(i)}, \overline{\alpha}_k^{(i)})$ depending upon whether $h_k(\alpha_k^{(i)})$ is positive or negative. Here we have excluded the remote possibility that ever exactly $q_k(\alpha_k^{(i)}) = 1$, in which case the iteration is terminated with $\alpha_k^{(*)} \equiv \alpha_k^{(i)}$. After the initial evaluation at $\alpha_k^{(0)} = 1$ the updating rule yields

$$(\underline{\alpha}_k^{(i)}, \overline{\alpha}_k^{(1)}) \equiv \begin{cases} (-\infty, 1) & if \quad q_k(1) > 1 \\ (1, +\infty) & if \quad q_k(1) < 1 \end{cases}. \tag{4.12}$$

From then on at least one of the bounds is always finite. Obviously the nested intervals $(\underline{\alpha}_k^{(i+1)}, \overline{\alpha}_k^{(i+1)})$ $(\underline{\alpha}_k^{(i)}, \overline{\alpha}_k^{(i)})$ must converge to a limit set $[\underline{\alpha}_k^{(*)}, \overline{\alpha}_k^{(*)}]$. In order to ensure that at least one of the two limiting bounds $\underline{\alpha}_k^{(*)}$ and $\overline{\alpha}_k^{(*)}$ represents a suitable $\alpha_k^{(*)}$ the trial values $\alpha_k^{(i)}$ must be appropriately restricted to the interior of $(\underline{\alpha}_k^{(i)}, \overline{\alpha}_k^{(i)})$. To this end we may express the reciprocal of any nonzero $\alpha_k^{(i)}$ as the affine combination

$$\frac{1}{\alpha_k^{(i)}} = \frac{1}{\overline{\alpha}_k^{(i)}} \frac{1}{\left[1 - \tau_k^{(i)}\right]} + \frac{1}{\underline{\alpha}_k(i)} \frac{1}{\left[1 - 1 / \tau_k^{(i)}\right]} \quad, \tag{4.13}$$

where the parameter $\tau_k^{(i)}$ is easily calculated as the ratio

$$\tau_k^{(i)} \equiv \frac{\left[1 - \alpha_k^{(i)} / \overline{\alpha}_k^{(i)}\right]}{\left[1 - \alpha_k^{(i)} / \underline{\alpha}_k^{(i)}\right]} \in [-\infty, +\infty] \quad . \tag{4.14}$$

These reciprocal expressions make sense even when one of the two bounds is infinite. Now we restrict $\alpha_k^{(i)}$ by imposing the condition

$$0 < \hat{\tau} \le \hat{\tau}_k^{(i)} \equiv \tau_k^{(i)} \left/ \frac{\left[q_k(\overline{\alpha}_k^{(i)}) - 1\right]}{\left[q_k(\underline{\alpha}_k^{(i)}) - 1\right]} \right. \le \frac{1}{\hat{\tau}} \quad , \tag{4.15}$$

where $\hat{\tau} \in (0, 1)$ is a constant. The choice $\hat{\tau}_k^{(i)} = 1$ yields simply the secant step from the interval bounds, which is feasible except when $\tau_k^{(i)} = 1$ and consequently $\alpha_k^{(i)} = 0$. In an attempt to avoid the numerical instabilities that arise when $\alpha_k^{(i)}$ becomes very small, we will impose the additional restriction that

$$\tau_k^{(i)} \notin [(1 - \tilde{\tau}), 1/(1 - \tilde{\tau})] \quad with \quad \tilde{\tau} \in [0, 1 - \hat{\tau}) \quad constant. \tag{4.16}$$

This condition is trivially satisfied if $\underline{\alpha}_k^{(i)} \overline{\alpha}_k^{(i)} > 0$ since then (4.10) and (4.14) imply $\tau_k^{(i)} < 0$. Otherwise the upper bound $\tilde{\tau} < (1 - \hat{\tau})$ ensures that (3.35) and (3.36) are always consistent and we derive for $\tilde{\tau} > 0$

$$\frac{1}{|\alpha_k^{(i)}|} = \frac{1}{|\overline{\alpha}_k^{(i)}|} \frac{1}{\left|1 - \tau_k^{(i)}\right|} + \frac{1}{|\underline{\alpha}_k^{(i)}|} \frac{1}{\left|1 - 1 \left/ \tau_k^{(i)}\right.\right|}$$

$$\le \left[\frac{1}{|\overline{\alpha}_k^{(i)}|} + \frac{1}{|\underline{\alpha}_k^{(i)}|}\right] (1 + \tilde{\tau}) \left/ \tilde{\tau}\right.$$

so that

$$|\alpha_k^{(i)}| \ge \frac{\tilde{\tau}}{4} \min\left\{|\underline{\alpha}_k^{(i)}|, |\overline{\alpha}_k^{(i)}|\right\} \quad . \tag{4.17}$$

The last inequality allows us to verify the Armijo condition (3.9) with $\hat{\alpha}_k$ being an earlier iterate. However, the following result holds even for the limiting choice $\tilde{\tau} = 0$ that excludes only $\alpha_k^{(i)} = 0$.

**PROPOSITION 3** If $\alpha_k^{(0)} = 1$ and subsequent $\alpha_k^{(i)} \ne 0$ are chosen such that (4.13), (4.14), and (4.15) hold for all $i > 0$, then they have a limit $\alpha_k^{(*)}$ that satisfies

$$\alpha_k^{(*)} \ne 0 \quad and \quad q_k(\alpha_k^{(*)}) = 1$$

or

$$\alpha_k^{(*)} = 0 \quad and \quad g_k^T G_k s_k = 0$$

unless exactly $q_k(\alpha_k^{(i)}) = 1$ for some $\alpha_k^{(*)} \equiv \alpha_k^{(i)}$ with $i$ finite.

Proof: Throughout the proof we exclude the last possibility so that we have an infinite sequence of distinct nonzero iterates $\alpha_k^{(i)}$. First let us check that any $\alpha_k^{(i)}$ of the form (4.13) with positive $\hat{\tau}_k^{(i)}$ as defined in (4.14) belongs to the interval $(\underline{\alpha}_k^{(i)}, \overline{\alpha}_k^{(i)})$. If this interval does not include the origin, it follows from (4.10) that $\tau_k^{(i)}$ is negative so that (4.13) represents in fact a convex combination of the reciprocals $1/\underline{\alpha}_k^{(i)}$ and $1/\overline{\alpha}_k^{(i)}$. Hence $1/\alpha_k^{(i)}$ lies between these values and taking the reciprocal again we find that (4.11) is indeed satisfied. On the other hand if the interval $(\underline{\alpha}_k^{(i)}, \overline{\alpha}_k^{(i)})$ straddles the origin, it follows from (4.10) that $\tau_k^{(i)}$ is positive so that the coefficients in the affine combination (4.13) have opposite sign. Consequently $1/\alpha_k^{(i)}$ lies outside the interval $\left[1/\underline{\alpha}_k^{(i)}, 1/\overline{\alpha}_k^{(i)}\right] \supset 0$ and taking the reciprocal verifies again (4.11) since the boundaries have opposite signs. Thus the sequences $\underline{\alpha}_k^{(i)}$ and $\overline{\alpha}_k^{(i)}$ are well defined and due to their monotonicity they must have limits $\underline{\alpha}_k^{(*)} \leq \overline{\alpha}_k^{(*)}$. By continuity of $h_k(\alpha)$ it follows from (4.10) that $h_k(\underline{\alpha}_k^{(*)}) \leq 0 \leq h_k(\overline{\alpha}_k^{(*)})$, which implies by (4.8) $\underline{\alpha}_k^{(*)} < \infty$ and $-\infty < \overline{\alpha}_k^{(*)}$. Consequently at least one of the limiting bounds is finite. Our final preparatory observation is that due to every $\alpha_k^{(i)}$ being equal to $\underline{\alpha}_k^{(i+1)}$ or $\overline{\alpha}_k^{(i+1)}$, the sequence $\alpha_k^{(i)}$ has at most two limit points, namely $\underline{\alpha}_k^{(*)}$ and $\overline{\alpha}_k^{(*)}$. Since the assertion follows immediately with $h_k(\underline{\alpha}_k^{(*)}) = 0 = h_k(\overline{\alpha}_k^{(*)})$ if the two bounds are equal we will assume from now on that they are distinct. The remainder of the proof has two main parts. First we show that if either $\underline{\alpha}_k^{(*)}$ or $\overline{\alpha}_k^{(*)}$ is in fact the only limit of the $\alpha_k^{(i)}$, then this value is a root of $h_k(\alpha)$. In the second part of the proof we exclude the possibility that both $\underline{\alpha}_k^{(*)}$ and $\overline{\alpha}_k^{(*)}$ are distinct limit points of the $\alpha_k^{(i)}$.

Suppose

$$\left\{\alpha_k^{(*)}\right\} \in \left\{\underline{\alpha}_k^{(*)}, \overline{\alpha}^{(*)}\right\} \neq \left\{\alpha_k^{(*)}\right\}.$$

Because the two limiting bounds are distinct, either of the sequences $\left\langle \underline{\alpha}_k^{(i)} \right\rangle_{i=0}^{\infty}$ and $\left\langle \overline{\alpha}_k^{(i)} \right\rangle_{i=0}^{\infty}$ must become stationary at a nonzero value for sufficiently large $i$. We may assume

$$\overline{\alpha}_k^{(*)} = \alpha_k^{(*)} \quad and \quad 0 \neq \underline{\alpha}_k^{(*)} = \underline{\alpha}_k^{(i)} \quad for \quad all \quad i \quad .$$

This does not represent any loss of generality since early iterates can be disregarded and the bounds may be interchanged by switching the sign of $\alpha$ and $s_k$ if necessary. Now we have to distinguish whether $\alpha_k^{(*)}$ is zero or not. The first case can only occur when $\underline{\alpha}_k^{(i)} \overline{\alpha}_k^{(i)} < 0$ throughout the iteration with all $\alpha_k^{(i)} = \overline{\alpha}_k^{(i+1)}$ being positive. This requires by (4.13) that all $\tau_k^{(i)}$ are less than 1 which implies by (4.15) and with the constancy of $\underline{\alpha}_k^{(i)} = \underline{\alpha}_k^{(*)}$

$$1 - q_k(\overline{\alpha}_k^{(i)}) \leq \left[1 - q_k(\underline{\alpha}_k^{(*)})\right] /\hat{\tau}.$$

Because the $q_k(\overline{\alpha}_k^{(i)}) = q_k(\alpha_k^{(i-1)})$ are also bounded above by 1, it follows from (4.2) that indeed $g_k^T G_k s_k = 0$ as asserted for the case $\alpha_k^{(*)} = 0$. Otherwise we have

$$\lim_{i \to \infty} 1/\alpha_k^{(i)} = 1/\overline{\alpha}_k^{(*)} \in R$$

and it follows from (4.13) that

$$0 = \lim_{i \to \infty} \left| \frac{1}{\alpha_k^{(i)}} - \frac{1}{\overline{\alpha}_k^{(i)}} \right| = \lim_{i \to \infty} \left| \frac{1}{\underline{\alpha}_k^{(i)}} - \frac{1}{\overline{\alpha}_k^{(i)}} \right| \frac{1}{\left[1 - 1/\tau_k^{(i)}\right]}. \tag{4.18}$$

Since the first factor on the right is bounded away from zero, the $\tau_k^{(i)}$ must tend to zero so that by (4.15)

$$\lim_{i \to \infty} \left| q_k(\overline{\alpha}_k^{(i)}) - 1 \right| \leq \lim_{i \to \infty} \left| q_k(\underline{\alpha}_k^{(*)}) - 1 \right| \tau_k^{(i)}/\hat{\tau} = 0 \quad .$$

By continuity of $q_k(\alpha)$ this implies $q_k(\alpha_k^{(*)}) = 1$ as asserted for the case $\alpha_k^{(*)} \neq 0$. Now we come to the third and most difficult part of the proof.

Suppose

$$\underline{\alpha}_k^{(*)} = \lim_{odd\ j\to\infty} \alpha_k^{(i_j)} \neq \overline{\alpha}_k^{(*)} = \lim_{even\ j\to\infty} \alpha_k^{(i_j)}$$

for some infinite index sequence $\mathcal{J} \equiv \langle i_j \rangle_{j=1}^\infty$. More specifically, we may define $\mathcal{J}$ such that it consists exactly of all *jump indeces* $i = i_j$ for which $(\alpha_k^{(i)} - \underline{\alpha}_k^{(*)})$ and $(\alpha_k^{(i-1)} - \overline{\alpha}_k^{(*)})$ have opposite sign. After deleting the first jump if necessary, we may write equivalently

$$\alpha_k^{(i_j)} = \underline{\alpha}_k^{(i_j+1)} \quad and \quad \alpha_k^{(i_j-1)} = \overline{\alpha}_k^{(i_j)} \quad if \quad j \quad odd \quad , \tag{4.19}$$

and

$$\alpha_k^{(i_j)} = \overline{\alpha}_k^{(i_j+1)} \quad and \quad \alpha_k^{(i_j-1)} = \underline{\alpha}_k^{(i_j)} \quad if \quad j \quad even \quad . \tag{4.20}$$

In other words the trial sequence $< \alpha_k^{(i)} >$ jumps to the left when $i = i_j$ with $j$ odd and to the right when $i = i_j$ with $j$ even. For later use we make the crucial observation that

$$\overline{\alpha}_k^{(i_j+1)} = \overline{\alpha}_k^{(i_j)} \quad if \quad j \quad odd \quad , \tag{4.21}$$

which simply means that the value of the right bound at the time of a jump to the left will remain unaltered until the next jump to the right. Clearly we have in analogy also

$$\underline{\alpha}_k^{(i_j+1)} = \underline{\alpha}_k^{(i_j)} \quad if \quad j \quad even \quad . \tag{4.22}$$

Because $\underline{\alpha}_k^{(*)}$ and $\overline{\alpha}_k^{(*)}$ can not both be infinite they are in fact both finite. Also one of them, say $\overline{\alpha}_k^{(*)}$, must be nonzero. Then (4.18) applies for all $i = i_j$ with even $j$ so that

$$\lim_{even\ j\to\infty} \tau_k^{(i_j)} = 0. \tag{4.23}$$

Similarly we can show that

$$\limsup_{odd\ j\to\infty} 1/|\tau_k^{(i_j)}| \leq 1. \tag{4.24}$$

When $\underline{\alpha}_k^{(*)}$ is also nonzero this follows exactly as for the even $j$ and the limit superior is in fact 0. The case $\underline{\alpha}_k^{(*)} = 0$ can only occur when $\underline{\alpha}_k^{(i)}\overline{\alpha}_k^{(i)} < 0$ throughout. Hence all $\alpha_k^{(i_j)} = \underline{\alpha}_k^{(i_j+1)}$ with odd $j$ must be negative, which requires by (4.13) that the corresponding $\tau_k^{(i_j)}$ are greater than one. Thus (4.22) is true whether $\underline{\alpha}_k^{(*)}$ is zero or not. Substituting (4.21) into (4.15) we find that

$$\frac{\left|q_k(\underline{\alpha}_k^{(i_j)}) - 1\right|}{\left|q_k(\overline{\alpha}_k^{(i_j+1)}) - 1\right|} \leq \frac{1/\hat{\tau}}{|\tau_k^{(i_j)}|} \quad if \quad j \quad odd \quad .$$

Similarly we have (4.15) and (4.22)

$$\frac{\left|q_k(\overline{\alpha}_k^{(i_j)}) - 1\right|}{\left|q_k(\underline{\alpha}_k^{(i_j+1)}) - 1\right|} \leq \frac{|\tau_k^{(i_j)}|}{\hat{\tau}} \quad if \quad j \quad even \quad .$$

Rewriting the last inequality in terms of an even j+1 and then multiplying it by the preceding one we can cancel the lower bounds to obtain finally

$$\frac{\left|q_k(\overline{\alpha}_k^{(i_j-1)}) - 1\right|}{\left|q_k(\overline{\alpha}_k^{(i_j+1)}) - 1\right|} \leq \frac{|\tau_k^{(i_j-1)}|}{|\tau_k^{(i_j)}|\hat{\tau}^2} \quad if \quad j \quad odd \quad .$$

Because of (4.23) and (4.24) this implies

$$\lim_{odd\ j\ \to\infty} \frac{\left| q_k(\overline{\alpha}_k^{(i_j-1)}) - 1 \right|}{\left| q_k(\overline{\alpha}_k^{(i_j+1)}) - 1 \right|} = 0$$

and hence

$$\lim_{even\ j\ \to\infty} \left| q_k(\overline{\alpha}_k^{(i_j)}) \right| = \infty$$

which contradicts the continuity of $q_k(\alpha)$ at $\overline{\alpha}_k^{(*)} \neq 0$. Thus we have lead the assumption that there are infinitely many jump indices to a contradiction, which completes the proof. ∎

In contrast to bracketing schemes based on nested bisection, the conditions (4.15) and (4.16) do not eliminate the possibility of Q-superlinear convergence. In fact for twice differentiable $g$, it can be shown that the secant step based on the last two iterates, will eventually always be acceptable, provided the slope of $h_k(\alpha)$ is nonzero at the root $\alpha_k^{(*)}$. Hence we have Q-superlinear convergence of order $\frac{1}{2}(1 + \sqrt{5}) \approx 1.62$. The Newton step would eventually also be acceptable but this is for us only of theoretical interest since derivatives are not available.

By a modification of the kind suggested by Bus and Dekker [7] one can ensure that the bracketing intervals $(\underline{\alpha}_k^{(i)}, \overline{\alpha}_k^{(i)})$ always converge to a singleton $\{\alpha_k^*\}$. However, here the exact value of the step multiplier $\alpha_k$ does not really matter as long as $q_k(\alpha_k)$ lies between $1 - \varepsilon$ and its reciprocal $1/(1 - \varepsilon)$. For the same reason the asymptotic rate of convergence may seem not very important. On the other hand we can expect that the number of function evaluations during a line-search will be relatively insensitive to the choice of $\varepsilon \in (0, \frac{1}{2})$ if the line-search algorithm is asymptotically fast. This was found to be true in our numerical experiments as reported in the next section.

Even for very smooth $g(x)$ the function $q_k(\alpha)$ may be quite volatile, especially if the search direction $s_k$ is nearly orthogonal to the gradient $\nabla\gamma_k$. Then one may have to perform five or more evaluations during a single line-search. This looks like a lot in the context of unconstrained optimization but seems acceptable for solving a general unvariate equation with some degree of accuracy. A line-search procedure based exclusively on the secant method itself was found to be particularly inefficient on the Rosenbrock function. On the first few iterations from the standard starting point it took five function evaluations to find an acceptable step multiplier.

Since the Rosenbrock vector function is quadratic, its values at $x_k, x_k + s_k$ and a third point $x_k + \alpha_k^{(1)}s_k$ provide sufficient information to find the root $\alpha_k^{(2)}$ of $q_k(\alpha) - 1$ exactly. For nonquadratic $g$ we may still use the $\alpha_k^{(2)}$ defined by the interpolating parabola as the next best value provided it satisfies the inequalities (3.35) and (3.36). If one of these conditions is violated one could enforce them by making a minimal adjustment to $\alpha_k^{(2)}$. We prefer instead to abandon the quadratic interpolation and define $\alpha_k^{(2)}$ by (3.33) with $\tau_k^{(2)}$ the closest element to $[q_k(\overline{\alpha}_k^{(i)}) - 1]/[q_k(\underline{\alpha}_k^{(i)}) - 1]$ in the union $[-\infty, 1 - \tilde{\tau}] \cup [1/(1 - \tilde{\tau}), \infty]$. This process is repeated until $1 - \varepsilon \leq q_k(\alpha_k) \leq 1/(1 - \varepsilon)$ for some $\alpha_k = \alpha_k^{(i)}$ or $|\alpha_k^{(i)}|$ have become "very small" in which case we set $\alpha_k = 0$.

Since the only effect of a trivial iteration $x_{k+1} = x_k$ is the update of $B_k$, the "very small" test on $\alpha_k^{(i)}$ should ensure that the resulting $y_k \equiv [g_k(x_k + \alpha_k^{(i)}s_k) - g(x_k)]/\alpha_k^{(i)} \approx G(x_k)s_k$ is reasonably accurate. During each parabolic interpolation one obtains an approximation $c_k^{(i)}$ to the second derivative of $g_k(x_k + \alpha s_k)$ with respect to $\alpha$. Since this m-vector represents a second order difference it is obtained with an error of size $\Delta g/[\alpha_k^{(i)}]^2$, when $\Delta g$ is the absolute error in evaluating *g*. Hence we may estimate the total error between $y_k$ and $G_k s_k$ by the product $|\alpha_k^{(i)}| \| c_k^{(i)} \|$ which will eventually start growing due to roundoff once $\alpha_k^{(i)}$ is nearly zero. Therefore, we consider the current trial multiplier $\alpha_k^{(i)}$ as very small if

$$|\alpha_k^{(i+1)}| \, \| c_k^{(i+1)} \| \geq |\alpha_k^{(i)}| \, \| c_k^{(i)} \| \geq |\alpha_k^{(i-1)}| \, \| c_k^{(i-1)} \| \quad .$$

In other words we terminate the line-search with $\alpha_k = 0$ if $q(\alpha_k^{(i)})$ was always less then $1 - \varepsilon$ and the error estimate for $y_k$ increased twice in a row. In the numerical experiments reported in the next section this condition was only met once before an $\alpha_k^{(i)}$ with acceptable $q_k(\alpha_k^{(i)})$ had been found. Even for very bad $B_k$ the resulting search-directions $s_k$ are quite unlikely to be nearly orthogonal to the gradient $g_k^T G_k$.

Closely related to the question of line-search termination is that of a stopping criterion for the Gauss-Broyden iteration itself. We will not discuss this difficult problem here and simply let the calculation run on for a sufficiently large number of iterations.

## 5 NUMERICAL EXPERIMENTS

All calculations were performed on a subset of the test functions collected by Moré et al [18]. The problems number 2, 3, 5, 7, 14, and 15 were excluded because even the Gauss-Newton version of our code failed to reach an acceptable solution. This was largely due either to very large residuals or singularity of the Jacobian at a least squares solution. Clearly, a step selection strategy based solely on the local linear models $g(x) \approx g_k + B_k(x - x_k)$ is unsuitable for the numerical treatment of such problems. On the helical value problem number 4 our Gauss-Broyden code got stuck during a line-search that crossed the coordinate axis where the function is by definition discontinuous.

In general $B_0$ was always initialized as the exact Jacobian $G(x_0)$ at the standard starting point $x_0$. Since this would lead to one step convergence on linear functions we used for problem 1 with m = 10 and n = 5 the initialization

$$B_0^T = \begin{bmatrix} 0 & 0 & \beta & 0 & 0 & 1 & 0 & 0 & 0 & 0 \\ -1 & 0 & 0 & 0 & 0 & 0 & 1 & 0 & 0 & 0 \\ 0 & 0 & 0 & 0 & 0 & 0 & 0 & 1 & 0 & 0 \\ 0 & 2 & 0 & 0 & 0 & 0 & 0 & 0 & 1 & 0 \\ 0 & 0 & 0 & 0 & 0 & 0 & -2 & 0 & 0 & 1 \end{bmatrix} \quad , \tag{5.1}$$

where the entry $\beta$ was set to 2, 20 and 200. Due to the linearity of the problem we can always perform an exact line-search at the cost of two function evaluations per iteration. Apart from the obvious quantities $\| g_k \|$ and $\alpha_k$ we also list the relative errors $\| y_k - B_k s_k \| / \| y_k \|$ and $\| B_k - G \|_F / \| G \|_F$ as well as the quantity *gap* $(G, B_k)$ which represents the gap [14] between the ranges of $G$ and $B_k$. The latter is calculated as $\sqrt{1 - \| P_k^T P \|_F^2}$ where $P$ and $P_k$ denote the orthogonal projection onto the ranges of $G$ and $B_k$ respectively. In contrast to the sequence of Frobenius norms $\| B_k - G \|_F$, which must be nonincreasing in the linear case, the gaps may go down and up. This can be observed on the 17-th iteration in Table 1. Since the correctness of the range appears so important for quasi-Gauss-Newton methods we tried two alternative updates that ensure at all steps a nonincrease of the gap in the linear case. Since neither modification yielded consistently better results we will not discuss them here.

| $k$ | $\|g_k\|$ | $\alpha_k$ | $\|y_k - B_k s_k\|/\|y_k\|$ | $\|B_k - G\|/\|G\|$ | $gap(B_k, G)$ |
|---|---|---|---|---|---|
| 1 | 5.00000000 | -1.2672 | 1.9106 | 2.2539 | 1.00000 |
| 2 | 4.03717094 | 0.0663 | 1.7359 | 2.0856 | 0.99730 |
| 3 | 4.03437164 | -0.3190 | 1.4131 | 1.9358 | 0.99915 |
| 4 | 3.79846789 | 0.4553 | 1.0500 | 1.8297 | 0.99567 |
| 5 | 3.39140183 | 0.1599 | 1.0155 | 1.7684 | 0.99992 |
| 6 | 3.32280158 | 0.0311 | 1.0706 | 1.7091 | 0.99941 |
| 7 | 3.32172759 | -0.1305 | 1.0148 | 1.6407 | 0.99985 |
| 8 | 3.31769304 | 0.7063 | 0.9642 | 1.5767 | 0.99973 |
| 9 | 3.28166117 | -1.1017 | 1.0343 | 1.5166 | 0.99990 |
| 10 | 3.22629317 | 0.4043 | 1.1445 | 1.4443 | 0.99987 |
| 11 | 3.22176847 | -1.1968 | 0.9638 | 1.3506 | 0.99863 |
| 12 | 3.20483746 | 1.9683 | 0.8137 | 1.2799 | 0.96363 |
| 13 | 3.14083984 | 6.8896 | 1.4058 | 1.2271 | 0.96711 |
| 14 | 2.86930847 | 5.8418 | 1.6220 | 1.0538 | 0.94171 |
| 15 | 2.60249754 | 1.2873 | 0.8345 | 0.7645 | 0.56483 |
| 16 | 2.40832998 | 1.8394 | 0.8216 | 0.6672 | 0.50654 |
| 17 | 2.27571879 | 2.1282 | 0.7740 | 0.5569 | 0.50629 |
| 18 | 2.24189568 | -2.5187 | 0.6316 | 0.4362 | 0.52893 |
| 19 | 2.24029299 | -5.7128 | 0.6411 | 0.3324 | 0.43266 |
| 20 | 2.23904686 | 2.8910 | 0.3400 | 0.1682 | 0.28346 |
| 21 | 2.23859101 | 19.9545 | 0.1199 | 0.0719 | 0.07403 |
| 22 | 2.23744899 | -25.6541 | 0.1012 | 0.0480 | 0.04983 |
| 23 | 2.23618698 | -16.0819 | 0.0333 | 0.0160 | 0.02785 |
| 24 | 2.23607172 | -56.9246 | 0.0090 | 0.0059 | 0.01060 |
| 25 | 2.23606928 | -199.7778 | 0.0090 | 0.0043 | 0.00822 |
| 26 | 2.23606814 | -21.7877 | 0.0031 | 0.0016 | 0.00246 |
| 27 | 2.23606814 | -9704.3572 | 0.0019 | 0.0008 | 0.00113 |
| 28 | 2.23606798 | 451.0730 | 0.0001 | 0.0000 | 0.00007 |
| 29 | 2.23606798 | -15.6860 | 0.0000 | 0.0000 | 0.00006 |

Table 1: Convergence of Gauss-Broyden to solution of linear problem with reasonable $B_0$, i.e. $\beta = 2$.

The results listed in Table 1 where obtained on the linear problem number 1 with $B_0$ given by (4.1) for $\beta = 2$. Then the relative error between $B_0$ and $G$ is greater than 1 and their ranges are very different as the initial gap has the largest possible value, namely 1. This means that some elements in the range of $B_0$ are exactly orthogonal to the range of $G$ and vice versa. Consequently the progress over the first fifteen items appears rather slow as the two relative errors and the range gap hover about 1. After that the iteration gradually starts rolling and in the end all three discrepancy measures tend to zero. The minimal residual achieved agrees with the one given in [18]. Despite the good agreement between $y_k$ and $B_k s_k$ during later iterations, the step multipliers $\alpha_k$ show no signs of converging to one, but rather oscillate between large positive and negative values. This observation supports our strategy of allowing for negative or arbitrarily large step multipliers.

During all our calculations the $\hat{\tau}$ and $\tilde{\tau}$ were both set to .01. On the run reported in Table 1 these constraints had no effect as the multiplier $\alpha_k^{(1)} = q_k(1)$ was always feasible with respect to (3.35) and (3.36). The first choice $\alpha_k^{(0)} = 1$ was never acceptable with respect to (3.3) and (3.8) for any tolerance $\varepsilon \in (0, \frac{1}{2})$. After 25 steps the numerical residual is attained with seven digits accuracy. Round-off effects become apparent on the 29th step, where for the first time the computed value of $q_k(\alpha(1))$ deviates from its theoretical value 1 by more than $\varepsilon = .1$.

Tables 2 and 3 list selected iterates from the same calculations with $B_0$ defined by (4.1) for $\beta = 20$ and $\beta = 200$ respectively. In these cases $B_0$ is not even roughly the same size as the exact Jacobian and it takes a long time until the $B_k$ come anywhere close to it. In the second case $\beta = 20$ the relative errors drop below one at about the 45nd iteration and final solution accuracy is attained ten steps later. In the third case $\beta = 200$ this effect occurs much later, namely the relative error drop below 1 at about the 382nd iteration. Again final solution accuracy is attained only ten steps later. Actually this behavior has nothing to do with the fact that we are considering a rectangular linear system.

On the linear problem with $m = 5 = n$ and $B_0^T$ reduced to the first 5 columns of (4.1), Broyden's method with perfect line search uses 224 iterations before reaching the solution seemingly exactly. This final step appears somewhat flukey and follows immediately after the first iteration at which the step size is exactly equal to 1. Throughout the whole calculations $\|B_k - G\|/\|G\|$ is only reduced to about 80. Clearly, the amazing success of the final step is related to the finite termination result of Burmeister [6] and Gay [10] which is based on full steps. However, the Gauss-Broyden method with $\alpha_k = 1$ at each step showed no signs of convergence on the linear problem with $m = 10, k = 5$ and $\beta = 200$. Hence we may conclude that on rectangular systems the algebraic properties of the Broyden update don't apply and we have to rely on its ability to eventually produce a close approximation to the actual Jacobian at the solution. To this end it appears necessary that $B_k$ be always at least a rough approximation to the exact Jacobians. When it seems that this property is lost, the resetting of $B_k$ to a divided difference approximation of $G_k$ might be advantageous. This approach is customarily for square systems, but has not been implemented in our code.

While the bad initializations on the linear functions may appear somewhat artificial similar effects may be observed on Problem 16, the Meyer function. There, the first column of the Jacobian is more than 1000 times larger than the others. Due to the large sensitivity of the residual with respect to the first variables the first components of the steps are comparatively small and the same is true for the corrections to the first column of the $B_k$. Hence the poor relative scaling of the variables severely limits the ability of the Broyden update to produce good Jacobian approximations. This can be seen from Table 4 where $e_k$ denotes the number of evaluations needed to reach the k-th iterate.

| $k$ | $\|g_k\|$ | $\alpha_k$ | $\|y_k - B_k s_k\|/\|y_k\|$ | $\|B_k - G\|/\|G\|$ | $gap(B_k, G)$ |
|---|---|---|---|---|---|
| 1 | 5.00000000 | -1.1243 | 1.7973 | 9.2585 | 1.00000 |
| 6 | 4.27451425 | -0.0313 | 1.4009 | 9.1366 | 0.98676 |
| 11 | 4.20264215 | 0.2181 | 1.4101 | 9.0012 | 0.98140 |
| 16 | 3.46229190 | 0.0922 | 2.1406 | 8.8648 | 0.98031 |
| 21 | 3.21066025 | -3.6776 | 4.3880 | 6.4065 | 0.99993 |
| 26 | 2.87148196 | -4.8225 | 5.2548 | 5.1898 | 0.87887 |
| 31 | 2.69641772 | 0.2124 | 1.1687 | 4.3128 | 0.62717 |
| 36 | 2.61409487 | 5.7640 | 3.4260 | 3.9309 | 0.40862 |
| 41 | 2.34684028 | -1.5404 | 1.6915 | 1.9908 | 0.42160 |
| 42 | 2.33209192 | -8.0293 | 2.9567 | 1.8415 | 0.37099 |
| 43 | 2.28346748 | -1.3822 | 1.6386 | 1.2817 | 0.37287 |
| 44 | 2.26604912 | -3.3455 | 2.3235 | 1.0516 | 0.30923 |
| 45 | 2.23839199 | -0.3729 | 0.2759 | 0.1613 | 0.19222 |
| 46 | 2.23712556 | -0.5830 | 0.0868 | 0.1038 | 0.17662 |
| 47 | 2.23697339 | -18.5478 | 0.0979 | 0.0963 | 0.16984 |
| 48 | 2.23648012 | -13.6616 | 0.1885 | 0.0858 | 0.15633 |
| 49 | 2.23612943 | -16.2366 | 0.0335 | 0.0159 | 0.02168 |
| 50 | 2.23607131 | 52.9474 | 0.0097 | 0.0053 | 0.00742 |
| 51 | 2.23606899 | -37.2689 | 0.0048 | 0.0030 | 0.00386 |
| 52 | 2.23606899 | -753.8577 | 0.0040 | 0.0020 | 0.00310 |
| 53 | 2.23606897 | 4114.0684 | 0.0023 | 0.0010 | 0.00133 |
| 54 | 2.23606798 | -80.5213 | 0.0001 | 0.0001 | 0.00009 |

Table 2: Convergence of Gauss-Broyden to solution of linear problem with bad $B_0$, i.e. $\beta = 20$.

| $k$ | $\|g_k\|$ | $\alpha_k$ | $\|y_k - B_k s_k\|/\|y_k\|$ | $\|B_k - G\|/\|G\|$ | $gap(B_k, G)$ |
|-----|-----------|------------|------------------------------|----------------------|----------------|
| 1 | 5.00000000 | -1.1042 | 1.7909 | 89.5551 | 1.00000 |
| 25 | 3.63904563 | -0.0248 | 1.3167 | 89.3747 | 0.94584 |
| 51 | 3.11536263 | -0.0029 | 1.1195 | 88.6210 | 0.99976 |
| 76 | 3.11466771 | -0.0044 | 1.1299 | 88.5854 | 0.96328 |
| 101 | 3.11400225 | -0.0086 | 1.1687 | 88.5484 | 0.86923 |
| 126 | 3.11354099 | -0.0144 | 1.2927 | 88.5070 | 0.77189 |
| 151 | 3.06934090 | 0.3060 | 1.4093 | 88.2872 | 0.66196 |
| 176 | 3.06826057 | 0.0552 | 1.6416 | 88.2222 | 0.58337 |
| 201 | 3.06635311 | 0.0913 | 2.4045 | 88.1170 | 0.52008 |
| 226 | 3.00206155 | 0.6699 | 4.1812 | 86.0688 | 0.51498 |
| 251 | 2.91677840 | 0.6182 | 5.0995 | 80.7244 | 0.46483 |
| 276 | 2.52430562 | -0.1082 | 2.2109 | 50.2837 | 0.54887 |
| 301 | 2.51899579 | -0.3669 | 2.3734 | 49.0077 | 0.50666 |
| 326 | 2.46703249 | -0.3053 | 2.3874 | 43.8077 | 0.48804 |
| 351 | 2.25121233 | -0.0047 | 1.1749 | 11.2685 | 0.81469 |
| 376 | 2.23811449 | -0.1738 | 3.6601 | 4.0857 | 0.41907 |
| 377 | 2.23779360 | -0.5048 | 6.0367 | 3.7435 | 0.40580 |
| 378 | 2.23691083 | -0.0303 | 1.7050 | 2.5934 | 0.45707 |
| 379 | 2.23684152 | -0.0949 | 2.6540 | 2.4788 | 0.40412 |
| 380 | 2.23667225 | -0.2852 | 4.3091 | 2.1761 | 0.37832 |
| 381 | 2.23622397 | -0.0156 | 1.1636 | 1.0109 | 0.42562 |
| 382 | 2.23619240 | -0.0516 | 1.5742 | 0.8667 | 0.31842 |
| 383 | 2.23613390 | -0.0319 | 1.0089 | 0.5054 | 0.25986 |
| 384 | 2.23610945 | -0.0182 | 0.4932 | 0.2278 | 0.16244 |
| 385 | 2.23610428 | -0.0160 | 0.1232 | 0.0568 | 0.05158 |
| 386 | 2.23610366 | 0.0804 | 0.0105 | 0.0137 | 0.02292 |
| 387 | 2.23610354 | -93.5910 | 0.0179 | 0.0129 | 0.02287 |
| 388 | 2.23608431 | 32.4474 | 0.0180 | 0.0101 | 0.01656 |
| 389 | 2.23608122 | 502.1454 | 0.0128 | 0.0060 | 0.01102 |
| 390 | 2.23606963 | 659.7123 | 0.0029 | 0.0019 | 0.00264 |
| 391 | 2.23606843 | 372.1550 | 0.0033 | 0.0015 | 0.00231 |
| 392 | 2.23606798 | -141.2999 | 0.0002 | 0.0001 | 0.00013 |

Table 3: Convergence of Gauss-Broyden to solution of linear problem with very bad $B_0$, i.e. $\beta = 200$.

| $k$ | 0 | 30 | 60 | 90 | 120 | 150 | 180 |
|-----|---|----|----|----|-----|-----|-----|
| $\|g_k\|$ | 41153.5 | 251.623 | 239.648 | 226.843 | 205.016 | 9.37994 | 9.377998 |
| $\alpha_k$ | .4360 | -.0003 | -.0004 | -.0008 | -.0032 | 0.035 | .0385 |
| $e_k$ | 0 | 186 | 282 | 378 | 470 | 559 | 653 |

Table 4: Slow convergence of Gauss-Broyden on the Meyer function

| $k$ | 0 | 1 | 2 | 3 | 4 | 5 | 6 | 7 | 8 |
|---|---|---|---|---|---|---|---|---|---|
| $\|g_k\|$ | 41153.4 | 41136.2 | 41132.4 | 41132.3 | 41131.9 | 501.81 | 9.4021 | 9.3780 | 9.377945 |
| $\alpha_k$ | .43607 | 1.0237 | .11643 | .30272 | 1 | 1 | 1 | 1 | 1 |
| $e_k$ | 0 | 14 | 26 | 43 | 58 | 62 | 66 | 70 | 74 |

Table 5: Fast Convergence of Gauss-Newton on the Meyer function.

After big initial reductions the method almost stalls between the 30th and 120th iteration. The next sixty iterations are again fairly productive and the minimal residual is attained with four digits accuracy. Since the exact Jacobian has a condition number of about $10^9$ this solution accuracy is not too bad. However, due to lengthy line-searches during the first 10 steps and the large number of overall iterations the computational expense is quite large, though comparable to the performance of the MINPACK routine LMDER1 as reported in [19]. The Gauss-Newton method with our line-search appears to be lucky since it converges in eight iterations at the expense of 74 function evaluations as listed in Table 5. Here every Jacobian evaluation is counted as $n = 3$ extra function evaluations. These results were obtained with the line-search tolerance $\varepsilon = .1$, but the less restrictive choice $\varepsilon = .49$ was found to save only a single function evaluation on this problem. In case of the Gauss-Broyden method the Jacobian errors $\|B_k - G_k\|/\|G_k\|$ were almost constant at .91 and the ratios $\|y_k - B_k s_k\|/\|y_k\|$ appeared to go through a five cycle without ever coming close to zero.

At first things look a lot better on the Kowalik and Osborne function 9 as reported in Table 6. Here the matrix error $\|B_k - G_k\|/\|G_k\|$ never exceeds a half and the range gap is always less than .4. The iteration proceeds quite well until it reaches the minimal residual with six figures accuracy on the 20th iteration. Until then the modules of all step-multipliers $\alpha_k$ is greater than a fifth. Subsequently, the step multipliers are cut down to a few thousands and there is clear evidence of a zig zagging motion. This can be seen in Table 7 which lists the steps $s_k \in \mathcal{R}^4$ in tenthousands for the iteration 20 to 29.

During these 10 iterations the relative errors $\|y_k - B_k s_k\|/\|y_k\|$ and $\|B_k - G_k\|/\|G_k\|$ as well as the range gap between $B_k$ and $G_k$ remain almost constant at values of .1, 0.2, and .009 respectively. So far we have not been able to explain the zig-zagging behavior. In an attempt to overcome this undesirable effect the line-search along $s_k$ was replaced by a *plane-search* over the affine variety $\{x_k + \alpha s_k + \tilde{\alpha} s_{k-1} : \alpha, \tilde{\alpha} \in \mathcal{R}\}$. While the plane-search lead to some improvement it did not eliminate the zig-zagging completely and made little difference on other problems.

In view of the results discussed above it seems clear that the Gauss-Broyden method cannot be expected to achieve a high solution accuracy, unless the approximating Jacobian is reinitialized periodically. However, especially when the number of variables is large, one might hope that a residual within a few percent of the optimal solution can sometimes be reached much cheaper than with methods that require the evaluation of the Jacobian at each iteration. We have tested this hypothesis on problem 18, the second function of Osborne. There $n = 11$ so that in terms of function evaluations every Gauss-Newton step is $13/2 = 6\frac{1}{2}$ times as expensive as a Gauss-Broyden step, provided the line-search involves two evaluations. As one can see in Figure 2, Gauss-Broyden is indeed initially ahead, but it stalls at a residual accuracy of 4 digits, which is surpassed by the Gauss-Newton scheme at the sixth iteration. In many nonlinear least squares problems the modeling

| $k$ | $\|g_k\|$ | $\alpha_k$ | $\|y_k - B_k s_k\|/\|s_k\|$ | $\|B_k - G_k\|/\|G_k\|$ | $gap(B_k, G_k)$ |
|---|---|---|---|---|---|
| 1 | 0.72891510d-01 | 0.3404 | 0.4621 | 0.0000 | 0.00000 |
| 2 | 0.44983118d-01 | 0.3462 | 0.6837 | 0.2288 | 0.14856 |
| 3 | 0.34658151d-01 | 1.0000 | 0.5306 | 0.4728 | 0.36324 |
| 4 | 0.25074788d-01 | 0.3708 | 0.7159 | 0.3430 | 0.31188 |
| 5 | 0.17997362d-01 | 1.0000 | 0.4336 | 0.2568 | 0.23578 |
| 6 | 0.17791993d-01 | 1.0000 | 0.5060 | 0.2226 | 0.08135 |
| 7 | 0.17696710d-01 | 1.0000 | 0.1858 | 0.2107 | 0.04016 |
| 8 | 0.17597603d-01 | 1.0000 | 0.2564 | 0.2222 | 0.04907 |
| 9 | 0.17548312d-01 | 2.2329 | 0.5171 | 0.2174 | 0.03634 |
| 10 | 0.17538912d-01 | 0.0476 | 0.9594 | 0.2105 | 0.04626 |
| 11 | 0.17538646d-01 | 0.4330 | 0.2239 | 0.2084 | 0.01235 |
| 12 | 0.17536220d-01 | 0.3063 | 0.4573 | 0.2095 | 0.00771 |
| 13 | 0.17536150d-01 | -0.6382 | 0.3932 | 0.2054 | 0.02001 |
| 14 | 0.17536140d-01 | -0.3649 | 0.1695 | 0.2051 | 0.02101 |
| 15 | 0.17536047d-01 | 0.1963 | 0.2472 | 0.2050 | 0.02289 |
| 16 | 0.17535892d-01 | 0.1499 | 0.0956 | 0.2051 | 0.00885 |
| 17 | 0.17535882d-01 | -0.0323 | 0.1879 | 0.2051 | 0.01797 |
| 18 | 0.17535881d-01 | 0.0720 | 0.0870 | 0.2051 | 0.00862 |
| 19 | 0.17535880d-01 | 0.0252 | 0.1556 | 0.2051 | 0.01427 |
| 20 | 0.17535879d-01 | 0.0264 | 0.0858 | 0.2051 | 0.00828 |
| 21 | 0.17535879d-01 | 0.0037 | 0.1430 | 0.2051 | 0.01304 |
| 22 | 0.17535879d-01 | 0.0069 | 0.0857 | 0.2051 | 0.00826 |
| 23 | 0.17535879d-01 | 0.0038 | 0.1346 | 0.2051 | 0.01227 |
| 24 | 0.17535879d-01 | 0.0034 | 0.0858 | 0.2050 | 0.00825 |
| 25 | 0.17535879d-01 | 0.0024 | 0.1284 | 0.2050 | 0.01172 |

Table 6: Gauss-Broyden stalls on Kowalik and Osborne with minimal residual 0.01753584.

| $k$ | $10^3 s_k^{(1)}$ | $10^3 s_k^{(2)}$ | $10^3 s_k^{(3)}$ | $10^3 s_k^{(4)}$ | $\alpha_k$ |
|---|---|---|---|---|---|
| 20 | -0.0769 | 1.3201 | 0.2998 | 0.4818 | 0.0264 |
| 21 | 0.1273 | -3.5416 | -0.7561 | -1.7379 | 0.0037 |
| 22 | -0.0785 | 1.3471 | 0.2883 | 0.5011 | 0.0069 |
| 23 | 0.1146 | -3.2525 | -0.7148 | -1.5991 | 0.0038 |
| 24 | -0.0803 | 1.3858 | 0.2907 | 0.5217 | 0.0034 |
| 25 | 0.1054 | -3.0367 | -0.6801 | -1.4960 | 0.0024 |
| 26 | -0.0816 | 1.4213 | 0.2966 | 0.5397 | 0.0015 |
| 27 | 0.0985 | -2.8715 | -0.6517 | -1.4174 | 0.0018 |
| 28 | -0.0826 | 1.4523 | 0.3034 | 0.5551 | 0.0008 |
| 29 | 0.0931 | -2.7382 | -0.6275 | -1.3540 | 0.0019 |
| 30 | -0.0835 | 1.4799 | 0.3103 | 0.5687 | 0.0005 |

Table 7: Zig-zagging motion of Gauss-Broyden near minimizer of Kowalik and Osborne function.

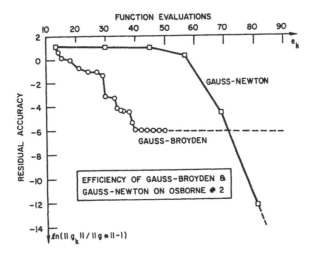

Figure 2

errors are so large that an iterate that lies only a few percent above the mathematical minimum would be entirely acceptable. Naturally the realization of such economies requires the development of a suitable stopping criterion.

Finally let us check the validity of Corollary 1 by considering the zero residual problem 12, the so called box-function. As shown in Table 8 the method converges clearly superlinearly with most step-multipliers $\alpha_k$ being equal to one. Here we used $\varepsilon = .4$ and the more restrictive choice $\varepsilon = .1$ was found to delay convergence considerably.

In Table 9 the overall performance of our Gauss-Broyden scheme is compared to that of the Gauss-Newton method with the same line-search and the MINPACK routine LMDER1. While the latter two methods reached the residual values given in [19], the Gauss-Broyden method achieved only on the Bard Function and the zero residual problems similar accuracies. On all other problems we have listed the ratio between the achieved and minimal residual in the last column of the table.

We conclude from the numerical results that nonlinear least-squares problems can be solved to moderate accuracy by a pure quasi-Gauss-Newton method, provided the solutions are nonsingular and have a smallish residual value. None of the test functions considered here involve a large number of variables and all have analytic Jacobians that are cheaply evaluated and factorized. Therefore, the Gauss-Broyden method always used more computing time even when it required fewer function evaluations. Obviously the situation would be radically different on really large problems. When m is much large than a, Householder QR factorization involves nearly $n^2m$ arithmetic operations. By comparison a rank one update can be incorporated into an existing QR-factorization for about 8mn arithmetic operations [8]. Hence the linear algebra overhead of the Gauss-Broyden method can be cheaper than that of Gauss-Newton when $m >> n >> 8$. However, on smaller problems like those in the test set of Moré et al one might as well refactorize the approximating Jacobian at every step even though that doubles the storage requirements.

| $k$ | $\|g_k\|$ | $\alpha_k$ | $\|y_k - B_k s_k\|/\|y_k\|$ | $\|B_k - G_k\|/\|G_k\|$ | $gap(B_k, G_k)$ |
|---|---|---|---|---|---|
| 1 | 0.32111584d+02 | 1.0000 | 0.0070 | 0.0000 | 0.00000 |
| 2 | 0.22733352d+00 | 2.3233 | 1.7875 | 0.3367 | 0.17102 |
| 3 | 0.52989427d-01 | 1.0000 | 0.1295 | 0.5066 | 0.65150 |
| 4 | 0.75816780d-02 | 1.9973 | 1.0665 | 0.5280 | 0.34620 |
| 5 | 0.14417961d-02 | 0.2298 | 1.0071 | 0.3434 | 0.69773 |
| 6 | 0.13647757d-02 | 1.0000 | 0.4542 | 0.2439 | 0.48954 |
| 7 | 0.94367371d-03 | 1.0000 | 0.4236 | 0.2410 | 0.28074 |
| 8 | 0.28010218d-03 | 1.0000 | 0.1493 | 0.2381 | 0.82475 |
| 9 | 0.48799239d-04 | 1.9676 | 0.9829 | 0.2380 | 0.65717 |
| 10 | 0.32155400d-05 | 0.0640 | 1.0224 | 0.1725 | 0.28887 |
| 11 | 0.31781913d-05 | 0.4311 | 0.9414 | 0.1640 | 0.12641 |
| 12 | 0.27543841d-05 | 1.0000 | 0.5456 | 0.1633 | 0.04869 |
| 13 | 0.10920408d-05 | 1.0000 | 0.3482 | 0.1632 | 0.03879 |
| 14 | 0.28141557d-06 | 1.0000 | 0.0136 | 0.1620 | 0.21894 |
| 15 | 0.37602296d-08 | 1.0000 | 0.0489 | 0.1620 | 0.22932 |
| 16 | 0.17426263d-09 | 1.0000 | 0.0032 | 0.1619 | 0.27395 |
| 17 | 0.54386588d-12 | 1.0000 | 0.0134 | 0.1619 | 0.27698 |

Table 8: Superlinear convergence of Gauss-Broyden on box function with zero residual.

| Problem | $m \times n$ | LM | GN | GB | Accuracy |
|---|---|---|---|---|---|
| Rosenbrock | $2 \times 2$ | 15/54 | 8/37 | 19/54 | $.2 * 10^{-20}$ |
| Powell sing. | $4 \times 4$ | 58/298 | 56/281 | 63/88 | $0.5 * 10^{-31}$ |
| Bard | $15 \times 3$ | 4/21 | 4/17 | 16/27 | $*1.0$ |
| Kowalik & Osb. | $11 \times 4$ | 15/82 | 12/64 | 19/51 | $*(1 + 10^{-5.6})$ |
| Meyer | $16 \times 3$ | 115/471 | 8/76 | 180/653 | $*(1 + 10^{-5.2})$ |
| Watson | $31 \times 6$ | 6/44 | 6/43 | 255/613 | $*(1 + 10^{-2.6})$ |
| Watson | $31 \times 9$ | 6/62 | 4/41 | 235/574 | $*(1 + 10^{-1.4})$ |
| Watson | $31 \times 12$ | 8/106 | 4/53 | 343/956 | $*(1 + 10^{-.57})$ |
| Box | $10 \times 3$ | 5/25 | 6/25 | 17/27 | $0.7 * 10^{-14}$ |
| Brown | $10 \times 10$ | 11/134 | 11/182 | 13/62 | $.33 * 10^{-15}$ |
| Osborne 1 | $33 \times 5$ | 14/93 | 6/41 | 20/40 | $*(1 + 10^{-4.9})$ |
| Osborne 2 | $65 \times 11$ | 11/148 | 12/91 | 16/40 | $*(1 + 10^{-2.5})$ |

Table 9: Comparison of Levenberg-Marquardt, Gauss-Newton, and Gauss-Broyden.

Apart from the plane-search and the two updates that ensure a nonincrease in the range gap for linear functions, various other modifications were tried. In particular we tested various implementations of the so-called "Bad-Broyden" update, which has the distinctive advantage of being invariant with respect to affine transformations on the range of the vector-function $g$. We also tried various diagonal and general weighting of the *good-Broyden* update in order to alleviate the effects of poor scaling on the domain. While some of these modifications lead to spectacular improvements on particular problems there were usually some disadvantages on others and the overall performance did not warrant the added complexity.

# 6 SUMMARY AND DISCUSSION

A generalization of Broyden's method to overdetermined nonlinear systems was motivated and implemented. Of particular concern was the design of an efficient line-search procedure in the absence of any reliable derivative information. The resulting Gauss-Broyden algorithm was found to solve all full-rank and small-residual problems in the test set of Moré et al with moderate accuracy. In agreement with Corollary 1, the rate of convergence was found to be Q-superlinear on zero residual problems.

Future research should concentrate on the development of a reliable stopping criterion. This could be combined with a strategy for resetting the Jacobian by differencing whenever little progress appears to be made. Also provisions must be made to deal with singularities on the way to or at the minimizer. Finally, an analysis of the zig-zagging observed on the Kowalik-Osborne problem might suggest an improvement of the update itself.

# ACKNOWLEDGEMENTS

The helpful comments of the referees lead to the elimination of many mistakes and a strengthening of Proposition 1.

References

[1] M. Al-Baali and R. Fletcher (1986), *An efficient line search for nonlinear least squares*. JOTA, Vol. 26, pp. 359-377.

[2] L. Armijo (1966), *Minimization of functions having Lipschitz continuous first partial derivatives*. Pacific J. of Math., Vol. 16, pp. 1-3.

[3] Y. Bard (1974), *Nonlinear Parameter Identification*. Academic Press, New York.

[4] S. K. Bourji, H. F. Walker (1987), *Least-Change Secant Updates of Nonsquare Matrices Research Report*. Dept. of Mathematics, Utah State University.

[5] C. G. Broyden (1970), *The convergence of a class of double rank minimization algorithms, Parts I and II*. J.I.M.A. Vol. 12, pp. 76-90 and 222-236.

[6] W. Burmeister (1975), *Zur Konvergenz einiger Verfahren der konjugierten Richtungen*. Talk VII, IKM, Weimer.

[7] J. C. P. Bus, T. J. Dekker(1975), *Two efficient algorithms with guaranteed convergence for finding a zero of a function*. ACM TOMS, Vol. 1,pp. 330-345.

[8] J. W. Daniel, W. B. Gragg, L. Kaufman, and G. W. Stewart (1976), *Reorthogonalization and Stable Algorithms for Updating the Gram-Schmidt QR Factorization*. Math. of Comp., Vol. 30, pp. 772-795.

[9] J. F. Dennis, R. B. Schnabel (1983), *Numerical Methods for Unconstrained Optimization and Nonlinear Equations*. Prentice Hall, Englewood Cliffs, N.J.

[10] D. M. Gay (1979), *Some convergence properties of Broyden's method*, SIAM J. Num. Anal. Vol. 16, pp. 623-630.

[11] A. Griewank (1987), *The global convergence of partitioned BFGS on problems with convex decompositions and Lipschitzian gradients*. Technical Memorandum No. 105, Mathematics and Computer Science Division, Argonne National Laboratory.

[12] A. Griewank (1986), *The 'global' convergence of Broyden-like methods with a suitable line search*. J. Austr. Math. Soc. Ser. B, Vol. 28, pp. 75-92.

[13] A. Griewank (1987), *The local convergence of Broyden-like methods on Lipschitzian problems in Hilbert spaces*. SIAM J. Num. Anal., Vol. 24, pp. 684-705.

[14] T. Kato (1976), *Perturbation theory for linear operators,* in Grundlehren der mathematischen Wissenchaften 132, Springer-Verlag, Berlin, Heidelberg, New York.

[15] C. L. Lawson, R. J. Hanson (1974), *Solving Least Squares problems*. Prentice Hall, Inc., Englewood Cliffs, N. J.

[16] P. Lindström and P.A. Wedin (1984), *A new line search algorithm for nonlinear least squares problems*. Mathematical Programming, Vol. 29, pp. 268-296.

[17] F. T. Luk and R. R. Gerber (1981), *A generalized Broyden's method for solving simultaneous linear equations*, SIAM J. Num. Anal., Vol. 18, pp. 882-891.

[18] J. J. Moré, B. S. Garbow, and K. E. Hillstrom (1981), *Testing Unconstrained Optimization Software*. ACM Trans. Math. Software, Vol. 7, pp. 17-41.

[19] J. J. Moré, B. S. Garbow, and K. E. Hillstrom (1981), *User Guide for Minpack-1*. Technical Report, ANL-80-74, Argonne National Laboratory.

[20] J. M. Ortega and W. C. Rheinboldt (1970), *Iterative solution of nonlinear equations in several variables*. Academic Press, New York, N.Y.

[21] M. J. D. Powell (1970), *A hybrid method for non-linear equations in Numerical Methods for Nonlinear Algebraic Equations*. edt. P. Rabinowitz, Gordon, and Breach, London.

[22] G. W. Stewart (1973), *Introduction to Matrix Computations*, Academic Press, New York.

[23] W. Warth and J. Werner (1977), *Effiziente Schrittweitenfunktionen bei unrestringierten Optimierungsaufgaben*. Computing, Vol. 19, pp. 59-72.

[24] P. Wolfe (1971), *Convergence conditions for ascent methods. II: Some corrections*. SIAM Rev. Vol. 13, pp. 185-188.

[25] G. Zoutendijk (1970), *Nonlinear programming, computational methods*. in **Integer and Nonlinear Programming** (J. Abadie edt.), pp. 37-86, North Holland, Amsterdam.

# Parallel Adaptive Algorithms for Multiple Integrals

ALAN GENZ, Computer Science Department, Washington State University, Pullman, WA 99164-1210

## 1. INTRODUCTION

Many large scale computations in the applied sciences require the numerical calculation of multiple integrals. These integrals often come in groups which are related by one or more varying parameters in the integrand, and have a common integration region. The application areas are diverse, but common areas are quantum chemistry and statistics. The number of integrals needed and the difficulty of the integrals will determine the scale of the computation.

Until recently there had not been much published research work in the area of parallel numerical calculation of integrals (the books by Davis and Rabinowitz [3] and Stroud [21] are good general references for numerical integration). The work that has appeared so far has mostly considered the computation of single integrals. A series of papers by Rice [14-17] consider algorithms for single one dimensional integrals on MIMD parallel computers, and a paper by de Doncker and Kapenga [5] describes implementation of some of these algorithms on MIMD computers. A report by Berntsen and Espelid [2] describes the implementation of a one dimensional parallel algorithm on hypercube architecture computers. A paper by Simpson and Yazici [20] discusses the implementation of an extrapolation method for integration over a triangle on a vector computer. The present author has studied [7,8] the implementation of a globally adaptive algorithm for multiple integrals on different types of parallel computers, and Berntsen [1] has extended this work.

The main purpose of this paper is to consider methods for parallelizing adaptive algorithms for multiple integrals and report on how successful the software that has implemented some of the methods has been, in terms of high processor utilization without loss of accuracy. We begin with a general description of adaptive algorithms for multiple integrals. Then, some background information about currently available parallel computers will be given. This is followed by a general discussion about different ways of introducing parallelism in adaptive algorithms. Finally we describe some details about actual implementation of adaptive algorithms on existing parallel machines and summarize some of the results that were obtained.

## 2. ADAPTIVE ALGORITHMS FOR MULTIPLE INTEGRALS

We consider integrals over an n-dimensional hyper-rectangular region $R = [a_1,b_1] \times [a_2,b_2] \times \cdots \times [a_n,b_n]$ in the standard form

$$I(f) = \int_{a_n}^{b_n} \int_{a_{n-1}}^{b_{n-1}} \cdots \int_{a_1}^{b_1} f(x_1,x_2,...,x_n) \, dx_1 \, dx_2 \cdots dx_n.$$

All adaptive algorithms for multiple integrals have as a major component an integration rule in the form

$$Q(f) = \sum_{i=1}^{N} w_i f(\mathbf{x}_i).$$

These rules are usually designed to provide good approximations to I(f) when f(**x**) has a good low degree polynomial approximation. With integration rules of this type N usually grows rapidly with the number of dimensions n.

If the error $E(f) = I(f) - Q(f)$ from such a rule is not small enough, then the approximation can usually be improved by using an appropriately chosen rule $Q'(f)$, with a larger number $N'$, of integrand values. Adaptive algorithms usually begin with a small number N, and then proceed in stages, adding points to the original Q in a way designed to decrease some error estimator as rapidly as possible. Because the convergence of integration rules is influenced by the presence of irregularities (peaks, oscillations, etc.) in the integrand, successful adaptive algorithms usually choose the new integration points in a way that takes account of the local behavior of the integrand in different parts of the integration region. It is in this sense that the algorithms "adapt" to the problem to be solved. Most adaptive algorithms use a sequence of finer subdivisions of the original integration region R, chosen to concentrate the integrand evaluation points in the subregions of R where coarser subdivisions did not yield good results.

In this paper we consider adaptive algorithms where a "fixed" basic rule Q(f) is used for any particular subregion. The rule is fixed in the sense that the points $\{x_i\}$ and weights $\{w_i\}$ have been chosen to be good for certain classes of functions (usually polynomials) for some standard region. When the rule is used in some other integration region the weights are scaled and the points transformed according to a linear transformation that takes the original standard region into the new one. The transformed rule retains the good properties that it had in the original region.

Associated with this basic rule we also assume that there is a basic error estimating rule $E_Q(f)$, with its own weights and points, that provides a reliable estimate for E(f). The points for Q and the error estimator are often the same or nearly the same. The cost of applying Q to f in some subregion consists of the time for transforming Q and $E_Q$, the time for evaluating f at the points for Q and $E_Q$ and the time for computing the weighted sums.

There are many different algorithms that have been proposed for subdivision of R in adaptive integration algorithms. We will restrict ourselves to subdivision algorithms that are defined in the following way. A subdivision of R is defined by $S_i = \{R_1, R_2, ..., R_{M_i}\}$, where $\cup R_j =$ R and $S_0 = \{R\}$. We assume $S_{i+1}$ is obtained from $S_i$ by taking one or more subregions $R_j$ from $S_i$ and dividing each of these subregions into two or more pieces. $S_{i+1}$ consists of the new pieces, together with the undivided subregions from $S_i$. Given a subdivision $S_i$, an integral estimate $I_i$, for I(f), is obtained by applying Q to each of the subregions in $S_i$ and summing the results. An estimate $E_i$, for E(f), is obtained in a similar way, using $E_Q$.

One method for modeling adaptive algorithms using subdivisions of this type associates the subdivision with a weighted digraph. The nodes of the digraph are the allowed subdivisions of R. The edges indicate a step in the adaptive algorithm from one subdivision to a finer subdivision. The weight for such an edge is the sum of the costs for determining the new subdivision, applying Q and $E_Q$ to the new subregions in the finer subdivision and updating the global estimates for the integral and error. The goal of an adaptive integration algorithm is to dynamically follow a low cost path through this graph to a node with an error estimate that is less than some user specified tolerance $\epsilon$.

A second model for adaptive algorithms of this type uses a rooted tree of subregions[19]. The root node of the tree is the initial integration region R. At the beginning of each stage in the adaptive algorithm, there is a leaf node in the tree for each subregion in the current subdivision of R. New leaves in the tree are produced when the adaptive strategy causes a subdivision of one or more of the current subregions. Global estimates for I(f) and E(f) are obtained by summing over the respective local estimates for the integral and error for each of the leaves in the tree.

Two main types of adaptive strategies have been used for serial algorithms: locally adaptive and globally adaptive. Locally adaptive algorithms are characterized by decisions to subdivide particular subregions which do not use results from other subregions. A popular locally adaptive strategy, described in terms of the subregion tree, divides the leftmost leaf subregion $R_j$ that has an error greater than $\epsilon V_j/V$, where $V_j$ is the volume of $R_j$ and V is the volume of R. Termination occurs when no leaf in the tree satisfies this condition. Locally adaptive algorithms have been very popular for one dimensional calculations, where good results are often obtained in a short time. However, for multidimensional integrals, it is often unknown at the beginning of a calculation whether a result accurate to within a given $\epsilon$ can be found in a reasonable amount of time. In this case locally adaptive algorithms can often spend most of some allotted amount of time working in a small part of the integration region R, trying to achieve an unreasonably small error. When the time runs out, all that is available globally is an inaccurate result. Although this can also be a problem with one dimensional integrals, much smaller times are usually involved.

Globally adaptive algorithms subdivide using information about all of the current subregions. A popular globally adaptive strategy[7,18] always subdivides the subregion with the largest error, until the global error sum is less than $\epsilon$, or a time limit has been reached. For difficult problems, early termination of this type of algorithm usually gives a much more accurate result than that obtained with a locally adaptive algorithm. However, globally adaptive algorithms usually require more working storage than locally adaptive algorithms, and globally adaptive algorithms may take more time to select subregions for subdivisions.

Either type of adaptive algorithm repeats these four main steps:
1) Choose some subregion(s) from a structured list of subregions.
2) Divide the chosen subregion(s) and apply the basic rule.
3) Update the subregion list.
4) Check for convergence.

## 3. PARALLEL COMPUTERS

For the purpose of discussing the parallel algorithms in this paper we will consider four general types of parallel computers: SIMD, shared memory MIMD, distributed memory MIMD and vector computers. The SIMD/MIMD classification is due to Flynn [6]. A typical SIMD (Single Instruction stream, Multiple Data stream) machine has a large number ($> 1000$) of simple processors arranged in an array. For a given instruction a selected set of active processors all execute that instruction on their own data. The main concern in programming machines of this type is to organize the data so that the set of active processors is as large as possible for each step in the computation. This is often difficult because of the constraint that all active processors must execute the same instruction.

MIMD (Multiple Instruction stream, Multiple Data stream) machines usually have a smaller number of more powerful processors, and the processors may all concurrently execute different instructions on different data. A typical shared memory MIMD machine has 2-30 processors which are connected through a bus to a common memory area. Examples are the Sequent Balance, Encore Multimax and Alliant FX computers. Distributed memory MIMD machines usually have a larger number of processors connected together in a network. Each processor will often have a sizable amount of local memory. Examples of distributed memory machines are the Intel iPSC and NCUBE machines. A typical program consists of a host program along with separate programs for the nodes. The programs run independently and communicate using the message passing subroutines. MIMD computers of this type are often more difficult to program than the shared memory machines because the programmer has to handle the communication between processors and local memories, and take more account of the cost of this communication. However, for certain applications, the larger feasible number of processors in distributed memory machines allows much more parallelism.

A vector computer uses pipelining to efficiently execute simple instructions on vectors of operands. Good examples are the CRAY-1S and CYBER-205 computers. These computers allow extremely fast computation, but only on low level arithmetic and some intrinsic function calculations for long vectors of operands.

Some computers now combine features of the four basic types of parallel computers. Further details about the hardware for the parallel computers briefly described here, and for other parallel computers, can be found in the report by Dongarra and Duff[4], and the recent review paper by Hwang[12].

## 4. PARALLELIZATION OF ADAPTIVE INTEGRATION METHODS

Parallel computation with adaptive integration algorithms can be done at one of three different levels:
  1) the integral,
  2) the subregion and
  3) the basic rule.
Discussion is given in the following subsections about parallelization at each of these levels for different types of parallel computers.

### 4.1. Parallelization at the Integral Level

Parallelization at this level may be appropriate in large scale calculations where a number of integrals need to be computed. It is most appropriate in cases where the integration region is different for each integral or the integrands differ enough to result in significantly different subdivisions of a common integration region for each integral. If similar integrands share a common integration region R, then the same subdivision of R might be good for the accurate computation of all of the integrals. The time finding this subdivision can be shared among the integrals and the parallelization should therefore be introduced at a lower level. For example, if the tree model is used for the adaptive algorithm, then each node represents one application of the basic rule. With many adaptive algorithms the tree is a binary tree, and such a tree has only one more leaf node than interior node. A consequence of this is that the cost for the (leaf)nodes in the final subdivision is almost the same as the cost of adaptively constructing that subdivision. Therefore, approximately half of the total computation time can be saved if only one of the integrands is used for constructing the subdivision tree and all of the other integrands are evaluated only in the subregions that make up the final subdivision.

Parallel calculations done at this highest level are large granularity calculations that are most efficiently done on MIMD computers, where is it is usually straightforward to arrange to keep all of the processors busy for most of the time that it takes to complete the job. Vector computers cannot efficiently be used for parallel work at this level because of the large granularity of the work. SIMD computers cannot be efficiently used at this level either, because of the different sequence and length of computations that different integrals might require.

## 4.2. Parallelization at the Subregion Level

If parallelization is considered at the subregion level, we see that many adaptive algorithms do not explicitly produce groups of subregions that could be handled concurrently. Usually the subregions are produced one at a time, from a stack or priority queue that is updated with each new set of results. The chosen subregion is also usually subdivided into a small number of pieces. If more than one integral is being computed over some common region then this need not be a problem, because different processors can be given different integrals to compute over the common selected subregion. For single integrals, two types of modification to these essentially serial algorithms allow some parallelism, and these are: i) to take more than one subregion at a time from the subregion list, or ii) to divide the one chosen subregion into a larger number of pieces. Either one of these modifications alone can result in an algorithm that loses efficiency. If the priority subregion has a large error compared to the other subregions in the list, then the first modification may waste processor time on relatively unimportant subregions. On the other hand, if the subregions in the list have similar errors, then modification ii) could divide some regions too finely, wasting time that could be used for more important work. In the next section we describe an algorithm that uses a mixture of these strategies to avoid some of the possible inefficiency.

MIMD type parallel computers are usually well suited for efficient work at this level, because each of the new subregions can be given to a different processor. When the integrands are simple enough to allow rapid application of the basic rule in a particular subregion, then the communication time required for distributed memory MIMD machines could be significant, and should be carefully accounted for. In general, adaptive algorithms for a single integral, can probably be modified to make efficient use of a modest number of processors (2-16). If there are more than this number of processors, then extra time associated with updating the subregion list, choosing a good selection of subregions and communication between processors and memory can contribute to loss of efficiency. Because of the relatively large granularity at this level, vector computers are not suitable for efficient parallel work here. SIMD computers are not suitable for single integral parallel work at this level because of the relatively small number of subregions that can be produced without significant loss of adaptivity. This is true for SIMD computers that are commercially available, where the number of processors is large, but is not an inherent problem with the SIMD architecture computers.

If there is a large number of similar integrals then an MIMD computer can be effectively used to apply the basic rule to the subregions provided by the adaptive algorithm. In this case the subregions need only be produced one at a time, because the parallelism is over the integrands. Some SIMD machines may also be used efficiently for this type of problem, particularly when the number of integrals is large.

### 4.3. Parallelization at the Basic Rule Level

The possible parallel computation of the basic rule $Q(f)$ can be decomposed into three steps:

      i) prepare vectors $W = (w_1, w_2, ..., w_N)$ and $X = (x_1, x_2, \cdots, x_N)$ ,

      ii) compute the vector $F = (f(x_1), f(x_2), \cdots, f(x_N))$ and

      iii) compute $Q(f) = W \cdot F$.

In a serial algorithm the steps are normally carried out together in a single loop which includes a function call for the integrand, but different parallel machines may often require very different code structures. If step i) is separate from step ii), then $N(n+1)$ extra memory locations are required.

Step i) is difficult to parallelize. This is due to the complicated structure of the set $\{x_i\}$ for many integration rules. If the rule is a nonproduct polynomial rule it can often be written as a sum of product rules. The points and weights used by product rules can be determined from some smaller sets using an indexing function, and step i) could be combined with step ii), but in some cases significantly increasing the computation time. Most number-theoretic rules and Monte-Carlo rules have a regular structure for $\{x_i\}$ that will allow straightforward parallelization of step i), but these rules are not often used with adaptive algorithms. Storage of the X and W vectors at compile time in appropriate data structures is a possible means of reducing the execution time at step i) but this strategy would often significantly increase the length of the code. Storage of the X and W vectors in arrays in an initialization step at the beginning of the adaptive algorithm can also significantly increase the memory requirements for large N. In any case, the amount of time needed for step i) will not depend on the integrand, so even when fairly extensive calculations are needed for step i) there will always be integrands that are difficult enough to make the computations for step ii) dominate the total time. These problems are the ones most likely to be considered for large scale calculations on parallel computers.

For SIMD machines, step ii) is an SIMD step consisting of the single instruction "compute f for different X values". As long as $N \approx kP$, when k is an integer for a P processor machine, step ii) is efficient for an SIMD machine where each processor can compute k function values. If the function f requires different computation times for different x values, then the work of the processors which already have accurate f values must be discontinued while the remaining processors finish, and this will result in loss of efficiency. Differences in computation times can occur if a series expansion is used to compute the function. In this case, more terms of the series may be required for some values of **x** than for others. On commercial SIMD computers P is usually large, with typical values in the thousands. For many integration problems, however, the most appropriate integration rule might have an N value that is small relative to P or not close to an integer multiple. The programmer could then be faced with the choice of using this rule and computing the integral without good processor utilization or using a more efficiently implementable rule (e.g. Monte-Carlo) that is not as accurate. It will not always be clear which of these choices is consistent with the overall aim of fast accurate calculation. This is often more of a problem for smaller values of the integral dimension n.

On MIMD machines a similar discussion applies, except that P is much smaller for the shared memory machines. The smaller P means that efficiently matching N with an integer multiple of P will not usually be a problem. Step i) will normally include some partitioning of X and W. For distributed memory machines step i) could also involve some distribution time, unless the points and weights required by a particular processor were prepared by the same processor. For MIMD machines the granularity at this level is small, particularly for simple integrands, and so in many problems it will be difficult to achieve maximum efficiency because of communication overheads. This problem will be more significant for distributed memory MIMD computers.

The efficient use of a vector machine will depend on whether the steps can or cannot be written as loops that vectorize. If the function evaluations in step ii) can be decomposed into simple combinations of the basic intrinsic FORTRAN functions, then it might be possible for step ii) to be programmed as a sequence of vectorizable loops and run efficiently on a CRAY-like machine. For more complicated integrands good vectorization will be difficult to achieve.

If step iii) is separated from the other two steps, it could be implemented as a cascade sum [11]. A procedure for this would usually be provided as part of a good parallel extended FORTRAN and would typically require $O(\log(N))$ (with a small constant) time, if N is not significantly larger than P. For many integration rules steps i) and ii) could be efficiently parallelized and it might appear that step iii) could dominate the computation, but the constant associated with the $O(\log(N))$ or the serial $O(N)$ term is usually so small that this will not happen.

Now consider the problem of rule evaluation for a large group of integrals in the form

$$Q(f_j) = \sum_{i=1}^{N} w_i f_j(\mathbf{x}_i), \quad j = 1, 2, ..., M.$$

Here, the most straightforward way to utilize parallel machines is to parallelize over the j index. As long as M is close to an integer multiple of P, good parallelism can be achieved with very little code modification on shared memory MIMD machines. For vector machines significant speedups could be achieved if the evaluation of the functions $f_j$ is vectorizable. On distributed memory MIMD machines this approach might have a host or master processor compute the rule sum and node processors computing the integrand values. In this case the communication times could be significant unless the integrands $f_j$ are relatively time consuming to compute.

For SIMD machines it might appear that parallelization is not possible because of the different integrals. However, the collections of integrals in large scale calculations often differ only with respect to some parameter(s). In this case high processor utilization could be achieved on SIMD computers.

## 5. PARALLELIZATION OF A GLOBALLY ADAPTIVE ALGORITHM

### 5.1. The Serial Algorithm

We now discuss in more detail the parallel implementation of a globally adaptive algorithm. This algorithm has been used in serial form in the NAG [13] library for several years. Parallelization was done at the subregion and basic rule levels, and efficiency studies were carried on a variety of parallel computers by the present author [8] and Berntsen [1]. The serial algorithm is an algorithm originally described by van Dooren and de Ridder [18] and modified by Genz and Malik [9]. The essential features of this algorithm will be described by following the outline for a generic adaptive algorithm given in section two. In step one the subregion with the largest error is chosen from a list of subregions. Two new subregions are obtained from this subregion by dividing it in half along the axis where the integrand has the largest fourth difference. In step two a polynomial degree seven basic rule is used in the two new subregions. In step three the new subregions, integral estimates and error estimates are added to the list, which is structured as an error keyed heap. The use of a heap allows a subregion list of length M to be updated in O(log(M)) time. In step four the global error and integral estimates are updated and checks are made for convergence and time and space consumption.

### 5.2. Parallelization at the Subregion Level

We consider parallelization at the subregion level for a P processor MIMD machine. A natural change to the serial algorithm is to select P/2 subregions at step one, taken consecutively from the top of the heap. These subregions are each divided into two pieces and the resulting P subregions used in step two. The algorithm can be initialized by subdividing the whole integration region into P equal parts.

When tests were done [Genz, 8] it was found that high processor utilization occurred for simple integrands on shared memory machines with 2-12 processors without attempting to parallelize steps one, three and four where the heap maintenance was done. However, the accuracy obtained for a fixed number of integrand evaluations was not as good as that obtained with the serial algorithm for some problems, because of wasted work on some subregions with relatively small errors. A small modification was made to the algorithm: as the subregions were removed from the heap, any subregion with an error estimate greater than four times the error estimate for the next subregion was divided into four pieces instead of two. This process was continued until there were P subregions ready for the basic rule application in step two. Further testing showed that this simple heuristic removed most of the inefficiency, without significantly complicating the algorithm.

In order to produce parallel code for the modified algorithm for distributed memory machines, separate programs had to be written for the host and node processors. Node programs were written to do the basic rule work. These programs waited for a subregion from the host, evaluated the basic rule on that subregion and sent the results back to the host. The host program contained most of the original code for the modified algorithm. This version of the parallel algorithm was found to have a high communication cost that dominated the processing time for simple integrals and low dimension (n < 5) [Genz, 8]. However, Berntsen and Espelid [2] showed how to reduce the communication cost by moving most of the host processor work to one of the network nodes and reorganizing the internode communication to minimize communication path lengths.

The Berntsen and Espelid algorithm actually used copies of the serial algorithm on all nodes except the new manager node, which produced an initial subdivision of R and collected global results. This form of the algorithm is really a parallelization at the integral level, because the initial problem is split at the beginning into separate subproblems. This was also discussed in [8], where the communication was organized so that the nodes were in a ring. In this paper test results showed that some nodes could waste time on unimportant subregions if the integrand had an isolated region of irregularity. For more difficult problems it is likely that this type of load imbalance between processors might occur often. A modification to the ring algorithm allowed nodes with difficult subregions to send one of their harder subregions to a neighbor node in order to provide some load balancing, but this increased the communication costs, and the hard subregions were not distributed rapidly enough. Another variation on this algorithm could initially assign each processor an error tolerance $\epsilon/P$, and processors that finished early can get more work from still working processors. This strategy is partially locally adaptive and has not been tested.

### 5.3. Parallelization at the Basic Rule Level

The globally adaptive algorithm has also been parallelized at the basic rule level [Berntsen, 1]. In this case, as was discussed in the previous section, the basic rule points and weights can be precomputed in an initialization step when the global algorithm is begun, and stored in an array of length N(n+1). The standard degree seven basic rule for this algorithm has $N \approx 2^n + 2n^2$, so this can be a considerable storage overhead. The transformation of the points and weights for a particular subregion can be done in parallel, along with the integrand evaluation. A second possibility is to do a complete generation of the points and weights for each subregion, but this cannot be done efficiently in parallel and so has a considerable overhead. Berntsen's results only showed good processor utilization with n > 6 and a reasonably complicated integrand (Bessel function), even though the test were done on a six processor Alliant FX/8. This shared memory MIMD computer has vector processors, so some speedup from vectorization should have occurred with the simplest test integrands.

The granularity at the basic rule level is probably too low for efficient use of current distributed memory MIMD machines, unless the integrand is very complicated, and this type of parallelization has not been tested for the globally adaptive algorithm on this class of computer. Some testing of parallelization at this level was also done [Genz, 7] using a 4096 processor ICL DAP. The timing results from this SIMD machine were not very good, but were somewhat limited. For simple integrands, it is also possible that a vector computer might be efficient at this level. Some of Berntsen's results show this; Gladwell [10] also has reported significant gains in efficiency when one dimensional algorithms are structured to allow for vectorization at this level.

The use of vectorization to speed up the algorithm for the case when there is more than one integral over the same region R has been investigated [Genz, 8], and software for a modified algorithm that allows for vectorization at this level is in the NAG [13] library (Mark 12). The original adaptive algorithm was modified to handle a vector of integrands by replacing each function evaluation in the original algorithm with a vector function call. This algorithm required additional storage in the subregion list for the integral and error estimates for each integrand; the sup norm was used to determine the error for each subregion. Simple tests showed that moderately good vectorization (speedups of 20-25 for 50-100 integrands) could be obtained when the vector integrand had components that were simple enough to allow for pipelining of their evaluation.

## 6. CONCLUDING REMARKS

The general discussion has suggested that there should be some parallel machines that will allow the efficient implementation of adaptive multiple integration algorithms, although a particular algorithm will not necessarily be efficiently implementable on all parallel machines. A more detailed description of implementation experiences with a particular globally adaptive algorithm on a variety of computer types and with parallelization at different levels has supported this conclusion. In the following table we summarize the discussion and reported results for parallelization at different levels with different computers.

| Expected Efficiency of Parallelization for Different Computers | | | | | |
|---|---|---|---|---|---|
| Level | Integrals | SIMD | Shared MIMD | Distrib. MIMD | Vector |
| Integral | Many | None | High | High | None |
| Subregion | One | None | High | Medium | None |
| Subregion | Many | Medium | High | High | None |
| Basic Rule | One | Medium | Medium | Low | Medium* |
| Basic Rule | Many | High | High | Medium | Medium* |

*For simple integrands.

In general we can see that shared memory MIMD computers are likely to be the most efficient computers for widest range of large scale problems which require the numerical calculation of integrals. With the present state of software tools and language extensions, these computers are also the easiest to program. However, because the number of processors on these computers is not very large the maximum speedup will never be high. For very large scale problems the extra software development required for the implementation of parallel algorithms on distributed memory MIMD computers and SIMD computers may be justified by the higher potential parallelism on these computers.

## REFERENCES

1.  J. Berntsen, Adaptive Multidimensional Quadratue Routines on Shared Memory Parallel Computers, University of Bergen Department of Informatics Report No. 29 (1987).

2.  J. Berntsen and T. Espelid, A Parallel Global Adaptive Quadrature Algorithm for Machines with Hypercube Architecture, University of Bergen Department of Informatics Report No. 27 (1987).

3.  P. J. Davis and P. Rabinowitz, *Methods of Numerical Integration*, Academic Press, New York, 1984.

4.  J. Dongarra and I. Duff, Advanced Computer Architectures, Argonne National Laboratory MCS Technical Memorandum No. 57 (1985).

5.  E. de Doncker and J. A. Kapenga, A Parallelization of Adaptive Integration Methods, *Numerical Integration*, P. Keast and G. Fairweather (Eds.), D. Reidel, Dordrecht, Holland (1987), pp. 207-218.

6.  M. Flynn, Some Computer Organizations and Their Effectiveness, *IEEE Trans. Comput. C-21* (1972), pp. 948-60.

7.  A. C. Genz, Parallel Methods for the Numerical Calculation of Multiple Integrals, *Comp. Phys. Comm. 26* (1982), pp. 349-352.

8.  A. C. Genz, The Numerical Evaluation of Multiple Integrals on Parallel Computers, *Numerical Integration*, P. Keast and G. Fairweather (Eds.), D. Reidel,

Dordrecht, Holland (1987), pp. 219-230.

9. A. C. Genz and A. A. Malik, An Adaptive Algorithm for Numerical Integration over an N-Dimensional Rectangular Region, *J. Comp. Appl. Math. 6* (1980), pp. 295-302.

10. I. Gladwell, Vectorization of One Dimensional Quadrature Codes, *Numerical Integration*, P. Keast and G. Fairweather (Eds.), D. Reidel, Dordrecht, Holland (1987), pp. 231-239.

11. R. Hockney and C. Jesshope, *Parallel Computers*, Adam Hilger, Bristol, U.K., 1981, pp. 178-192.

12. K. Hwang, Advanced Parallel Processing with Supercomputer Architectures *Proc. IEEE 75* (1987), pp. 1348-1379.

13. Numerical Algorithms Group Limited, Mayfield House, 256 Banbury Road, Oxford OX2 7DE, United Kingdom.

14. J. R. Rice, Parallel Algorithms for Adaptive Quadrature, Convergence, *Proc. IFIP Congress '74*, North Holland, New York, 1974, pp. 600-604.

15. J. R. Rice, Parallel Algorithms for Adaptive Quadrature II, Metalgorithm Correctness, *Acta Informatica 5* (1975), pp. 273-285.

16. J. R. Rice, Parallel Algorithms for Adaptive Quadrature III, Program Correctness, *ACM TOMS 2* (1976), pp. 1-30.

17. J. R. Rice and J. M. Lemme, Speedup in Parallel Algorithms for Adaptive Quadrature, Purdue University Computer Science Technical Report CSD-TR 192 (1976).

18. P. van Dooren and L. de Ridder, An Adaptive Algorithm for Numerical Integration over an N-Dimensional Rectangular Region, *J. Comp. Appl. Math. 2* (1976), pp. 207-217.

19. H. D. Shapiro, Increasing Robustness in Global Adaptive Quadrature through Interval Selection Selection Heuristics, *ACM TOMS 10* (1984), pp. 117-139.

20. R. B. Simpson and A. Yazici, An Organization of the Extrapolation of Multidimensional Quadrature for Vector Processing, *Parallel Computing 4* (1987), pp.175-188.

21. A. H. Stroud, *Approximate Calculation of Multiple Integrals*, Prentice-Hall, New Jersey, 1971.

# A Comparison of Hypercube Implementations of Parallel Shooting

Herbert B. Keller and Paul Nelson*

Department of Applied Mathematics

California Institute of Technology

Pasadena, California 91125

## 1. INTRODUCTION

In this work we consider issues relating to the use of (discrete) parallel shooting [1] for the numerical approximation of linear two-point boundary-value problems with separated boundary conditions,

$$y' = Ay + f, \tag{1a}$$

$$B_a y(a) = \beta_a, \quad B_b y(b) = \beta_b, \tag{1b, c}$$

upon a hypercube. By a hypercube we intend a distributed memory MIMD computer with communication between processors by message passing via a communication network having the topology of an $n$-dimensional cube, with the vertices considered as processors and the edges as communication links. At this writing several versions of such machines are commercially available, but our experience has been with the Mark II Hypercube at Caltech [2-7], under the Crystalline Operating System [8]. This background may be reflected in some of the assumptions underlying our analyses, although we have tried to minimize such specifics. References 3, 6, 7 and 9, and other works cited therein, provide more detailed

---

* Permanent Address: Departments of Nuclear Engineering and Mathematics; Texas A&M University; College Station, TX 77843-3133.

discussions of considerations relating to scientific computation upon computers having a hypercube architecture.

At this early juncture it seems appropriate to consider the circumstances under which one might reasonably wish to consider using a hypercube, or any other type of computer having a high degree of explicit user-accessible parallelism, to solve a linear two-point boundary-value problem. It seems to be a general feature (cf. Section IV F of [7]) that sufficient problem size is the only requirement for efficient use of concurrent processors. This requirement matches need, because, as has been emphasized by Geoffrey Fox [10], the basic justification for considering computers of novel architecture must be regarded as the solution of problems not otherwise (reasonably) soluble, not merely faster solution of problems presently soluble. This view of the appropriate "problem scaling" is at the heart of the recent well-publicized hypercube work at Sandia [28]. Numerical solution of (linear or nonlinear) two-point boundary-value problems on basically sequential computers is a highly developed field, with a number of well tested codes available for this specialized purpose [11-14]. However, most of these codes seem to be tailored to problems such that the vector of dependent variables ($y$) has dimension on the order of tens. As the computational cost tends to increase with the cube of this dimension, problems of the form (1) having dimensionality on the order of hundreds or larger would seem to be candidates for one type of problem not reasonably soluble on most current sequential computers. Such problems can easily arise via application of the method of lines to partial differential equations. For another source of such problems, a quadrature approximation to the three-dimensional integral appearing in the integro-differential (linear) transport equation can give rise to systems (1) having dimensionality on the order of 100-1000, even for steady-state situations involving only spatial dimension.

The computational cost of solving (1) also can be expected to increase, although not so rapidly, with increasing length, where the latter is measured by the number of mesh points necessary to obtain the required accuracy by means of the particular underlying discrete approximation. We shall therefore focus upon problems (1) that are either *long*, in the sense just defined, or *large*, in the sense of the preceding paragraph. This restriction will simplify our analysis, in that it permits neglect of lower order terms in these parameters in estimating computational times. We shall also see that, in keeping with the general observation of the preceding paragraph, it is in precisely one or both of these situations that the methods we consider are most efficient.

The parallel shooting method perhaps should be considered a class of methods, with the particular method within this class depending upon the precise selection of "shooting" points from among the mesh points. The two extreme methods from this class are the underlying finite difference method itself, for which each mesh point is a shooting point, and ordinary superposition, for which only the right endpoint ($b$) is a shooting point. Finite-difference methods *per se* are of some interest, but there are some additional special considerations for them that make our analysis inapplicable; we do note these, at the appropriate point. On the other hand, we do not consider ordinary superposition to be of interest, because of its well known [1] susceptibility to instability. We primarily focus upon implementations involving dynamic selection of the shooting points, although we shall, where appropriate, indicate variations that would result from *a priori* selection of the shooting points (e.g. as in finite-difference methods).

In Section 2 following we review the parallel shooting process, primarily for the purpose of establishing notation and terminology. For subsequent considerations it is convenient to consider this process as divided into two phases, "integration" and "solution." In Section 3 we consider two different types of decompositions, by which is intended mappings of subproblems (or tasks) onto the processors. In the *domain* decomposition,[1] each processor is assigned responsibility for the computations associated with the fundamental matrices corresponding to a particular subinterval of $[a, b]$, whereas in the *column* decomposition, each processor is assigned the computations corresponding to certain columns of the fundamental matrices associated with *each* shooting subinterval. Section 4 is devoted to timing estimates for each of the four phase/decomposition combinations. In Section 5 we use these estimates to discuss the relative advantages of using a domain decomposition for both phases, as opposed to using a column decomposition for both phases. In Section 6 we study the efficiency question for these two *strategies,* where a strategy is efficient if

---

[1] By "domain decomposition" we shall exclusively mean decomposition within the domain of the independent variable. Under a sufficiently broad interpretation of this term even what we term "column decomposition" could be considered an instance of domain decomposition. For examples of approaches to domain decomposition in this broad sense, including some that might have application to parallel solution of two-point boundary-value problems, see [15].

the reduction in computational time, compared to a single processor implementation, is approximately proportional to the number of processors used. Some concluding remarks are given in Section 7.

Throughout the following we denote the dimensionality of the differential system (1a) by $n$. Further we assume that the given vectors $\beta_a$ and $\beta_b$ have respective dimensions $k_1$ and $k_2 := n - k_1$, where $1 \leq k_1 \leq n$. We also assume that the known $k_1 \times n$ ($k_2 \times n$) matrix $B_a$ (respectively, $B_b$) has rank $k_1(k_2)$, as these are necessary conditions (see p. 14 of [16]) for existence of a unique solution of the linear two-point boundary-value problem (1). Finally, throughout we also assume that (1) indeed does have a unique solution. Necessary and sufficient conditions for this to hold, which serve to assure that it does indeed hold for a nontrivial class of problems, are given in Section 2 of [16].

## 2. PARALLEL SHOOTING

We consider a compact (i.e. single step) finite difference approximation to (1), on a mesh having points $x_j = a + jh$, $j = 0, \ldots, m$, where $h = (b - a)/m$. The particular (single-step) finite difference approximation is not important for our considerations, but for convenience we shall discuss the box scheme (centered Euler method, midpoint rule),

$$y_{j+1} - y_j = h[A_{j+1/2}(y_{j+1} + y_j)/2 + f_{j+1/2}], \tag{2a}$$

$$B_a y_0 = \beta_a, \quad B_b y_m = \beta_b. \tag{2a, b}$$

Here $A_{j+1/2} := A(a + (j + 1/2)h)$, $f_{j+1/2} := f(a + (j + 1/2)h)$ and $y_j$ is the approximation to $y(x_j)$ thus defined.

Let $\{j_i : i = 1, \ldots, s\}$ be a strictly increasing sequence such that $j_1 > 0$ and $j_s = m$. The $\hat{x}_i := x_{j_i}$ are the *shooting points*. (Selection of these is discussed at the beginning of the next section.) If $n$ is the dimension of the differential system (1a) let the $n \times n$ matrices $\Phi_i$ be defined by

$$\Phi_i := \prod_{j=j_{i-1}}^{j_i-1} (I - hA_{j+1/2}/2)^{-1}(I + hA_{j+1/2}/2). \tag{3}$$

Here $j_0 := 0$, and we assume $h$ is sufficiently small so that the $I - hA_{j+1/2}/2$ are nonsingular. Similarly let

$$\varphi_i := (I - hA_{j_i-1/2}/2)^{-1}(I + hA_{j_i-1/2}/2)\tilde{y}_{j_i-1-j_{i-1}}$$

$$+ h(I - hA_{j_i-1/2}/2)^{-1}f_{j_i-1/2}, \ i = 1, \ldots, s, \qquad (4a)$$

where

$$\tilde{y}_0 := 0 \qquad (4b)$$

and

$$\tilde{y}_{i+1} = (I - hA_{i+1/2}/2)^{-1}(I + hA_{i+1/2}/2)\tilde{y}_i + h(I - hA_{i+1/2}/2)^{-1}f_{i+1/2}. \qquad (4c)$$

Then it is readily shown that

$$\hat{y}_i = \Phi_i\hat{y}_{i-1} + \varphi_i, \ i = 1, \ldots, s, \qquad (5a)$$

where $\hat{y}_i := y_{j_i}$.

In the present notation, the boundary condition $(1b, c)$ may be written as

$$B_a\hat{y}_0 = \beta_a, \ B_b\hat{y}_s = \beta_b. \qquad (5b, c)$$

Equations (5) then comprise a system of $n(s + 1)$ linear equations for $n(s + 1)$ unknowns, where $n$ is as above. This is the essence of the parallel shooting procedure, as adapted to linear problems. First, one computes the $\Phi_i$ and $\varphi_i$, by (3) and (4), respectively, then the $\hat{y}_i$ are obtained by solving the linear system (5). This yields approximations to the solution of (1) at the shooting points. Approximations at additional points, if desired, can then be obtained by initial-value techniques, but we shall not consider this matter in the following.

In the following we shall assume, as indicated notationally above, that the step-size for the underlying finite-difference approximation is both independent of position (i.e. value of the independent variable) and *a priori* known. The first of these assumptions is merely a matter of convenience in notation. Our results readily extend to (*a priori* known) variable meshes. However, the second assumption evades the important question of how to select step sizes so as to attain a desired accuracy. Ultimately one would wish to have algorithms employing adaptive selection of variable step sizes. The problem of maintaining load balance between processors (see the next section) in an environment of an inhomogeneous time varying mesh is both important and difficult. From the viewpoint of these considerations our work perhaps is best viewed as providing an optimistic estimate of the utility and efficiency of (two particular implementations of) parallel shooting on a hypercube for numerical solution of linear two-point boundary-value problems.

We note parenthetically that if a suitable constant step size can be found, and additionally the data ($A$ and $f$) of (1a) are piecewise constant, then it may be possible to effect significant savings in the computations indicated by (3) and (4). We do not pursue this matter here, because our primary interest is in linear problems only as the core of an iterative solver for nonlinear two-point boundary-value problems, and in this context the data of (1a) hardly ever will be piecewise constant. However, there are important applications for which such considerations may be significant.

## 3. PHASES AND DECOMPOSITIONS

It is convenient to consider the parallel shooting process, as described in the preceding section, to be segmented into two phases. In the *integration phase* the basic objective is to compute the $\Phi_i$ and $\varphi_i$, by means of (3) and (4). An auxiliary task that must be accomplished in this phase is the dynamic selection of the shooting points, so as reasonably to control error propagation. In the following we assume that the latter is accomplished by taking each shooting subinterval (i.e. each $[\hat{x}_{i-1}, \hat{x}_i]$) as large as is consistent with a requirement of the form

$$\|\Phi_i\| \leq M, \; i = 1, \ldots, s. \tag{6}$$

Here $M$ is some given (i.e. user selected) parameter, and $\|\cdot\|$ is a matrix norm that must be selected so as to be inexpensively computable. A more rigorous control on error propagation would perhaps be provided by a condition number criterion,

$$\|\Phi_i^{-1}\| \cdot \|\Phi_i\| \leq M, \; i = 1, \ldots, s,$$

but experience [17] has shown that the additional assurance of stability thereby provided is not worth the additional computational effort required to compute the matrix inverses needed to invoke such a criterion.

The *solution phase* of parallel shooting consists of the solution of the linear algebraic system (5). We note that the coefficient matrix of this system has the block bidiagonal ("almost block diagonal," "staircase") form shown in Figure 1. Several algorithms specialized to such systems have appeared in the literature [18-23]. We shall suppose that some version of the alternate-row-and-column-elimination method of Varah [20] is used, as this method seems both highly stable and economical of storage. Our conclusions would not appear to be sensitive to the particular solution algorithm to be used for (5).

Within each of the two phases one must select some *decomposition,* by which is intended a mapping of tasks onto processors. In general terms a decomposition should be selected so as to minimize communication overhead (relative to computation) and to balance the computational workload, in the sense of keeping as many processors as possible computationally occupied at any given time. One obvious possibility is *domain decomposition,* in which a given processor would be assigned the task of all computations associated with mesh points lying in some *a priori* determined subset, which we always take as a subinterval of the underlying interval. For the integration phase such domain decomposition would require that the endpoints of these subintervals be shooting points, in order that the processors could execute the calculations implied by (3) and (4) in parallel. This may introduce several more shooting points than would be required by the criterion (6), as we shall discuss in Sections 5 and 6. However, in the solution phase this corresponds to a decomposition such that each processor is assigned to (tasks associated with) a set of contiguous columns in the coefficient matrix of Figure 1. Such a "contiguous column" decomposition would tend to provide poor load balance (in this phase), in that at most two processors could be involved in the elimination calculations at any given time, i.e. under such a decomposition the solution phase would be essentially serial in nature.

$$\begin{pmatrix} B_a & & & & & \\ -\Phi_1 & I & & & & \\ & -\Phi_2 & & I & & \\ & & \ddots & & \ddots & \\ & & & -\Phi_s & & I \\ & & & & & B_b \end{pmatrix}$$

Figure 1 – The block bidiagonal structure of the coefficient matrix of the linear system (5).

In numerical linear algebra it has been found advantageous to use other types of decompositions to overcome similar problems. For example, Fox [24] has considered a variety of decompositions by "blocks" of (not necessarily contiguous) rows and columns, while Geist [25] has considered a variety of row decompositions, and Davis [26] studied several types of column decompositions. It seems likely that some approaches of this type might be able to alleviate the potential difficulty with domain decomposition that was noted in the preceding paragraph. In order to explore this possibility, we shall consider *column decompositions* in which processor $k(0 \leq k \leq p-1)$ is assigned responsibility for all elimination (and other) calculations on entries of $I_{j_i-1/2}$ or $\Phi_i$ in columns having indices equal to $k(\bmod\ p)$. Such a decomposition is termed a "wrap mapping" by Davis [26], in the context of column decompositions, and a "scatter" decomposition by Fox [24], in the context of block decompositions. A similar decomposition in the integration phase would assign processor $k$ the responsibility for all calculations in (3) leading to matrix elements having column index equal to $k(\bmod\ p)$.

The two types of decompositions we have just described (i.e. domain decomposition by subintervals, and column decomposition by mod $p$ "scattering" or "wrapping," where $p$ is the number of processors) can be described in terms of the general decomposition procedure of Fox [7]. According to this procedure one first identifies a *dataset* ("world"), which is composed of individual *members*. The dataset then is partioned into *cells*, with each processor assigned the responsibility for computational tasks associated with one cell. Each member may itself have several *components*. For the domain decomposition the members are the mesh points, a cell consists of a set of contiguous mesh points, and the members are the (finite-difference approximations to) the values of the scalar components of the vector $y$, at the particular mesh point. As viewed in terms of the coefficient matrix of Figure 1, this leads to a decomposition as indicated schematically in Figure 2. On the other hand, for the column decomposition, the members are the scalar components of the vector $y$, a cell consists of all such components having indices that are congruent modulo $p$, and the components of a member consists of its approximate values, over all mesh points. The corresponding decomposition, in terms of the coefficient matrix of Figure 1, is depicted schematically in Figure 3.

Other types of decompositions might have application to numerical solution of two-point boundary-value problems by parallel shooting, but for present purposes we limit considerations to the two just described in detail. In each of the two phases of parallel

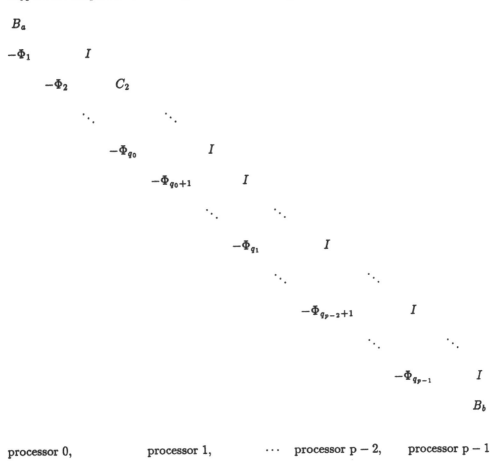

processor 0,      processor 1,    ⋯ processor p − 2,   processor p − 1

Figure 2 – The domain decomposition, in terms of the coefficient matrices of Figure 1 $\left(\sum_{i=0}^{p-1} q_i = s\right)$.

shooting, we thus have a choice of each of these two decompositions. In the present work we restrict our considerations to the strategies that use the same one of these decompositions in each of the two phases. As a step toward generating guidelines for selecting appropriate decompositions, we now turn to the task of obtaining timing estimates for these four combinations of phase and decomposition.

$$
\begin{pmatrix}
B_a^1 & B_a^2 & B_a^p & B_a^{p+1} & B_p^{kp} & & & & & \\
-\Phi_1^1 & -\Phi_1^2 & -\Phi_1^p & -\Phi_1^{p+1} & -\Phi_1^{kp} & I^1 & I^{kp} & & & \\
& & & & -\Phi_2^1 & -\Phi_2^{kp} & & I^{kp} & & \\
& & & & & & -\Phi_{q_{p-1}}^{kp} & I^1 & I^{kp} & \\
& & & & & & & & B_b^1 & B_b^{kp} \\
0 & 1 & p-1 & 0 & p-1 & 0 & p-1 & p-1 & 0 & p-1
\end{pmatrix}
$$

$D_1^\ell = \ell$th column of matrix $D_i$, $n$ assumed equal to $kp$, for some integer $k$.

Figure 3 – The column decomposition, in terms of the coefficient matrix of Figure 1.

## 4. TIMING ESTIMATES

Our estimates will be expressed in terms of the following parameters, which have already been introduced:

$m =$ number of mesh points;

$n =$ dimension of the vector of dependent variables (i.e. $y$) in the differential system (1a);

$p =$ number of processors used;

$s =$ number of shooting points used.

We shall use the following guidelines in obtaining timing estimates:

a) Estimates will be made in terms of a unit of time equal to "the time taken to perform a typical single precision floating point calculation."

b) Terms that are subdominant for large $m, n$ and $p$ will be neglected. That is, if there are contributions of the form $f(s)g_1(m, n, p)$ and $f(s)g_2(m, n, p)$, then the latter is neglected in comparison to the former if

$$
\lim_{\substack{(m,n,p) \\ \to (\infty,\infty,\infty)}} \frac{g_2(m, n, p)}{g_1(m, n, p)} = 0.
$$

c) Factors on the order of unity will be neglected.

d) It will be assumed that the time required to communicate one floating point number from one processor to a neighboring processor is on the order of one unit, as defined in a).

(e) Each processor will be assumed to have an infinite amount of associated memory.

f)  It will be assumed that $m - s = \gamma m$, where $\gamma$ is of order unity; further it will be assumed that most shooting intervals contain approximately the same number of mesh points. (The assumption $\gamma \approx 1$ does invalidate our analysis near the limiting case of parallel shooting that is comprised of the underlying finite-difference method itself.)

In guideline a) we are following in the spirit of Fox [7]. As emphasized in [7], this unit of time is not precisely defined, because there are a variety of such operations. However, timing differences for these can be expected to involve factors on the order of unity, and thus distinctions are blurred within the spirit of guideline c). Guidelines b) and c) are based upon the idea that our basic interest is in studying *trends* for large values of $m, n$, and $p$. The reasons for limiting attention to large $m$ and $n$ already have been explained in the Introduction. The justification for restricting attention to large $p$ is simply that the original objective underlying development of hypercube was the solution of significant problems by a network of low-performance (and hence low cost) processors; for such a system a high level of overall performance can be expected only for a "massively parallel" system consisting of a relatively large number of processors.

Guideline d) is quite appropriate for the current generation of Caltech Hypercubes [7], at least within the framework of guideline c). However, this perhaps is the single point within our analysis that is most sensitive to the details of a particular machine, and its associated operating system. We refer to the recent work of McBryan and Van de Velde [27] for a very thorough detailed discussion of such matters.

Guideline e) is an idealization to avoid considerations associated with use of external memory, particularly communication times between external devices and nodes. The actual memory required per node is on the order of $(s/p)n^2$. If $s/p$ is on the order of ten, then this is $10n^2$. The Mark II Hypercubes at Caltech have 256 Kbyte memories, as does the Intel iPSC, so that $n$ on the order of a hundred is quite feasible on these machines, without use of external storage. The Mark III Caltech Hypercubes have 4 megabyte memories, thus similarly allowing $n$ on the order of a thousand. It is clear from these facts that even the current generation of hypercubes permit consideration of problems of the size targeted in the Introduction, without violating the spirit of guideline e).

The first part of guideline f) is what precludes our estimates from including finite-difference methods. For such methods one has $m - s = 0$, as opposed to $O(m)$, which would significantly change some of the estimates given in this and the following two sections. The latter part of guideline f) is somewhat restrictive, and perhaps should be viewed in the same spirit as that of an *a priori* known mesh, as discussed in Section 2.

With the underlying groundrules thus clarified, let us now turn to the task of actually obtaining timing estimates. Additional assumptions, more specific to particular computational procedures, will be described at the appropriate points.

**Integration Phase/Domain Decomposition**

Within the phase/decomposition combination we have the following computations, and associated timing estimates:

1) Formation of the matrices $I - hA_{j+1/2}/2$: $(m/p)n^2$;

2) Computation of the $LU$ decompositions of these matrices; $(m/p)n^3$;

3 Use of these $LU$ decompositions to effect the computations indicated in (3): $(m - s)n^3/p \approx (m/p)n^3$;

4) Determination of the shooting points, via (6): $(m/p)n^2$;

5) Formation of the $f_{k+1/2}$, as needed in (4c): $(m/p)n$;

6) Calculation of the $\varphi_i$, by (4): $(m - s)n^2/p$.

Here we have assumed that the various processors are assigned subintervals of approximately the same lengths (and hence the same number of mesh points), which maintains load balance. In the first two of these counts we have also assumed that the matrices $I - hA_{i+1/2}/2$ are all distinct. This is motivated by our basic interest in ultimately using parallel shooting for linear problems as the core of a nonlinear two-point boundary-value problem solver based upon Newton (or other) iteration. In such a context the $A_{i+1/2}$ would be (iteration dependent and) more-or-less continuously variable. However, if one is interested in linear problems *per se*, it might be appropriate also to consider the fact that the piecewise homogeneous problems (i.e. $A$ piecewise constant) constitute an important subclass of such problems. For this subclass, $p$ in 1) and 2) should be replaced by the number of such homogeneous regions.

The times required for steps 1), 4), 5), 6) and 7) are subdominant. Thus the total time required for the integration phase, under the domain decomposition, is of the order of

$$T_{ID} = mn^3/p. \tag{7}$$

### Integration Phase/Column Decomposition

The computational steps remain the same as for the domain decomposition, but now there are communication requirements, with corresponding additional times. The steps, and their associated appropriate time estimates are:

1) Formation of the matrices $I - hA_{i+1/2}/2$: $mn^2/p$;

2) Computation of the $LU$ decompositions of these matrices: $m(n^3/p + n^2 \log p)$;

3) Calculation of the $\Phi_k$ by (3): $m(n^3/p + n^2 \log p)$;

4) Invocation of the shooting-point criterion (6): $m(n^2/p + n \log p)$;

5) Formation of the $f_{i+1/2}$, as needed in (4c): $mn$;

6) Computation of the $\varphi_i$ via (4): $mn^2$;

In step 2) the $LU$ factorization of the inverse of the $I - hA_{i+1/2}/2$ might be computed, for example, by the "pivot-by-row eliminate-by-column" approach of Davis [26]. A "pivot-by-column eliminate-by-column" approach would appear more natural, although maintenance of load balance during elimination would thereby be somewhat less certain. In any event, the first term in the timing estimate for step 2) comes from the elimination computations, and the second from communciation of the elimination columns. The "$\log p$" term in 2) (as in 3) and 4)) stems from communications; $\log p$ is the *diameter* of a hypercube of $p$ processors, where this is defined as the maximum number of communication links between any two processors. The computational and communication times associated with determining and using the pivots are subdominant. (Pivoting may not be required here, but it is always safer and, because of the subdominance just noted, it does not change our considerations to assume that it is performed.) Similarly, the terms in 3) arise respectively from carrying out the computations in (3), and from communicating the $I + hA_{i+1/2}/2$ and the $LU$ decompositions of the $I - hA_{i+1/2}/2$ to the various processors, as necessary to carry out these computations. We are using guidelines e) to assume that these are communicated once, and then stored in the memory of each individual node.

The timing estimates for the other steps are obtained similarly, but those for steps 2) and 3) are dominant. Thus the total time required is of order

$$T_{IC} = mn^3/p + mn^2 \log p. \tag{8}$$

As a practical matter, computational steps 3) and 6) probably would be combined, by assigning some node the responsibility for the calculations (4), to be carried out in parallel with those for (3). In this case the time required for these combined steps would be of order $mn^2(n+1)/p + mn^2 \log p \approx mn^3/p + mn^2 \log p$, and thus the timing estimate is not changed, to dominant orders. This strategy would introduce some load imbalance, in that the node assigned the calculations associated with the inhomogeneous term would be required to form the $f_{i+1/2}$, step 5) above, which has no counterpart in the $\Phi$ calculations. However, again this has only a subdominant effect on the timing estimate.

**Solution Phase/Domain Decomposition**

As mentioned above, we use the alternating-row-and-column elimination method of Varah [18] for this phase. The primary computational cost stems from the elimination, which now must be done sequentially, and thus requires time of order $(s+1)n^3$. Communication between processors assigned to adjacent subintervals requires an additional time of order $pn^2$, provided there is a direct communication link between such processors, which always can be arranged. However, as $p \leq s$ for domain decomposition, this is a subdominant term. Thus the total time required is of order

$$T_{SD} = (s+1)n^2 \approx sn^3, \tag{9}$$

where we have used guideline c), along with the obvious fact that $s \geq 1$.

**Solution Phase/Column Decomposition**

Use of the column decomposition permits the alternating row and column elminations to be carried out in parallel, so that the time required for these eliminations is of order $(s+1)n^3/p$. The dominant communication time comes from transmittal of the elimination columns, and is of order $(s+1)n^2 \log p$. (Here we have assumed that the fraction of the boundary conditions imposed at either end remains bounded away from 0 as $n$ varies, i.e. that $\alpha < k_1/n < 1 - \alpha$, for some fixed positive $\alpha$.) Thus the time required is of order

$$T_{SC} = (s+1)n^3/p + (s+1)n^2 \log p \approx sn^3/p + sn^2 \log p. \tag{10}$$

## 5. STRATEGIES

As stated previously, here we consider only the strategies in which the same decomposition is used in each phase. From (7) and (9) it follows that the time required for a *pure domain strategy,* by which we mean use of a domain decomposition in both the integration and solution phases, is of order

$$T_D = mn^3/p + sn^3. \tag{11}$$

Likewise, it follows from (8) and (10) that the time required for a similarly defined *pure column strategy* is

$$\begin{aligned} T_C &= mn^3/p + mn^2 \log p + sn^3/p + sn^2 \log p \\ &\approx mn^3/p + mn^2 \log p, \end{aligned} \tag{12}$$

where we have used the (obvious) fact that $m \geq s$. We shall compare these two strategies on the basis of their respective time requirements, according to (11) and (12), for the same values of $m, n$,

$N :=$ number of nodes (processors) *available for use,* and

$S :=$ number of shooting points called for by the criterion (6).

In considering the comparison the reader is cautioned to keep in mind limitations stemming from the fact that the underlying timing estimates (11) and (12) are valid *only* for large values of $m, n$ and $p$ (hence $N$).

Now it is not necessarily the case that one will have $p = N$, for either of the above strategies, or $s = S$, for the pure domain strategy. However, let us momentarily consider the nominal situation such that all of these equalities do hold. Then (11) and (12) show that the two strategies perform approximately equivalently if $Sn = m \log N$, or $m/S = n/\log N$. This suggests that it will be convenient to describe our comparison of the two strategies in terms of the parameters

$$\xi := n/\log N,$$

and

$$\eta := m/S.$$

In (11) and (12) the parameters $m$ and $n$ are determined by the problem and associated accuracy requirements. However, $p$ and $s$ are variable, within the constraints

$$p \leq N, \tag{13a}$$

$$p \leq s, \text{ (domain decomposition only)}, \tag{13b}$$

$$p \leq n, \text{ (column decomposition only)}, \tag{13c}$$

$$s \geq S. \tag{13d}$$

The significance of the first three of these constraints is clear. The last is a stability (accuracy) constraint.

Our first problem therefore is to determine $p$ and $s$ so as to minimize (11) and (12), subject to (13). Let us denote the respective solutions of these minimization problems by $(p_D, s_D)$ and $(p_C, s_C)$. Then it is routine to show that

$$p_D = N, \ s_D = S, \text{ for } S \geq N, \tag{14a}$$

$$p_D = N, \ s_D = N, \text{ for } \sqrt{m} \geq N \geq S, \tag{14b}$$

$$p_D = \sqrt{m}, \ s_D = \sqrt{m}, \text{ for } N \geq \sqrt{m} \geq S, \tag{14c}$$

$$p_D = S, \ s_D = S, \text{ for } N \geq S \geq \sqrt{m}, \tag{14d}$$

and

$$p_C = N, \ s_C = S, \text{ for } N \leq n, \text{ and} \tag{15a}$$

$$p_C = n, \ s_C = S, \text{ for } N \geq n. \tag{15b}$$

With this preliminary optimization problem now solved, we proceed to describe the basic results of our comparison, as shown in Figs. 4. We now discuss these figures in some detail.

The upper bound $\eta \leq m/2$ in Figs. 4 simply serves to avoid trivialities, by insuring that $S \geq 2$, in the region of the $\xi\eta$-plane considered, and thus that we are conisdering a problem for which ordinary shooting (or superposition) is inadequate. Likewise, the lower bound, $\eta \geq 2$, serves to insure that guideline f) holds, with $\gamma \geq \sqrt{2} - 1$. This can be seen by checking each individual case in (14) and (15); for (14b) and (14c) it is necessary to use the (obvious) fact that $m \geq 2$, and for the other cases we actually find $\gamma \geq 1/2$.

Conversely, if only $m \geq 4$, then the portion of the $\xi\eta$-plane indicated in Figs. 4 is nonempty. Note also that if $N > m/2$, then the horizontal line $\eta = m/N$ falls below the portion of the $\xi\eta$-plane depicted in Figs. 4.

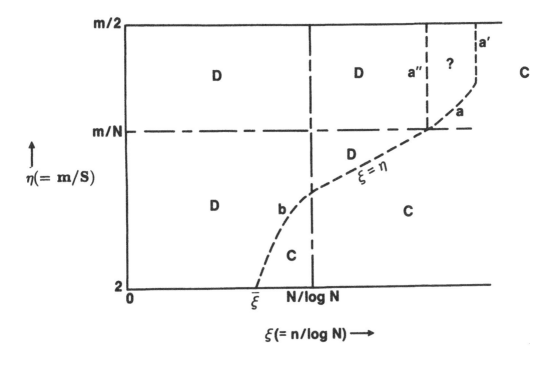

--- : **Boundary curve between regions of preference**
D: **Region in which the domain strategy is preferable**
C : **Region in which the column strategy is preferable**
a: $\xi = \eta\,[1 + \eta\,(\eta/m - 1/N)]^{-1}$, $m/N \leq \eta \leq \sqrt{m}$,
a': $\xi = \sqrt{m}\,(2 - \sqrt{m}/N)^{-1}$, max $\{m/N,\ \sqrt{m}\} \leq \eta$ and $m \leq N^2$,
a": $\xi = m/N$, max $\{m/N,\ \sqrt{m}\} \leq \eta$ and $m \geq N^2$,
b: $\eta = \xi \cdot \log N\,[1 - \xi \cdot \log N/N + \log(\xi \log N)]^{-1}$
?: **Domain strategy preferable if $m \leq N^2$, else column strategy preferable.**

**Figure 4a - Regions of preference between the domain and column strategies (for $m > N^2/\log N$).**

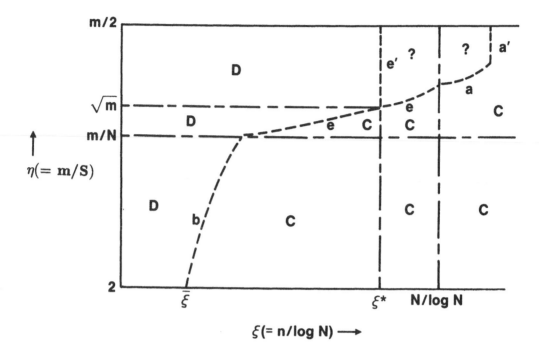

--- : **Boundary curve between regions of preference**
D: **Region in which the domain strategy is preferable**
C : **Region in which the column strategy is preferable**
a, a', b: **See Fig. 4b**
e: $(\xi/m)(\log N)\, \eta^2 - \eta\,[1 + \log(\xi \log N)] + \xi \log N = 0$
e': $\xi = \xi^*$, **where** $2\,\xi^* \log N = \sqrt{m}\,[1 + \log(\xi^* \log N)]$
?: **Domain strategy preferable if** $N/\log N \geq \xi^*$, **else column strategy preferable**

**Figure 4b - Regions of preference between the Domain and Column strategies (for $m < N^2/\log N$).**

Let us consider the region of the $\xi\eta$-plane portrayed in Figs. 4, and lying to the right of the vertical line $\xi = N/\log N$, and below the horizontal line $\eta = m/N$. Here we have $p_D = N$, $s_D = S$, $p_C = N$ and $s_C = S$, so we are in the "nominal" case, as discussed above. It follows that the boundary between the regions of preference for the two strategies lies along the line $\xi = \eta$, with the domain strategy preferred to the upper left of this line, and the column strategy preferred to its lower right. If

$$m/N > N/logN,$$

then the communication costs associated with the pure column strategy are sufficiently large so that this line actually intersects the region presently under discussion (if it exists), as illustrated in Fig. 4a. In the contrary case these costs are sufficiently small so that the region under discussion lies entirely within the region of preference for the pure column strategy, as shown in Fig. 4b.

Now consider the region to the right of $\xi = N/\log N$, but above $\eta = m/N$. Again we have $p_C = N$ and $s_C = S$, but now the values of $p_D$ and $s_D$ are as in (14b-d). For present purposes these perhaps can be most conveniently expressed as

$$(p_D, s_D) = (S, S), \text{ for } \eta \leq \sqrt{m},$$
$$(p_D, s_D) = (\sqrt{m}, \sqrt{m}), \text{ for } \sqrt{m} \leq \eta \text{ and } \sqrt{m} \leq N,$$

and

$$(p_D, s_D) = (N, N), \text{ for } \eta \geq \sqrt{m} \text{ and } \sqrt{m} \geq N.$$

Upon using these in (11) and (12), we find that the bounding curve between the two regions of preference is that labelled a-a' or that labelled a'' in Figs. 4, according respectively as $m \leq N^2$ or $m \geq N^2$. (In the case of Fig.  4b, it *must* be a-a'.) In writing the $\xi\eta$-equation for this curve, as for other segments of the boundary curves, we select $m$ and $N$ as parameters. The reasons are that this is permissible (i.e. they do not determine either $\xi$ or $\eta$), and it is convenient, both in that the vertical and horizontal lines that naturally form segments of the boundary curves already use these as parameters, and in that $m$ and $N$ are usually *a priori* known, so that their use as parameters permits ready construction of the boundary curves. In simplified terms, what is happening in this region is that $S < N$, so that optimization of the pure domain strategy requires, in the integration phase, either use of fewer processors than available, or use of more shooting subintervals than required for accuracy alone, or both. In either case the pure domain strategy is "penalized" in this region, relative to the nominal case discussed above.

Similar calculations show that the boundary curve in the region to the lower left of the point $(\xi, \eta) = (N/\log N, m/N)$ is that designated "b" in Figs. 4. To the upper left of this point it is either the curve "e" or "e-e'" in Fig. 4b. Here the pure column strategy is penalized, relative to the nominal case, because $n < N$, and therefore it cannot actually use all available processors. (We are not giving "credit" for possible alternative productive use of idle processors.)

We explicitly mention two of the interesting qualitative and semiquantitative conclusions that it is possible to draw from Figs. 4.

1. For given values of $m$ and $N$, the pure column strategy always is preferable for differential systems so large that $n \geq \sqrt{m} \log N \max\{\sqrt{m}/N, (2 - \sqrt{m}/N)^{-1}\}$. Similarly, if $\bar{\xi} = \bar{\xi}(N, S)$ denotes the root (in $\xi$) of

$$2 - \xi \log(N)/N + \log(\xi \log N) = \xi \log N,$$

then the pure domain strategy always is preferred for problems sufficiently small so that $n \leq \bar{\xi} \log N$. Between these two extremes, the pure domain strategy tends to become relatively better as the number of mesh points required increases (e.g. as the error tolerance decreases.)

2. If a sufficiently large number of nodes is available, then the pure column strategy always is preferable (to the pure domain strategy).

A limited use can be made of Figs. 4 in terms of selecting the best strategy to use for a specific problem. The parameters $n$ and $N$ will be known, which locates the vertical line $\xi = N/\log N$, and the operating point along some vertical line in Figs. 4. If $m$ can then be estimated (e.g. from the accuracy requirement, and experience with similar problems), then the horizontal line $\eta = m/N$ and the boundary curve can be constructed. In the two extreme cases described in 1 above, this is sufficient to determine the preferred strategy; otherwise, it will be necessary to estimate $S$, in order to select a strategy.

Perhaps it is appropriate here to recall that, in consideration of the various approximations appearing in the guidelines of the preceding section, a definite preference for one strategy over the other is warranted only if the relevant operating point is far removed from the boundary curves of Figs. 4. A choice sometimes also can be effected by knowing when the two strategies are good in some absolute sense, as opposed to the question of when one is good relative to the other, which has occupied us in the present section.

## 6. EFFICIENCY

### Generalities

The speedup of a parallel algorithm on $N$ processors is

$$\text{speedup} = T(1)/T(N),$$

where $T(p)$ is the time required on $p$ processors. (This definition is attributed by Seitz [5] to W. Ware.) The efficiency is then defined as the speedup per processor,

$$\varepsilon := T(1)/NT(N). \tag{16}$$

Efficiencies in the range 80-90% usually are sought, and 50% typically is considered the minimum acceptable.

For parallel shooting we have, in the spirit of the timing estimates of Section 4,

$$T(1) = mn^3 + Sn^3. \tag{17}$$

For the domain strategy the value of $T(N)$ is given by combining (11) and (14), and likewise for the column strategy by combining (12) and (15). This is the approach we shall use in studying the parametric dependence of efficiency in the following.

Clearly a good parallel algorithm necessarily has efficiency near unity; however, this alone is not sufficient for such an algorithm to be regarded as good. For that it is necessary (and sufficient) additionally to require that a serial implementation of the algorithm compare favorably with other serial algorithms for solving the same problem. Although there is no clear cut method of choice for solving linear two-point boundary-value problems, it is at least arguable that parallel shooting is reasonably competitive with other methods; therefore we shall restrict our considerations to those of efficiency, as defined above.

### The Domain Strategy

For this strategy the efficiency is given by

$$\varepsilon = \frac{m/N + S/N}{m/p_D + s_D},$$

where $p_D$ and $s_D$ are given by (14). Perhaps the first noteworthy fact is that the efficiency is (to the present degree of approximation) independent of the size of the underlying differential system.

It is instructive to use this expression to show how the efficiency varies with increasing $N$, for a fixed problem (i.e. for given $m, n$ and $S$). For sufficiently large $N$, one eventually is in one of the cases (14c) or (14d). In either case, simple calculation shows that the corresponding efficiency is inversely proportional to $N$. Thus, in these regions an increase (by one) in the *dimension* of the hypercube will lead to a 50% reduction in the efficiency! In order to attain efficiencies as targeted above, one clearly does not wish to go far into such regions. That is, we have found that, for a given problem, there is an upper bound on the number of nodes that can be used while attaining reasonable efficiencies.

We can actually go a bit further, and estimate this maximal efficient number of nodes. In the case (14a) and (14b) we find, respectively,

$$\varepsilon(N) = \frac{m+S}{m+SN}, S \geq N, \tag{18a}$$

and

$$\varepsilon(N) = \frac{m+S}{m+N}, \sqrt{m} \geq N \geq S. \tag{18b}$$

Given that $N$ is large compared to unity, it is clear that if $S$ is comparable to $m$, then the subject strategy cannot be efficient in either of these regions. But if $S \ll m$, then these become respectively

$$\varepsilon(N) = m/(m + SN),$$

and

$$\varepsilon(N) = m/(m + N).$$

In the first case the efficiency is approximately 50% or more if $N \leq m/S$, and in the second case it is approximately 50% or more if $N \leq \sqrt{m}$. We can summarize these results by the assertion that

$$N \leq \min\{\sqrt{m}, m/S\} \tag{19}$$

is a necessary condition for the domain strategy to attain approximately 50% efficiency in application to a give problem. Thus in order to use a 32 node hypercube efficiently, one would need a problem with $m$ on the order of 1000, and for a 128 node system, on the order of 15,000 mesh cells would be required for efficient operation. Given such situations, the maximum number of shooting intervals permitting efficient operation would be approximately 30 and 100, respectively.

In the situation presently under consideration there are different natural notions of computational grain (or atomic task) for each of the two phases. For the integration phase (and the domain strategy) this is the computations associated with an integration step across a given cell (which requires time on the order of $n^3$). There are $m$ such grains within this phase, and thus the workload (i.e. number of grains) per processor (in unit of grains per processor) during this phase would be

$$g_{ID} := m/N,$$

in the idealized case that these tasks could be uniformly distributed across all available processors (which they are in our model of the domain strategy). Similarly, for the solution phase (and the domain strategy) the natural notion of computational grain is the computation associated with a given shooting interval (and the *grain size*, in units of computational time, is again $n^3$). Accuracy requires at least $S$ such grains within this task (efficiency may dictate more). Thus the (problem dictated) number of grains per processor during this phase would be

$$g_{SD} := S/N,$$

again in the idealized case that only this many shooting intervals actually are used and that the corresponding grains are uniformly distributed across all available processors. (The latter is far from the actual case for the domain strategy). Figure 5 shows typical regions of efficiency, in terms of these two idealized workloads, for a target efficiency $\varepsilon_0 > 1/2$.

Perhaps the single most important observation from Fig. 5 is that, in order to maintain efficiency as $N$ (the number of available nodes) increases, $g_{ID}$ must increase proportionally

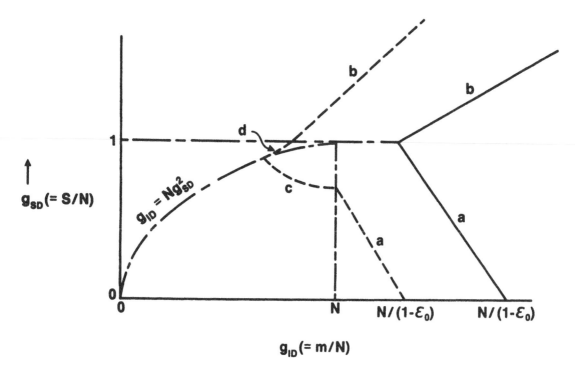

a: $g_{SD} + (1 - \varepsilon_0)\, g_{ID} = \varepsilon_0 N$

b: $(\varepsilon_0 N - 1)\, g_{SD} = (1 - \varepsilon_0)\, g_{ID}$

c: $g_{SD} + g_{ID} = 2\, \varepsilon_0 \sqrt{N}\, \sqrt{g_{ID}}$

d: $(\varepsilon_0 N - 1)\, g_{SD}^2 = g_{ID}\, (g_{SD} - \varepsilon_0)$

—: Efficiency for grain-size combinations to the right of this curve if $N >$ $(2\, \varepsilon_0 - 1)^{-1}$

---: Efficiency for grain-size combinations to the right of this curve if $N <$ $(2\, \varepsilon_0 - 1)^{-1}$

Figure 5 - The efficient region in grainsize space for the domain strategy and target efficiency $\varepsilon_0 > 1/2$.

to $N$. This is in contrast to $g_{SD}$, which must only remain constant in order to retain efficiency with increasing number of nodes. The latter behavior is more typical [7]. This scaling requirement for the grain size in the integration phase suggests that the domain strategy probably is *not* suitable for application on systems with large numbers of nodes (unless each processor also is extremely powerful). Another view of this observation is that, in order for the domain strategy to remain efficient as $N$ increases, $m$ must increase as $N$. This requirement also can be seen from the efficiency condition (19).

**The Column Strategy**

For this strategy the efficiency is

$$\varepsilon = \frac{(m+S)/N}{(m/p_C + m \log p_C/n)},$$

where $p_C$ is given by (15). In constrast to the pure domain strategy the dimension of the underlying differential system (i.e. $n$) does appear in the expression for the efficiency of the pure column strategy; indeed, as we see below, this parameter plays a crucial role in efficiency considerations for this strategy.

For sufficiently large $N$ we are in the case (15b), whence

$$\varepsilon(N) = \frac{1}{N} \frac{1 + S/m}{N(1/N + \log n/n)}.$$

From this expression it again follows that the efficiency ultimately declines rapidly with increasing dimension of the hypercube, provided the other parameters remain fixed. Thus again, for a given problem there is an upper bound on the number of nodes that can be used while retaining reasonable efficiencies.

Again we can bound the maximal efficient number of nodes. In the case (15a) we have

$$\varepsilon(N) = \frac{1 + S/m}{1 + N \log(N)/n}, N \leq n.$$

As $S/m \leq 1$, it follows that

$$N \log N \leq 3n$$

is a necessary condition for the pure column strategy to attain an efficiency of 50% or greater. Thus for a 32 node hypercube one would need a problem of dimensionality on

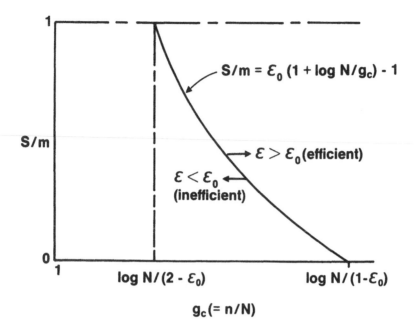

**Figure 6 - Efficient regions for the pure column strategy and arbitrary
target efficiency $\mathcal{E}_0$**

the order of $n \geq 50$ for efficient operation under this strategy, while for a 128 node system $n \geq 300$ would be required.

For the column decomposition the ideally uniformly distributed workload per processor is proportional to the number of columns per (available) processor,

$$g_C := n/N.$$

The efficiency can be expressed as

$$\varepsilon = \frac{1 + S/m}{1 + \log(N)/g_C}, N \leq n.$$

Figure 6 shows typical regions of efficiency, in terms of the parameters $g_C$ and $S/m$, for an arbitrary target efficiency $\varepsilon_0$. Note that, in order to maintain a given level of efficiency as $N$ increases, $g_c$ must increase proportionally to $\log N$, for fixed $S/m$. This is a substantially less rapid growth than required (for $g_{ID} := m/N$) with the pure domain strategy, but still is somewhat short of the objective of maintaining efficiency at fixed grain size, as $N$ increases.

## 7. CONCLUDING REMARKS

Perhaps the appropriate way of using the estimates presented here to select a strategy for parallel shooting on a hypercube is as follows. First use the results of Section 6 to determine which (if either) of the stategies meets some selected target efficiency. If only one of the strategies is efficient for the situation under consideration (and estimates of the corresponding absolute execution time indicate feasibility), then obviously that is the strategy that should be implemented. If both strategies are feasible, then the results of Section 5 should be used to help select the strategy of choice. If neither is efficient, then some alternate strategy should be sought. (See below for one possibility.)

If we use this procedure to consider the merits of the two strategies for increasing $m$ (number of mesh points), $n$ (dimension of the differential system) and $N$ (number of processors), then the following conclusions result:

1. For sufficiently large $m$, and target efficiencies $\varepsilon_0 > 1/2$, the pure domain strategy is always efficient. On the other hand the pure column strategy is inefficient if $n < N \log(N)/(1 - \varepsilon_0)$, but is efficient otherwise. In the former situation the pure domain

strategy clearly is preferable. In the latter case the relative preference of the two strategies is governed by Fig. 4a. In the limit $m \to \infty$ the boundary curve between regions of preference in this figure consists only of the segment labelled "b" and the appropriate portion of the line $\xi = \eta$. For sufficiently large $m$, other parameters remaining fixed, the operating point always is located above this boundary, and thus again the pure domain strategy is preferable. Thus for a fixed problem and hypercube, the pure domain strategy always should be used if the accuracy requirements are sufficiently stringent. A simple intuitive explanation for this is that, under these circumstances, the effort in the integration phase strongly dominates that in the solution phase, and in the integration phase the domain decomposition always is more efficient than the column decomposition, because of a lesser communication requirement.

2. For suffciently large $n$, other parameters being equal, Fig. 6 shows that the pure column strategy always is efficient. On the other hand, the efficiency of the pure domain strategy is independent of $n$. Nonetheless, Figs. 4 show that the pure column strategy always is preferable. The serial bottleneck faced in the solution phase by the domain strategy ultimately (with increasing $n$) makes it less efficient than the column strategy, although it cannot *per se* make it inefficient.

3. For large $N$, other parameters being equal, Figs. 5 and 6 show that *both* strategies ultimately become inefficient. If one must use one of the two strategies here, then the column strategy probably is preferable, because its efficiency degenerates only as $1/\log N$, whereas that of the domain strategy varies as $1/N$.

Except in situations such that the desired accuracy dictates large $m$ (i.e. point 1 above prevails) it would appear that the column strategy is perhaps overall the best of the two possibilities considered here. Nonetheless, the pure domain strategy does appear to be reasonably efficient for a substantial class of problems, and further it undeniably is the easier of the two to implement. For these reasons we presently are undertaking an implementation of the pure domain strategy. We hope to present elsewhere a comparison of the performance of this implementation with the theoretical estimates presented here.

Inasmuch as the domain decomposition is most efficient during the integration phase, while the column decomposition is most efficient during the solution phase, it is natural to consider a *hybrid strategy*, in which domain decomposition is used in the integration phase and column decomposition in the solution phase. Of course such a strategy will involve communication costs associated with redistributing the results of the integration phase as appropriate to a column decomposition. We hope also to present elsewhere a theoretical study of this strategy, similar to that presented here for the two pure strategies, along with comparisons of the latter two strategies with the hybrid strategy.

Acknowledgments    This research was partially supported by the U.S. Department of Energy, under Contract No DE-AA03-76-SF00767. The work of Paul Nelson also was partially supported by a Faculty Development Leave from Texas Tech University, and by Contract No. DE-FG05-87ER25042 between Texas Tech and the Department of Energy. A (slightly) modified version of this work is to appear in the Proceedings of the 1986 ODE Conference, which will be published in the journal *Applied Mathematics and Computation*.

## REFERENCES AND NOTES

1. H.B. Keller, *Numerical Solution of Two-Point Boundary Value Problems,* SIAM/CBMS Regional Conference Series In Applied Mathematics, Vol. 24, SIAM, Philadelphia, 1976.

2. E. Brooks, et. al., "Pure Gauge $SU(3)$ Lattice Theory on an Array of Computers," *Phys. Rev. Letters* 52 (1984), 2324-2327.

3. G. Fox, "Concurrent Processing for Scientific Calculations," in *Proceedings COMP-CON '84 Conference,"* pp. 70-73, IEEE Computer Society (1984).

4. C.L. Seitz, "Experiments with VLSI Ensemble Machines," *J. VLSI Comput. Syst.* 1 (1984), 3.

5. C.L. Seitz, "The Cosmic Cube," *Comms. ACM* 28 (1985), 22-33.

6. G.C. Fox, "Are Concurrent Processors General Purpose Computers?," *IEEE Trans. Nucl. Sci.* NS-32 (1985), 182-186.

7. G.C. Fox, "The Performance of the Caltech Hypercube in Scientific Computations," in Supercomputers: Algorithms, Architectures and Scientific Computation, F.A. Matsen and P. Tajima, eds., Univ. Texas Press (1986).

8. J. Patterson, "Caltech-JPL Concurrent Computation Project Introductory User's Guide," Caltech Concurrent Computation Project memo. Hm-159, March 1985. (Esp. Section 7).

9. G.C. Fox and S.W. Otto, "Algorithms for Concurrent Processors," *Physics Today* 37, No. 5 (May 1984), 50-59.

10. G.C. Fox, "On the Sequential Component of Computation," Caltech Concurrent Computation Project memo. Hm-130, December 11, 1984.

11. R.M.M. Mattheij and G.W.M. Staarink, "An Efficient Algorithm for Solving General Linear Two Point BVP," *SIAM J. Sci. Stat. Comp.* 5 (1984), 745-763.

12. U. Ascher, J. Christiansen and R.D. Russell, "Collocation Software for Boundary Value ODE's," *ACM Trans. Math. Software* 7 (1981), 209-229.

13. See also various papers in *Codes for Boundary-Value Problems in Ordinary Differential Equations,* Eds. B. Childs et al., Springer-Verlag (1979).

14. Uri M. Ascher and R.M. Mattheij, *Numerical Solutions of Boundary Value Problems for Ordinary Differential Equations,* Prentice-Hall, 1988.

15. *Proceedings of the First International Symposium on Domain Decomposition Meth-*

*ods for Partial Differential Equations,* Roland Glowinski, Gene H. Golub, Gérard A. Meurant and Jacques Periaux, Eds., SIAM, Philadelphia, 1988.

16. H.B. Keller, "Accurate Difference Methods for Ordinary Differential Systems Subject to Linear Constraints," *SIAM J. Numer. Anal.* 6 (1969), 8-30.

17. F.T. Krogh, "Workshop Selection of Shooting Points," pp. 159-163 of Ref. 13.

18. J.M. Varah, "On the Solution of Block-Tridiagonal Systems Arising from Certain Finite-Difference Equations," *Math. Comp.* 26 (1972), 859-869.

19. A.B. White, "Numerical Solution of Two-Point Boundary-Value Problems," Ph.D. thesis, 1974, California Institute of Technology, Pasadena, California.

20. J.M. Varah "Alternate Row and Column Elimination for Solving Certain Linear Systems," *SIAM J. Numer. Anal.,* 13 (1976), 71-75.

21. J.C. Diaz, G. Fairweather and P. Keast, "FORTRAN Packages for Solving Certain Almost Block Diagonal Linear Systems by Modified Alternate Row and Column Elimination," *ACM Trans. Math. Software* 9 (1983), 358-375.

22. J.C. Diaz, G. Fairweather and P. Keast, "COLROW and ARCECO: FORTRAN Packages for Solving Certain Almost Block Diagonal Linear Systems by Modified Alternate Row and Column Elimination (Algorithm 603)," *ACM Trans. Math. Software* 9 (1983), 376-380.

23. J.K. Reid and A. Jennings, "On Solving Almost Block Diagonal (Staircase) Linear Systems," *ACM Trans Math Software,* 10 (1984), 196-201.

24. G. Fox, "Square Matrix Decompositions-Symmetric, Local, Scattered," Caltech Concurrent Computer Project memo. Hm-97, August 13, 1984.

25. G.A. Geist, "Efficient Parallel LU Factorization with Pivoting on a Hypercube Multiprocessor," Oak Ridge National Laboratory Technical Report ORNL-6211, October 1985.

26. G.J. Davis, "Column LU Factorization with Pivoting on a Hypercube Multiprocessor," Oak Ridge National Laboratory Technical Report ORNL-6219, November 1985.

27. O.A. McBryan and E.F. Van de Velde, "Hypercube Algorithms and Implementation," *SIAM J. Sci. Statist. Comput.,* 8 (1987), S227-S287.

28. J.L. Gustafson, G.R. Montry and R.E. Benner, "Development of Parallel Methods for a 1024-Processor Hypercube," *SIAM J. Sci. Statist. Comput.,* 9 (1988), 609-638.

# An Asymptotic Induced Numerical Method for the Convection-Diffusion-Reaction Equation

Jeffrey S. Scroggs*
Institute for Computer Applications in Science and Engineering
NASA Langley Research Center
Hampton, Virginia
and
Danny C. Sorensen†
Math and Computer Division
Argonne National Lab.
Argonne, Illinois

A parallel algorithm for the efficient solution of a time dependent reaction convection diffusion equation with small parameter on the diffusion term will be presented. The method is based on a domain decomposition that is dictated by singular perturbation analysis. The analysis is used to determine regions where certain reduced equations may be solved in place of the full equation. Parallelism is evident at two levels. Domain decomposition provides parallelism at the highest level, and within each domain there is ample opportunity to exploit parallelism. Run-time results demonstrate the viability of the method.

## 1 INTRODUCTION

In this paper, a new approach to solving partial differential equations which model fluid flow is discussed and demonstrated. The algorithm is appropriate for modeling laminar transonic flow, such as through a duct of variable width. The method is an asymptotics-

---

*Research conducted while in residence at the Center for Supercomputing Research and Development, University of Illinois supported in part by the National Science Foundation under Grant No. US NSF PIP-8410110, the U.S. Department of Energy under Grant No. US DOE-DE-FG02-85ER25001, the Air Force Office of Scientific Research under Grant No. AFOSR-85-0211, the IBM Donation to CSRD, and by the Applied Mathematical Sciences subprogram of the Office of Energy Research, U.S. Department of Energy by Lawrence Livermore National Laboratory under Contract No. W-7405-Eng-48. Research was also partially supported by NASA Contract No. NAS1-18605 while in residence at ICASE.

†Work supported in part by the Applied Mathematical Sciences subprogram of the Office of Energy Research, U.S. Department of Energy under Contracts W-31-109-Eng-38, DE-AC05-840R21400.

induced numerical method suitable for parallel processors which represent the state of the art in scientific computers. The contents of this paper concentrate on a description of the method and computational results. The complete theoretical basis for the algorithm has been developed in [32] and will appear separately.

Competition between convection, diffusion, and reaction is crucial to the understanding of fluid flow. When modeling transonic flow, except in regions of rapid variation such as in shocks and boundary layers, convection and reaction dominate over diffusion. A novel aspect of this method is the use of asymptotic analysis to exploit these physical properties, providing the theoretical basis for a domain decomposition. The analysis identifies the following two types of subdomains: regions where the solution is smooth, where a reduced equation may be solved; and regions of rapid variations, such as in a neighborhood of a shock, where the full equation must be solved. Domain decomposition provides large-grain parallelism. The domain decomposition is independent of the choice of numerical schemes for the subdomains; thus, schemes may be chosen which are a source of smaller-grain parallelism. Even though large grain parallelism is not exploited in the implementation, significant speedups are demonstrated. In addition to dictating the domain decomposition, asymptotics also provides a means of approximating solutions to the problem. In this way, a set of simplified problems is obtained that is better conditioned for numerical computations; hence, they may be solved by conventional techniques. The use of asymptotic analysis to precondition the computations is a new aspect of this method.

The techniques presented herein are applicable to Computational Fluid Dynamics (CFD) in the transonic and supersonic regimes, in physical settings such as laminar flow through a nozzle (duct) and laminar flow around airfoils and other bodies. The gasdynamic equations, including viscous effects, are used as a model in these settings. Except for very simple geometries and boundary conditions there is no analytic solution to these gasdynamic equations, and a numerical solution is difficult to obtain. For these reasons new algorithms are usually developed and tested on a more tractable canonical equation. The convection-diffusion-reaction equation

$$\frac{\partial u}{\partial t} + A(x,t,u)\frac{\partial u}{\partial x} - \epsilon\frac{\partial^2 u}{\partial x^2} - r(x)u = 0, \tag{1}$$

is such a canonical equation and will be the focus of this paper. The flows considered in this paper are not reacting fluids. Here, the reaction term arises from the effects of a variable cross sectional area in a duct. When the equation is nondimensionalized [25], the diffusion coefficient $\epsilon$ is inversely proportional to the Reynolds number. Based on free-stream conditions in transonic flow, the Reynolds number for this problem is large. Asymptotic analysis exploits the smallness of the positive parameter $\epsilon$ and involves study of the solution as $\epsilon$ tends to zero ($\epsilon \downarrow 0$). This equation contains many of the properties that make the gasdynamic equations difficult to solve; namely, it is capable of modeling rapid variations such as shocks and boundary layers. The method is capable of obtaining solutions to (1) when the shock is not stationary, which extends Howes' studies [9,10] into the time-dependent regime.

Asymptotic analysis gives qualitative and quantitative information as $\epsilon \downarrow 0$. The numerical method presented here exploits the analysis to determine an accurate solution for small positive $\epsilon$. The method is in the spirit of matched asymptotic expansions [20,16], but it is not a numerical implementation of matched asymptotics. The asymptotic analysis involves the derivation of analytic upper and lower bounds on the solution, and is performed in the style of Howes [13,12,11]. Initially, bounds are discussed which are valid only in certain subregions. Then the bounds are combined to form a global *a priori* error bound.

Another novel feature of the method is the availability of extensive error information in the form of both *a priori* error bounds and reliable *a posteriori* error estimates. Reliable *a posteriori* error estimates are obtained using the error analysis which accompanies the numerical schemes used to solve the sub-problems. In addition, *a priori* error bounds are provided through the use of asymptotic analysis. The error analysis is based on the physical mechanisms associated with the problem; hence, it is based on accurate information (see [24]), not on the truncation of a Taylor series of a poorly behaved function.

The method is an iterative technique. A linearized version of the original problem is solved in each step of the iteration. Theorems establishing the convergence of the method are presented, the proofs will appear in a subsequent paper. Computational experiments show that in just a few steps of the iteration, the solution to the nonlinear equation may be obtained. The iterative algorithm as well as the theorems associated with it are novel.

In the next section, some of the ideas behind multiple scales asymptotic analysis are discussed. In addition, an introduction into how the asymptotic analysis and the numerical analysis are blended to form a computational method is presented. In Section 3 the problem is presented. Asymptotic analysis specific to this problem is discussed with the theorems supporting the method in Section 4. The iteration and method for detection of the subdomain boundary is discussed in Section 5. The numerical schemes used in the method are presented in Section 6. A global error approximation is presented in Section 7. The method is stated in algorithmic form in Section 8. In Section 9 computational results on an Alliant FX/8 are presented.

## 2 MULTIPLE SCALES

Many problems of scientific interest have multiple scales. These problems are characterized by the presence of distinguishable physical mechanisms, each associated with a temporal or spatial gauge or scale. When modeling a shock in a duct, for example, the width of the duct provides one scale, and the thickness of the shock layer provides another. The resolution of these scales is frequently required to determine the physics of interest. Asymptotic analysis provides analytic tools to identify and utilize the multiple scales. The relative importance of any two physical processes in a given domain may be measured by the ratio of the corresponding scales; thus, the various scales may be ranked by a set of dimensionless parameters, the ratios of scales. When the ratio of two scales is a large or a small number, then it often happens that one of the competing mechanisms is dominant in most of the domain. For example, in laminar duct flow with large Reynolds number the effects of viscosity may be ignored except in a neighborhood of the shock and boundary layers. The scales of the various competing processes (and, therefore, the relative magnitudes of the dimensionless parameters) usually change as the phenomenon evolves. Consider the behavior of the solution of the nonlinear parabolic equation,

$$P[u] := u_t + uu_s - \epsilon u_{ss} - ru = 0, \tag{2}$$

where $\epsilon$ is a small positive parameter. This equation may be used as a model for shocks and boundary layers. For example, if $r(x) = -A'(x)/A(x)$, where $A(x)$ is the width of the duct of Figure 1, then this equation is associated with the flow through the duct [10]. There

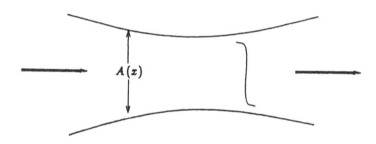

Figure 1 Variable width duct. (From [32].)

are (at least) two sets of scales appropriate when modeling shocks—the scales associated with the original variables $(x, t)$, and the scales appropriate in a small neighborhood of the shock (these are discussed in Section 4).

The most easily tractable multiple-scale problems are those in which there are only a small number of widely separated groups of scales and the motion on the fastest scales has little influence on the smooth part of the solution. An identifying feature of this class is the presence of local regions in which the solution undergoes rapid variation. Such regions are called boundary or internal layers, when located in the neighborhood of a boundary or in the interior of the domain, respectively. These are the problems that are most natural for multitasking because it is easy to break up the domain according to the regions of different local behavior. The method presented here is appropriate for this class of multiple-scale problems.

The decomposition into domains is accomplished using a symbiosis of numerics and asymptotics. The asymptotic analysis identifies the regions where diffusion is negligible.

In these regions, it is sufficient to solve a reduced equation. Solving this reduced equation can significantly reduce the work in the numerical method, and/or increase the potential for parallelism. For example, this allows the use of the method of characteristics to obtain a good approximation for the solution of (2). The numerics provides a means of solution in the subdomains, and also a feedback mechanism. The numerical scheme can expose regions of unexpected behavior, confirming or correcting the asymptotics-induced subdomain boundaries. This decomposition permits the use of locally refined meshes, allowing the concentration of computational effort in the regions where it is needed most.

There is much literature on multiple-scales problems. Analytic methods for multiple-scales problems are discussed in the books [3,16,19,34]. The theory of multiple-scales analysis is discussed in [17,6,21]. The books [2,23] discuss both the techniques and the theory behind them. Finally, numerical techniques for multiple-scales problems are discussed in the papers [22,26,33]. This list is meant only as an introduction to the literature, and not as a complete list.

## 3   THE NONLINEAR PROBLEM

The method will be described and demonstrated by solving (2) on the domain

$$D := \{(x,t)|0 \leq x \leq b, 0 \leq t < T\}, \tag{3}$$

subject to

$$u(x,0) = \gamma(x), \quad 0 < x < b; \tag{4}$$

$$u(0,t) = \alpha(t), \quad 0 < t < T; \text{ and} \tag{5}$$

$$u(b,t) = \beta(t), \quad 0 < t < T. \tag{6}$$

The portion of the boundary along which the data is specified is denoted by

$$\Pi := \{(x,t)|0 \leq x \leq b, \ t = 0\} \bigcap \{(x,t)|0 \leq t < T, \ x = 0, b\}.$$

For the sake of simplicity, it is assumed that all boundaries are inflow boundaries, that is, $\alpha(t) \geq \alpha_0 > 0$ and $\beta(t) \leq \beta_0 < 0$. The boundary data is assumed to be compatible; thus,

$$\alpha(0) = \gamma(0), \quad \text{and} \quad \gamma(b) = \beta(0). \tag{7}$$

The coefficient of the forcing term $r(x)$ is bounded with bounded derivatives. In addition, it is assumed that the solution to the reduced equation

$$P_0[\check{u}] := \check{u}_t + \check{u}\check{u}_x - r\check{u} = 0 \tag{8}$$

has continuous derivatives at $(x,t) = (0,0)$, and $(x,t) = (b,0)$. This last restriction prevents the formation of corner layers, and may be expressed as

$$\frac{d\alpha}{dt} + \gamma\frac{d\gamma}{dx} - r\gamma = 0, \quad \text{for } (x,t) = (0,0); \tag{9}$$

$$\frac{d\beta}{dt} + \gamma\frac{d\gamma}{dx} - r\gamma = 0, \quad \text{for } (x,t) = (b,0). \tag{10}$$

Under these conditions, the solution to (2) is uniquely defined [4].

## 4   ASYMPTOTIC ANALYSIS

Asymptotic analysis is employed to identify the dominant physics, creating an efficient and accurate numerical method. The analysis in the neighborhood of a shock is outlined here.

Shocks form in regions of merging characteristics (see Figure 2). Since the boundary

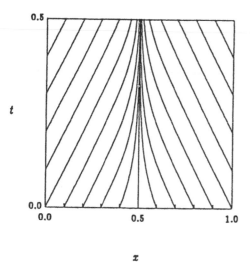

**Figure 2** Characteristics of steady-state solution $u = -\tanh[(.5 - x)/2\epsilon]$, for $\epsilon = .05$. (From [32].)

conditions imposed on the problem are inflow conditions on both the $x = 0$ and $x = b$ boundaries, the characteristics are traveling in the direction of increasing $x$ from $x = 0$, and in the direction of decreasing $x$ from $x = b$. These will merge (become asymptotically close) somewhere inside $D$, forming a shock. The merging of the characteristics stabilizes the shock, and keeps it from dispersing.

Since the behavior of $u$ as $\epsilon \downarrow 0$ is of interest, it is natural to first study the solution of the reduced equation (8). Weak solutions $\breve{u}$ are sought for (8) with boundary data (4-6). In order that $\breve{u}$ be uniquely defined, it is necessary to impose an entropy condition [18]. Suppose that $\breve{u}$ has a single shock. That is, suppose $\breve{u}$ is the solution to (8) subject to (4-6) that is discontinuous only along a curve $(x, t) = (\Gamma(t), t)$. For small $\epsilon$, this curve lies in the shock-layer region of the solution to the full problem. The size of this region tends to zero as $\epsilon \downarrow 0$. Analytic methods for choosing $\Gamma$ are discussed by Whitham [36], Kevorkian and Cole [16], and others. The path of the discontinuity is an analytic tool needed only for the theory. Since $\Gamma$ is not needed for the computations, methods for choosing $\Gamma$ will not be discussed here.

The initial and boundary data are assumed to be smooth; thus, the shock does not exist at $t = 0$. Rather, $\Gamma$ is assumed to be undefined for $t < t^\Gamma$, where $t = t^\Gamma$ is the time $\breve{u}$

becomes discontinuous. It is natural to describe $\breve{u}$ in terms of the following functions:

$$\breve{u}(x,t) = \begin{cases} \breve{u}_0(x,t) & \text{for } 0 < t \le t^\Gamma \\ \breve{u}_1(x,t) & \text{for } x < \Gamma^{-1}(t) \text{ and } t \ge t^\Gamma \\ \breve{u}_2(x,t) & \text{for } x > \Gamma^{-1}(t) \text{ and } t \ge t^\Gamma. \end{cases}$$

The shock speed for $\breve{u}$ is

$$(\breve{u}_1(\Gamma(t),t) + \breve{u}_2(\Gamma(t),t))/2), \tag{11}$$

so the entropy condition may be expressed as

$$\breve{u}_1(\Gamma(t),t) > (\breve{u}_1(\Gamma(t),t) + \breve{u}_2(\Gamma(t),t))/2) > \breve{u}_2(\Gamma(t),t). \tag{12}$$

Under these conditions,

$$\mu(t) = \breve{u}_1(\Gamma(t),t) - \breve{u}_2(\Gamma(t),t) \ge 0. \tag{13}$$

The regions where $\breve{u}$ is a good approximation to $u$ are defined by presenting functions which bound the difference $\breve{u} - u$. These bounds are small except in an asymptotically small neighborhood of the shock. The bounds, based on Howes [11], are reflected in the following theorem.

THEOREM 1. (Howes) Let $\breve{u}$ be the solution to $P_0[\breve{u}] = 0$ and $u$ the solution to $P[u] = 0$ on $D$, each satisfying the the boundary data (4-6). The solution to the reduced equation, $\breve{u}$, is possibly discontinuous on the curve $(x,t) = (\Gamma,t)$. Assume that the boundary data satisfy the following smoothness conditions: the data (4-6) satisfy the compatibility conditions (7),(9-10), and $\alpha$, $\beta$, $\gamma$, with their first and second derivatives are all bounded. Then for $\epsilon$ small enough

$$|u - \breve{u}| = O(\mu \exp[-f^2(x,t)/\epsilon^{1/2}]) + O(\epsilon)$$

when the derivatives of $\breve{u}$ are continuous across $\Gamma$, and

$$|u - \breve{u}| = O(\mu \exp[-f^2/\epsilon^{1/2}]) + O(\epsilon^{1/2}\delta \exp[-f/\epsilon^{1/2}]) + O(\epsilon)$$

in the more general case when the derivatives of $\breve{u}$ are not continuous across $\Gamma$. Here $f(x,t)$ is a distance function between $(x,t)$ and $(\Gamma,t)$, and $\delta$ is an upper bound on the difference of the normal derivative of $\breve{u}$ across $\Gamma$.

It is now reasonable to utilize the theorem to make the definitions of the subdomains more precise. The internal layer is the following neighborhood of $\Gamma$:

$$D_{IL} = \{(x,t)|(x,t) \in D, |x - \Gamma^{-1}(t)| \le \Delta(t)\}. \tag{14}$$

Here $\Delta(t) \le K\eta(t)\epsilon^{1/4}\ln^{1/2}\epsilon$ is the width of the internal layer at time $t$ ($K$ is a constant independent of $\epsilon$). The outer region is the complement of $D_{IL}$ with respect to $D$, that is,

$$D_{OR} = \{(x,t)|(x,t) \in D, |x - \Gamma^{-1}(t)| > \Delta(t)\}. \tag{15}$$

The upper bound on the size of the internal layer is based on the $\exp[-f^2/\epsilon^{1/2}]$ term in the error bounds of Theorem 1.

Theorem 1 motivates a preconditioning for the problem in $D_{OR}$. The theorem states that (8) may be solved in place of (2). In addition, the theorem provides an error bound if diffusion (artificial or implicit in the numerical scheme) is incorporated into the solution process of either (8) or (2). Thus, the numerical method for $D_{OR}$ may be chosen from the wide variety of methods designed for hyperbolic equations [1,35,8,28].

The solution in the outer region is used to provide boundary data for the problem in the internal layer. (This is justified in Section 5.1.) Thus, it is possible to have boundary data for the internal-layer problem which is perturbed from the exact solution. The effects of this perturbation are that the height and location of the shock can vary by the same magnitude as the perturbation itself. This results in an error of magnitude $O(1)$ as $\epsilon \downarrow 0$ in an asymptotically small neighborhood of the shock, and is reflected in the error bound of the following theorem [32].

THEOREM 2. Let $u$ and $v$ be solutions to (2) on the domain $D_{IL}$ with their boundary values satisfying the following smoothness conditions: the data are bounded with bounded derivatives. Assume that the curve defining $\partial D_{IL}$ is smooth. [1] Let

$$\psi(x,t) = u - v, \quad \text{for } (\text{x},\text{t}) \in \partial\mathrm{D}_{\mathrm{IL}}.$$

Then for $\epsilon$ and $\Delta$ small enough

$$|u - v| = O(\psi) + O((1 + f^2/\epsilon^2)^{-1}),\qquad(16)$$

where $f(x,t)$ measures the distance from $(x,t)$ to $(\Gamma,t)$.

The theorem is the source for a second preconditioner. Namely, a scaled and translated coordinate system based on (16). Let $f = |x - \Gamma(t)|$. Setting $\tilde{x} = (x - \Gamma)/\epsilon$, the second term in the right hand side of (16) will be large when $\tilde{x} = O(1)$ as $\epsilon \downarrow 0$. An analogue of $\tilde{x}$ will be used for the spatial coordinate in $D_{IL}$, and will be described in Section 6.2. The use of this scaled and translated coordinate system creates a better conditioned numerical problem, thus it is a preconditioning.

The local error bound of Theorem 1 is now used to form a global *a priori* error bound. The bound, as presented in Theorem 3 below, is sharp in $D_{OR}$; however, the bound reflects the possibility of shock displacement in the internal-layer region.

THEOREM 3. Let $u$ be the solution to (2) satisfying (4-6). Suppose $v$ is obtained by first solving (8) in $D_{OR}$ subject to (4-6), then solving (2) on $D_{IL}$ with boundary data $v$ on $\partial D_{IL}$. Assume the compatibility conditions (7),(9-10) obtain, and that the data (4-6) are bounded with bounded derivates. If $E = \|u - v\|_1$, then for $\epsilon$ small enough

$$E = O(\epsilon)$$

---

[1] The curve is continuous with a continuous tangent.

in $D_{OR}$, and
$$E = O(\epsilon^{1/4} \ln^{1/2} \epsilon)$$
in $D_{IL}$.

The computational results were much stronger than the theorem suggests. Both the magnitude of $\Delta$, and the magnitude of the error in the internal-layer subdomain were smaller in the computational results. Thus, the *a priori* error bounds of the asymptotic analysis did not reflect all of the accuracy and behavior of the computational algorithm.

Asymptotics identified two subdomains and provided preconditioners for the problems within the subdomains. The preconditioner for the full equation in $D_{IL}$ is the use of the local scale $\tilde{x} = (x - \Gamma)/\epsilon$ dictated by Theorem 2. This scale allows the diffusion to be modeled accurately, hence the grid is fine enough to resolve the shock. It is reasonable to use this scaling in the method, because computationally the internal-layer subdomain is of width $O(\epsilon)$. The preconditioning in the outer-region subdomain $D_{OR}$ is to solve (8) in place of (2), and was justified in Theorem 2. First, the domain decomposition and preconditionings are combined with a functional iteration to form the computational method. The particular numerical methods discussed herein are not new; however, their combination to form this method is.

## 5   DISCUSSION OF THE METHOD

An iteration is formed by linearizing the reduced problem. Each step of the iteration requires the solution of (8) in the outer-region subdomain, and (2) in the neighborhood of a shock. This is discussed in Section 5.1. Once the iteration has been described, the boundary detection scheme is presented. The method assumes no *a priori* information about the location of the internal-layer boundary, and is supported by theory. In Section 5.3, convergence of the method is presented. Numerical details of the method will be presented in Section 6.

### 5.1   Iteration

In general, each step of the iteration requires the solution of a linear convection-reaction equation in the outer-region subdomain, followed by the solution of a nonlinear convection-diffusion-reaction equation in the internal layer. The convection-reaction equation

$$U_t^{k+1} + U^k U_x^{k+1} - rU^{k+1} = 0 \tag{17}$$

is formed by lagging the convection coefficient of (8). The boundary of the internal-layer subdomain is allowed to change during the iterations. Thus, denote the outer-region subdomain for iterate $U^{k+1}$ by $D_{OR}^k$, and denote the complement of $D_{OR}^k$ with respect to $D$ by $D_{IL}^k$. That is, $U^{k+1}$ is obtained by solving (17) in $D_{OR}^k$, then solving (20) in $D_{IL}^k$.

The solution of (17) in $D_{OR}^k$ is obtained via a modification of the method of characteristics to account for the forcing term $rU^{k+1}$. The characteristic transformation $(x,t) \to (\xi, \tau)$

is defined by setting $t = \tau$, and by solving

$$\frac{\partial x^k}{\partial \tau} = U^k(x^k(\xi, \tau), \tau), \tag{18}$$

with initial conditions

$$x^k(0) = \xi, \quad \text{for } b > \xi > 0;$$
$$x^k(y_a^{-1}(\xi)) = 0, \quad \text{for } \xi < 0;$$
$$x^k(y_b^{-1}(\xi)) = 0, \quad \text{for } \xi > b;$$

where, $(\xi, \tau) = (y_a(\tau), \tau)$ is the image of the curve $(x, t) = (0, t)$, and $(\xi, \tau) = (y_b(\tau), \tau)$ is the image of the curve $(x, t) = (b, t)$. Utilizing this transformation, it is a simple task to solve

$$U_\tau^{k+1} = r U^{k+1} \tag{19}$$

in place of (17) along the characteristics defined by (18). This transformation becomes singular in a neighborhood of a shock, hence cannot be applied in the internal-layer subdomain. This fact is the basis of the procedure used to determine $\partial D_{IL}^k$.

Solutions to the reduced equation are poor approximations to the solution of the full equation in regions of large gradients, such as in the internal-layer subdomain. Thus, the full equation is solved in the internal layer at each iteration. The equation

$$U_t^{k+1} + U^{k+1} U_x^{k+1} - \epsilon U_{xx}^{k+1} - r U^{k+1} = 0 \tag{20}$$

is solved in the internal-layer subdomain for each $k$. Boundary data for the internal-layer subdomain is provided by the solution of (17) in the outer region. This is justified by observing that for $\epsilon$ small enough, the boundary of the internal-layer subdomain will be an inflow boundary.

## 5.2   Boundary Detection

It is desirable to be able to compute the location of the internal-layer subdomain during the course of the iteration. This has the advantage of requiring less *a prior* information. In addition, if the initial guess provides a poor approximation to the location of the internal-layer subdomain, the method will be able to correct the location of the boundary in the course of the iteration.

The method used to locate the internal-layer subdomain boundary is based on properties of the transformation used to solve (17). The transformation used to solve (17) will become singular (or nearly singular) in the region where characteristics merge (see Figure 2). Thus, the Jacobian

$$J^k = \partial x / \partial \xi, \tag{21}$$

of the transformation (18) will be asymptotically small in a neighborhood of the shock, while it is $O(1)$ in the outer-region subdomain. (For more details on the relationship between the magnitude of $J^k$ and the nature of the transformation, see [31].) This is the measure used to locate the boundary of the internal-layer subdomain. The size of the

Jacobian is monitored via the solution of an ODE along each characteristic path of interest. Combining the partial of (18) with respect to $\xi$ and the partial of (21) with respect to $\tau$, the equation

$$\frac{\partial J^k}{\partial \tau} = J^k \frac{\partial U^k}{\partial x}(x(\xi, \tau), \tau) \tag{22}$$

may be derived. The Jacobian is determined by solving this equation subject to $J^k(x_0, t_0) = J_0^k$ for $(x_0, t_0) \in \Pi$. The behavior of the solution inside the domain $D$ is of primary interest; therefore, it is sufficient to monitor $\hat{J}^k := J^k(x_1, t_1)/J_0^k$ in place of $J^k$ to determine the boundary. The ratio is determined using (22) with $\hat{J}^k$ in place of $J^k$, subject to $\hat{J}^k(x_0, t_0) = 1$ for $(x_0, t_0) \in \Pi$. The curve on which $\hat{J}^k$ becomes nearly singular, that is, where

$$\hat{J}^k(x, t) = TOL, \tag{23}$$

for some small number $TOL$, is the boundary of the internal layer subdomain for iteration $k$. The subdomains separated by (23) will, in general, be different than the subdomains used in the theorems. Thus, the subdomains used in the numerical method are

$$\hat{D}_{OR}^k = \{(x, t) | (x, t) \in D, \ \hat{J}^k(x, t) \geq TOL\}, \tag{24}$$

and,

$$\hat{D}_{IL}^k = \{(x, t) | (x, t) \in D, \ \hat{J}^k(x, t) < TOL\}. \tag{25}$$

The theory applies provided

$$D_{IL} \subseteq \hat{D}_{IL}^0 \subseteq \hat{D}_{IL}^1 \subseteq \ldots \subseteq \hat{D}_{IL}^{k-1} \subseteq \hat{D}_{IL}^k; \tag{26}$$

however, convergence was observed when this relation failed, thus the constraint (26) is not a necessary condition for the computational method to converge.

Heuristics, based on both accuracy and efficiency, are used to choose $TOL$. If $TOL$ is too small, then accuracy will suffer. This is because the internal-layer subdomain will be too small, and the data provided at the boundary of the internal-layer subdomain will have large perturbations as compared to the desired solution. If $TOL$ is too large, the internal-layer subdomain will be too large, and the computational mesh will be refined in regions where the solution is smooth, creating excess work.

## 5.3  Convergence

An advantage to this method is the availability of extensive error information. A global error bound based on Theorem 3 will be presented in this section. First, convergence of the iteration is established by showing the iteration is a contraction mapping. For more details on these results, see [32].

The convergence of the iteration (17) to a solution of (8) in the outer region will be shown by comparing successive iterates, then establishing a lower bound on the latest time at which the iteration is a contraction. For the sake of the theorem, the boundary of the internal-layer subdomain is assumed to be stationary from iteration to iteration

$(\hat{D}_{OR}^{k-1} = \hat{D}_{OR}^{k} = D_{OR})$. With the analysis that follows, this theorem provides a lower bound for the largest time at which the iteration will converge:

THEOREM 4. Let $\{U^k\}_{k=1}^{\infty}$ be the set of successive iterates of (17) in the subdomain $D_{OR}$ satisfying (4-6) with initial guess $U^0$. Assume $U^0$ satisfies (4-6) and is Lipschitz continuous on $D$. The boundary data are assumed to satisfy the compatiblity conditions (7),(9-10) and to have bounded first and second derivatives. Let

$$\delta = \sup_{D} |U^k - U^{k-1}|.$$

Then

$$|U^{k+1} - U^k| < \delta C e^{-\lambda t}(e^{Rt} - 1) \qquad (27)$$

for $(x,t) \in D_{OR}$. Here $C$, $\lambda$ and $R$ are known positive constants.

This theorem provides an upper bound on the latest time for which the iteration converges. Apply the infinity norm to (27) to obtain

$$\|U^{k+1} - U^k\|_\infty \le \hat{C}\|U^k - U^{k-1}\|_\infty.$$

Then the following corollary provides the conditions for convergence.

COROLLARY 5. Suppose that the conditions of Theorem 4 hold. Let $T_{max}$ be the largest positive number such that

$$\hat{C} = \sup_{0 \le t \le T_{max}} C e^{-\lambda t}(e^{Rt} - 1) \le 1.$$

If the bound on time in (3) satisfies $0 < T < T_{max}$, then the iteration in $D_{OR}$ defined by (17) is a contraction mapping; therefore, the sequence of iterates converges to $v = \lim_{k\to\infty} U^k = U^\infty$, which is a solution of (8) on $D_{OR}$ satisfying (4-6).

A statement of an *a priori* error bound for the method is presented in Corollary 6 below. As with Theorem 3, the bound is sharp in $D_{OR}$; however, the bound is crude in the region of the shock.

COROLLARY 6. Let $u$ be the solution to (2) satisfying (4-6). Suppose each iterate $U^k$ is obtained by first solving (17) in $D_{OR}$ subject to (4-6), then solving (20) on $D_{IL}$ with boundary data $U^k$ on $\partial D_{IL}$. Assume that the compatibility conditions (7),(9-10) hold, and that the data (4-6) are bounded with bounded derivates. Suppose $0 < T < T_{max}$, and let $v = U^\infty$. If $E = \|u - v\|_1$, then for $\epsilon$ small enough

$$E = O(\epsilon)$$

in $D_{OR}$, and

$$E = O(\epsilon^{1/4} \ln^{1/2} \epsilon)$$

in $D_{IL}$. Here $\Delta(t)$ of (14)-(15) has size $O(\epsilon^{1/4} \ln^{1/2} \epsilon)$.

As with Theorem 3, the computational results reflect that both $\Delta$ and the error in the internal-layer subdomain are smaller than the corollary suggests.

## 6 NUMERICAL DETAILS

The asymptotic analysis has provided a means to precondition the numerical problems. Because the sub-problems are well conditioned, the choice of numerical schemes may be made from a variety of standard methods. This is not usually the case. The class of problems for which the new algorithm is applicable are notoriously difficult to solve, and only a small number of schemes could be employed for its solution (prior to preconditioning). Since the sub-problems are well conditioned, numerical schemes used in the method presented here can be chosen based on criteria such as efficiency or the potential to exploit parallelism. It is possible that a better choice of numerical schemes for the subproblems could be made in a more complicated situation.

### 6.1 Schemes in the Outer Region

The method of characteristics is used to solve the hyperbolic PDE in the outer-region subdomain. This method allows the exploitation of physically motivated parallelism. In addition, the method of characteristics allows handling of the free boundary at $\partial \hat{D}^k_{IL}$ in a straightforward manner.

The method is not limited to using the method of characteristics for the problem in $\hat{D}^k_{IL}$. For example, the schemes for hyperbolic conservation laws [1,8,35] might be modified to account for the term $rU^k$ and used. If this where done, then the method for detection of $\partial \hat{D}^k_{IL}$ could be based on the gradients instead of monitoring the Jacobian.

The method of characteristics scheme involves laying down a characteristic coordinate system, updating solution values, and monitoring the Jacobian. To update the solution and monitor the Jacobian requires a negligible computational cost. The discrete version of (18) is used to determine the characteristic coordinate system. Thus, for each iteration $k$, a set of characteristics $\{x^k_i\}_{i=1}^{I_\beta}$ are computed and used as the computational grid. Here, the superscript $k$ is the iteration number, and the subscript $i$ identifies the characteristic. To determine a characteristic,

$$\frac{dx^k_i}{dt} = U^k(x^k_i, t) \tag{28}$$

is solved for each $i = 1$ to $I_\beta$. All of $\Pi$ is an inflow boundary for $D$, thus initial conditions

$$x^k_i(\tau_i) = \xi_i,$$

are specified along all of $\Pi$. On the $(x,t) = (0,t)$ portion of $\Pi$, the initial condition is $\xi_i = 0$ at $\tau_i = il$, for $i = 1$ to $I_\alpha$. On the $(x,t) = (x,0)$ portion, the initial condition is $\xi_i = (i - I_\alpha - 1)h$ at $\tau_i = 0$, for $i = I_\alpha + 1$ to $I_\gamma$. And on the $(x,t) = (b,t)$ portion, the initial condition is $\xi_i = b$ at $\tau_i = (i - I_\gamma - 1)l$, for $i = I_\gamma + 1$ to $I_\beta$. Here $l$ and $h$ are the increments for time and space, respectively. The locations of characteristics are desired at time increments of $\Delta t$. It is assumed that $\tau_i$ is some integer multiple of $\Delta t$; thus, the same temporal points are used for all characteristics for all iterations. Each characteristic is obtained on a $\Delta t$ interval using the Trapezoid rule to solve (28). The Trapezoid rule is solved via a Newton Iteration. The computed value of $x^k_i(t_j)$ is denoted $x^k_{i,j}$.

As discussed in Section 5.2, $\hat{J}^k$ is monitored along each characteristic to determine the boundary of the internal-layer subdomain at each iteration. The solution to (22) along characteristic $i$ is

$$\hat{J}^k(x_i^k(t), t) = \exp[S_i^k(t)], \tag{29}$$

where

$$S_i^k(t) = \int_{\tau_i}^t U_x^k(x_i^k(\tau), \tau) d\tau. \tag{30}$$

Since computations of (30) are better conditioned than those of (29), $S_i^k$ is monitored in place of $\hat{J}^k$. Denote the computed value of $S_i^k(t_j)$ by $S_{i,j}^k$. The integral is determined using the right-hand rectangle rule

$$S_{i,j+1}^k = S_{i,j}^k + \Delta t U_x^k(x_{i,j+1}^k, t_{j+1}) \tag{31}$$

with the initial value $S_i^k(\tau_i) = 0$. The monitoring is performed at a minimal cost. It is necessary to keep only the most recent value of $S_i^k$, thus minimal storage is required for this technique. In addition, the values of $U_x^k(x_{i,j+1}^k, t_{j+1})$ are saved from the Newton iteration, hence very little computational cost is required.

The criteria used to determine the boundary will now be made more precise. The boundary $\partial \hat{D}_{IL}^k$ is defined by $\hat{J}^k = TOL$, hence a characteristic is considered to be in the outer region as long as

$$S_{i,j}^k \geq \ln(TOL).$$

Let $t = t_{\hat{j}} = \hat{j}\Delta t$ be the first time this inequality is violated for characteristic $i$ during iteration $k$. Then the point $(x, t) = (x_{i,\hat{j}}^k, \hat{j}\Delta t)$ is considered to be inside $\hat{D}_{IL}^k$, and characteristic $i$ is considered to be incident with the boundary at the point $(x, t,) = (x_{i,\hat{j}-1}^k, (\hat{j} - 1)\Delta t)$, and is *flagged* as being part of $\hat{D}_{IL}^k$.

After $x_{i,j}^k$ has been determined, the solution $U^{k+1}$ is computed by solving (19) subject to (4-6) using the right-hand rectangle rule

$$U_{i,j+1}^{k+1} = [1 + \Delta t \, r(x_{i,j+1}^k)] U_{i,j}^{k+1}. \tag{32}$$

This formula is used until either $j = T/\Delta t$, or until characteristic $i$ enters $\hat{D}_{IL}^k$, whichever happens first. This formula has minimal computational requirements; however, the iteration requires the storage of the most recent iterate for a portion of $D$.

## 6.2   Schemes In the Internal Layer Region

The subproblem in the internal-layer subdomain requires the solution of a parabolic PDE subject to boundary data provided by the solution in the outer region. There are two major aspects of the computations in the internal-layer subdomain—mesh generation, and the difference technique. The mesh follows the shock, and has been scaled; therefore the variation in the solution is resolved on the new coordinate system. Thus, the mesh provides a preconditioning for the problem in the internal-layer, and the computations are not overly sensitive to the particular difference scheme used to solve the partial differential equation. Russell's Modified Method of Characteristics (MMC) [30], an explicit/implicit finite difference method, was chosen to solve the equation due to the regularity of the linear algebra problems which it generates. As with the schemes in the outer-region subdomain, other methods (see [7,5]) could be employed for the solution in $\hat{D}_{IL}^k$. It is necessary that the boundary of the internal-layer subdomain be identified with respect to the grid in the internal-layer subdomain. This is described first. Then the finite difference technique is reviewed.

The base of the internal-layer subdomain is identified by finding which characteristic (or set of characteristics) has the lowest time at which it is flagged as being part of the internal-layer subdomain boundary. Denote the computed value of $t^\Gamma$ by $\hat{t}^\Gamma$. For $t > \hat{t}^\Gamma$, the base is taken to be the region between the two outermost characteristics. This could result in non-flagged characteristics being part of the base, but this is not a problem. Once the base characteristics have been located, it is simple to identify whether a flagged characteristic is on the left or on the right boundary of $\hat{D}_{IL}^k$. A flagged characteristics with an index of lower value than a base characteristic is on the left boundary, and a flagged characteristics with an index of higher value than a base characteristic is on the right boundary.

A description of the method used to locate the left and right boundaries is facilitated by first introducing the coordinate system. The computations will be done on a scaled and translated coordinate system with temporal variable $t^* = t/\epsilon$. (The spatial variable will be described later). The temporal grid is the set of points $\{t_n^*\}$, where $t_n^* = n\Delta t^*$ for $n = \hat{t}^\Gamma/(\epsilon\Delta t^*)$ to $T/(\epsilon\Delta t^*)$. These points are a refinement of $\{t_j\}$, the temporal points on which the characteristics are known. The left and right boundaries of the internal-layer subdomain at time $t_n^*$ are denoted by $L_n$ and $R_n$, respectively. Algorithm 1 describes the method used to determine $R$; the procedure used to determine $L$ is symmetric, and hence will not be described.

It is now appropriate to describe the coordinate system on which the computations in the internal layer are based. Denote the middle of the internal-layer subdomain at time $t$ by $M(t) = [R(t) + L(t)]/2$. Then the spatial coordinate in the internal layer is

$$x^* = [x - M(t)]/\epsilon. \tag{33}$$

The temporal coordinate is also scaled, $t^* = t/\epsilon$. Equation (20), may be written as

$$\bar{U}_{t^*}^{k+1} + \left(\frac{dM}{dt} - \bar{U}^{k+1}\right)\bar{U}_{x^*}^{k+1} - \bar{U}_{x^*x^*}^{k+1} - \epsilon\bar{r}\bar{U}^{k+1} = 0. \tag{34}$$

Here, $\bar{U}^{k+1}(x^*, t^*) = U(M + \epsilon x^*, \epsilon t^*)$, and $\bar{r}(x^*) = r(M + \epsilon x^*)$. This is the form of the equation solved in the internal layer.

Equation (34) on the grid defined above is solved using the first step of the MMC. The reader should refer to [27,29,30] for a more complete description of the MMC; however, a review of the first step of the scheme is presented here for the sake of completeness. Denote the the spatial points of the computational grid by $x_i^* = i\Delta x^*$, for $i = -I^{IL}$ to $I^{IL}$. Since $\Delta x^*$ is a constant and the width of $\hat{D}_{IL}^k$ varies with time, the number of grid points also varies with time, and $I^{IL} = I^{IL}(t)$. The MMC involves approximation of the

Do $n = \hat{t}^{\Gamma}/(\epsilon \Delta t^*)$ to $T/(\epsilon \Delta t^*)$
   If $(t_n^* \neq t_j/\epsilon)$ for any $j$
      then $R_n := R_{n-1}$
   otherwise
      $j := \epsilon t_n^*/\Delta t$
      $H :=$ number of characteristics incident with the right boundary at $t_n^*$
      If $H = 0$
         then $R_n := R_{n-1}$
      If $H \geq 1$
         then
            $i :=$ index of right-most characteristic incident
                 with the right boundary
            $R_n := x_{i,j}^k$

**Algorithm 1** Determination of $R_n$.

convective term $\bar{U}_{t^*}^{k+1} + \left(\frac{dM}{dt} - \bar{U}^{k+1}\right)\bar{U}_{x^*}^{k+1}$ by a backward Euler approximation along the subcharacteristics of (34) using the following formula

$$\bar{U}_{t^*}^{k+1}(x^*, t^*) + \left(\frac{dM(t^*)}{dt} - \bar{U}^{k+1}(x^*, t^*)\right)\bar{U}_{x^*}^{k+1}(x^*, t^*) \tag{35}$$

$$\simeq [\bar{U}^{k+1}(x^*, t^*) - \bar{U}^{k+1}(\check{x}^*, t^* - \Delta t^*)]/\Delta t^*.$$

Here $\check{x}^* = x^* + \Delta t^*\left(\frac{dM(t^* - \Delta t^*)}{dt} - \bar{U}^{k+1}(x^*, t^* - \Delta t^*)\right)$ and $\bar{U}^{k+1}(\check{x}^*, t^* - \Delta t^*)$ are determined by linear interpolation between spatial grid points. This linear interpolation is performed element by element, thus there is ample opportunity to exploit parallelism within the method here. Once this quantity has been calculated, the second derivative is approximated using a centered difference. Hence, the full formula in the internal layer is

$$\left[1 - \epsilon \Delta t^* \bar{r}(x^*) + \frac{2}{(\Delta x^*)^2}\Delta t^*\right]\bar{U}^{k+1}(x^*, t^*) - \tag{36}$$

$$\frac{\Delta t^*}{(\Delta x^*)^2}(\bar{U}^{k+1}(x^* + \Delta x^*, t^*) + \bar{U}^{k+1}(x^* - \Delta x^*, t^*)) = \bar{U}^{k+1}(\check{x}^*, t^* - \Delta t^*).$$

There is a domain of dependency requirement on the method [29], which requires the absolute value of the partial with respect to $x^*$ of the convection coefficient to be bounded by $1/\Delta t^*$. Since $M$ is independent of $x^*$, this requirement is that

$$|\bar{U}_{x^*}^{k+1}(x^*, t^*)| < 1. \tag{37}$$

Extra inner iterations may be necessary to step the solution between the temporal grid lines, $t_n^*$.

## 7 ERROR ANALYSIS

In this section the global error analysis for the method is presented. This analysis couples the theorems quoted earlier with the error approximations of the numerical schemes. Let $u$ be the exact solution of (2). Let $v = U^\infty$ be the final iterate defined by solving (17) then (20) until convergence. Let $\tilde{u}$ be the numerical approximation to $v$ using the schemes described previously.

First, the error induced by the numerical schemes will be studied. In the continuum, the method of characteristics obtains $v$ in $D_{OR}$ exactly; however, the numerical scheme used to solve the ODEs is only first order, and

$$x_i^k(t_j) = x_{i,j}^k + \delta,$$

for some $\delta = O(\Delta t)$. Using the Mean Value Theorem,

$$\tilde{u}(x_{i,j}^k, t_j) = v(x_{i,j}^k, t_j) = v(x_i^k(t_j), t_j) - \delta v_x(\rho, t_j),$$

for some $\rho$ between $x_{i,j}^k$ and $x_i^k(t_j)$. Since $v$ is smooth in the outer-region subdomain, $v_x$ is bounded and

$$\|U - \tilde{u}\|_1 = O(\Delta t) \tag{38}$$

for $(x, t) \in D_{OR}$.

The error induced by the numerical schemes in the internal-layer subdomain involves the error at the boundary of the subdomain and the error introduced by the difference scheme used in $D_{IL}$. The boundary conditions at $\partial D_{IL}$ are obtained via first order interpolation between characteristics. This interpolation introduces an error the same size as the spacing between the characteristics. Namely,

$$|v - \tilde{u}| = O(h + \Delta t) + O(l + \Delta t)$$

for $(x, t) \in \partial D_{IL}$, where $h$ and $l$ are the spatial and temporal spacing of the characteristics along $\Pi$, respectively. The first step of the MMC[2] provides the computed solution inside $D_{IL}$. This is a first order approximation. The solution is smooth in the local coordinate system used in $D_{IL}$; thus, the MMC will be truly first order accurate. Combining this with the error induced at the boundary, the solution satisfies $|U - \tilde{u}| = O(\nu)$ for $(x, t) \in D_{IL}$, where $\nu = \max(h + \Delta t, l + \Delta t, \Delta t^*, \Delta x^*)$. Corollary 6 dictates an internal-layer subdomain of size $O(\epsilon^{1/4} \ln^{1/2} \epsilon)$; thus, the contribution of the numerical schemes for $D_{IL}$ to the $L_1$ error is

$$\|v - \tilde{u}\|_1 = O(\nu \epsilon^{1/4} \ln^{1/2} \epsilon) \tag{39}$$

for $(x, t) \in D_{IL}$. Since $1 >> \nu$, this term will have a negligible contribution to the error.

Combining the estimates (38) and (39) with Corollary 6, the following global error estimate follows.

---

[2]The complete MMC is a predictor-corrector method. Only the predictor step is used here. When the corrector step is used, the method has much greater accuracy.

THEOREM 7. Suppose that the conditions of Corollary 6 hold. Let $u$ be the exact solution to (2) satisfying (4-6). Suppose $\tilde{u}$ is the final iterate obtained by applying the numerical schemes described in this section on the iteration defined by (17) and (20). If $E = \|u - \tilde{u}\|_1$, then for $\epsilon$ small enough

$$E = O(\epsilon) + O(\Delta t)$$

in $D_{OR}$, and

$$E = O(\epsilon^{1/4} \ln^{1/2} \epsilon) + O(\Delta t)$$

in $D_{IL}$.

## 8 OUTLINE AND IMPLEMENTATION OF THE ALGORITHM

### 8.1 Algorithm

As a summary, the numerical method is outlined in Algorithm 2 below. The parameters such as $TOL$ and $\Delta t$ are assumed to have been provided by the user. Several steps are not mentioned. For example, if the domain of dependency requirement is violated in the MMC, it may be necessary to reset $\Delta t^*$. However, Algorithm 2 shows the major computational requirements.

The algorithm requires an initial guess. In the theory, the initial guess must satisfy the boundary conditions (4-6), and must be Lipschitz continuous; however, for the computational method, it is only required that some approximation technique which is consistent with the effects of the terms of (2) be used to determine $U^0$. The MMC was chosen for the initial guess. The computations were done on the original coordinates, $x$, and $t$. In order that the domain of dependency requirement be met, artificial diffusion was added instead of restricting the size of $\Delta t$.

Use of artificial diffusion could lead to some ill effects, especially in the location of the shock. The size of the diffusion coefficient effects the speed of a nonsteady shock. Thus, using artificial diffusion may result in computing the location of a nonsteady shock incorrectly. This will be shown in the Model Problem II. However, this inaccuracy is acceptable because the method is not sensitive to errors in the initial guess, as long as $U^0$ is continuous.

### 8.2 Parallelism

Parallelism may be exploited at several levels in the implementation of Algorithm 2. Consider first the parallelism which may be exploited in the solves for the characteristics. Characteristic $x_i^k$ is obtained at discrete points in time by solving (18) using a Newton iteration. These solves may be scheduled asynchronously, or they may be grouped as vectors. Grouping characteristics and assigning the spatial location of each characteristic in a group to a component of a vector allows the exploitation of vector processing capabilities.

Do $k = 1$ till converged

    I. Solve in $\hat{D}_{OR}^k$

        A. Do $i = 1$ to $I_\beta$

            1. $j := \tau_i/\Delta t$

            2. $x_{i,j}^k := \xi_i$

            3. Do while $j < T/\Delta t$ and $S_i^k \leq \ln(TOL)$

                a. $j := j + 1$

                b. Step $x_i^k$ from $t_{j-1}$ to $t_j$

                c. Compute solution value using (32)

                d. Update size of $S_i^k$ with (31)

    II. Solve in $\hat{D}_{IL}^k$

        A. Determine $\partial\hat{D}_{IL}^k$

            1. Find base (determine $\hat{t}^\Gamma$)

            2. Find $R$ using Algorithm 1

            3. Find $L$ (symmetric to $R$)

        B. Set boundary values along $\partial\hat{D}_{IL}^k$

            1. Initial conditions

            2. Right boundary values

            3. Left boundary values

        C. Discretize

            1. $t^* := \hat{t}^\Gamma/\epsilon$

            2. Determine $\Delta t^*$ to satisfy (37) as a subdivision of $\{t_j\}$

            3. $t^* := t^* + \Delta t^*$

            4. Do while $t^* \leq T/\epsilon$

                a. $t^* := t^* + \Delta t^*$

                b. Obtain $U^k$ at $t^*$

                    i. Form right hand side of (36)

                    ii. Solve implicit portion of (36)

                c. If (37) is violated, go to step C.1.

**Algorithm 2** Computational method.

Once the location of the characteristic is known for a time step, the value of $\ln(\hat{J}^k)$ may be approximated using (31).

For a large number of processors, a parent process could spawn a task for each characteristic, and allow them to all execute in parallel. In turn, the task solving a particular characteristic could then spawn two tasks, one for the monitoring of the Jacobian, and one for the updating of $U^{k+1}$. To avoid extra computations, the child task monitoring the Jacobian would need a means of interrupting the parent task computing the characteristic; however, no communication from the task computing $U^{k+1}$ to the parent task is needed. This is reflected by the data dependency graph in Figure 3. The parent routine is

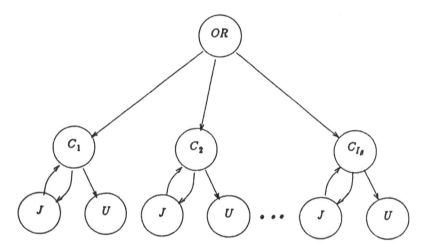

**Figure 3** Data dependency for characteristic solves. (From [32].)

labeled $OR$. The tasks which compute characteristics are labeled $C_i$, for $i = 1$ to $I_\beta$. Each characteristic task spawns two processes, one for the Jacobian monitor, and one to obtain the value of the solution. These are labeled $J$ and $U$, respectively. The data dependency graph for $\hat{D}^k_{OR}$ is a subgraph of the dependency graph for the whole problem.

For a smaller number of processors, since a large number of characteristics will be computed, the exploitation of the medium grain parallelism outlined above is not needed. Thus, for the implementation on the Alliant FX/8, a single task performs the Newton iteration to determine the new location of $x^k_i$. Then the same task updates the values of $S^k_i$ and $U^{k+1}_i$ at the new characteristic location.

Other parallelism evident from the description of Algorithm 2 is reflected in Figure 4. Nodes are labeled with the corresponding step number of Algorithm 2. The only sequential step is the location of the base of $\partial \hat{D}^k_{IL}$, and this is a small portion of the overall computations. The major portion of the computations are the characteristic solves in the outer region and the discretization for the internal layer. More efficient methods could

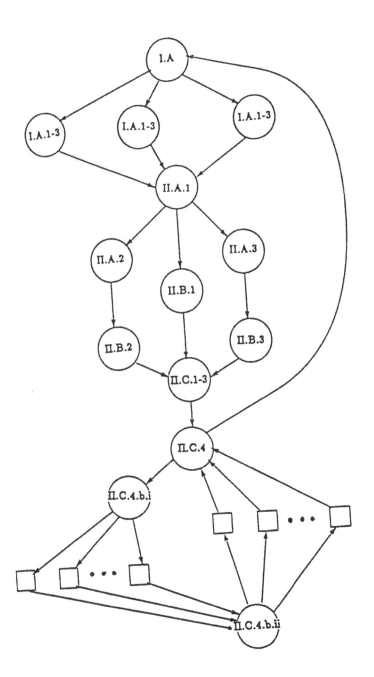

**Figure 4** Data dependency for algorithm

possibly be applied in the internal layer, potentially providing more parallelism.

A less obvious source of parallelism is the exploitation of another type of domain decomposition to form a pipeline out of the outer ($k$) iteration. To obtain $U^2$ up to time $t = T_1$ requires the knowledge of $U^1$ for $0 < t \leq T_1$, but not for $t > T_1$. Thus, $U^1$ for $T_1 < t \leq T_2$ can be computed at the same time that $U^2$ is determined for $0 < t \leq T_1$. In general, subdivide the domain $D$ in the temporal direction as $0 < T_1 < T_2 \ldots < T_N < T$. While $U^1$ is being computed for $T_k \leq t \leq T_k + 1$, the computations for $U^2$ through $U^k$ could be taking place, thus forming a pipeline based on the temporal subdivision of the domain.

### 8.3   Implementation

Algorithm 2 was implemented on an Alliant FX/8 using a package called Schedule [14]. This package provides a common user interface to the parallel capabilities of a variety of shared memory parallel computers. All of the synchronization required to enforce the data dependencies is automatically provided by the Schedule Package once the graph has been specified correctly. Moreover, there are no machine dependent statements within the user code. All such machine dependencies are internal to Schedule. This provides for transportability of the code between the various machines Schedule has been ported to. The implementation is meant as a demonstration of the viability of the method. Thus, not all of the available parallelism has been exploited. Even so, significant speedups were achieved, and will be discussed in the section on the experiments.

A useful feature of Schedule is the automatic generator of a data flow graph associated with a computation. In Figure 5, the graph Schedule produced for an older version of

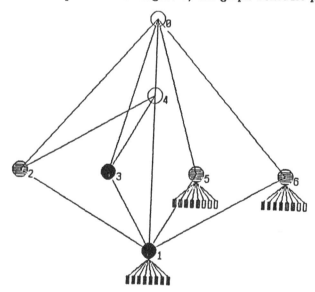

Figure 5 Dependency graph generated by Schedule. (From [24].)

the code [24] is shown. The nodes represented by circles form the static portion of the data dependency graph (the portion of the graph which does not change for different input data). The rectangular nodes associated with static nodes represent dynamically spawned processes which, in this case, are each a characteristic grid line. The graph represents a snapshot of computation shown midway through its execution. The black nodes represent computed processes, the hatched nodes represent active processes, and the white nodes represent processes waiting to execute. Information from this graph and various statistics available with it have been used to improve load balance in this algorithm.

The matrix equation in step II.C.4.b.ii is a symmetric positive definite tridiagonal matrix of about 300 unknowns. Since the number of unknowns is small, the Linpack tridiagonal solver, sptsl, [15] was parallelized instead of using a more complicated block scheme. The scheme in sptsl uses two dual sweeps of the tridiagonal matrix. The first dual sweep is for the forward elimination, and begins at both the top and bottom of the matrix, then works inward. The backward substitution sweep begins at the middle of the matrix, then works outward in both directions. Thus, the computations can be parallelized for a maximum speedup of two. The implementation used for this method had a speedup of about 1.5 on 2 to 8 processors (computational elements or CEs) of the FX/8 over the compiler-optimized version of the sequential code.

## 9  EXPERIMENTS

Two problems are solved, each demonstrating different features of the algorithm. Model Problem I has a steady shock and no forcing term, with an exact solution being known. It demonstrates that the behavior of the error in the computations is the same as the theoretical error estimates of the theorems in the outer regions. Model Problem II has a nonsteady shock. It is used to demonstrate that the method is not sensitive to the initial guess.

### 9.1  Model Problem I

The first model problem will demonstrate that the error tends to zero as $\epsilon \downarrow 0$. With no forcing term, (2) is Burgers' equation,

$$u_t + uu_x - \epsilon u_{xx} = 0. \tag{40}$$

The initial and boundary conditions are that

$$u = -2\tanh(x/\epsilon),$$

for $(x, t) \in \Pi$. Under these conditions, the exact solution to (40) is $u = -2\tanh(x/\epsilon)$, hence the computed solution may be compared easily to the exact. The $L_1$ error is presented for runs with different values of $\epsilon$ in Table 1. The error was measured at time $t = .1$, and remained constant for the remainder of the computational domain. Only one pass of the domain decomposition was needed for this problem. The results presented in Table 1 indicate that the error in the computational method is $O(\epsilon)$.

**Table 1** $L_1$ error

| $\epsilon$ | Error |
|------------|-------|
| $10^{-2}$ | $5.0 \times 10^{-02}$ |
| $10^{-3}$ | $2.2 \times 10^{-03}$ |
| $10^{-4}$ | $2.3 \times 10^{-04}$ |

## 9.2   Model Problem II

The second model problem has a nonsteady shock, and shows the effect of the forcing term on the location of the shock. In addition, this method demonstrates that the method is not overly sensitive to the accuracy of the initial guess. The equation solved is

$$u_t + uu_s - \epsilon u_{ss} - 4\sin(2\pi x)u = 0,$$

subject to the initial guess

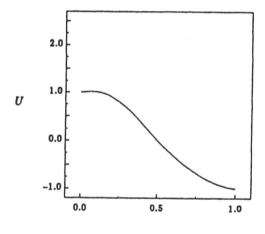

**Figure 6** Initial guess for iteration. (From [32].)

$$u(x, 0) = \cos(\pi x), \quad \text{for } 0 < x < 1 = b$$

(see Figure 6). Then the boundary conditions are

$$u(0, t) = 1, \ u(1, t) = -1, \quad \text{for } T > t > 0.$$

For the experiments presented here, $\epsilon = .001$. If there were no forcing term ($r = 0$), then this problem would have a steady shock develop in the center of the domain; however, the forcing term is $r = 4\sin(2\pi x)$. This represents a duct of width $A(x) = \exp[2\cos(2\pi x)/\pi]$, which has the general shape of the one in Figure 1. The effects of the shape of the duct on the location of the shock are reflected by the internal-layer subdomain having a different location as time increases. The movement of the internal layer subdomain can be seen in Figure 7, where the characteristics of the outer-region solution and internal-layer subdomain after four passes of the domain decomposition are shown.

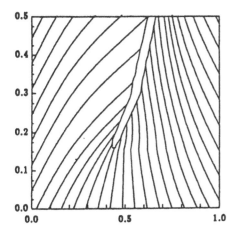

**Figure 7** Flow lines and domain decomposition. (From [32].)

The initial iterate is obtained using the Modified Method of Characteristics on a rectangular grid with 100 spatial points. Figure 8 shows the initial iterate, $U^0$, at time $t = .3$. The use of artificial diffusion resulted in the location and the width of the shock being wrong for the initial iterate. Comparing the initial iterate, $U^0$, with $U^3$ in Figure 9, these errors may be seen. The initial guess has the wrong amplitude, and the shock is slightly to the left and wider than the shock in $U^3$.

The succession of internal-layer subdomains may be seen in Figures 10-13. One of the manifestations of the convergence of the iteration is that the internal-layer subdomain has a much smoother boundary after convergence than before. The boundary for $U^4$ is smooth up to approximatedly time $t = .3$.

Timings for computing the initial guess and the first iteration are presented in Table 2. Speedup is the execution time for multiple CEs divided by the time for one CE. Efficiency is the speedup divided by the number of CEs. The data indicates that significant speedups were attained, even though a significant portion of the parallelism was not exploited. For example, the pipelining described in Section 8.2 was not used.

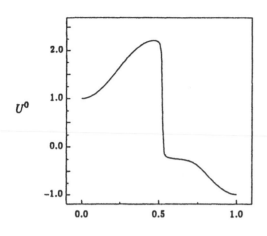

**Figure 8** Initial Guess at $t = .3$. (From [32].)

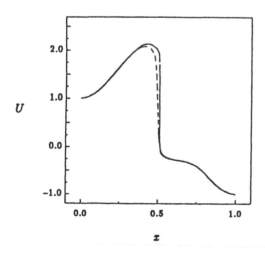

**Figure 9** $U^0$ (dotted) and $U^3$ at $t = .3$. (From [32].)

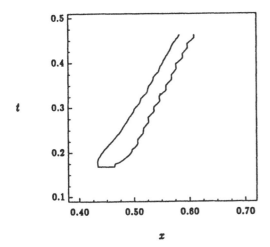

**Figure 10** Internal-layer boundary for $U^1$. (From [32].)

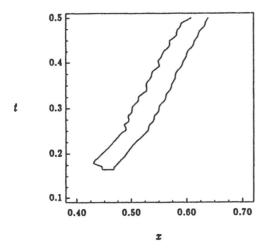

**Figure 11** Internal-layer boundary for $U^2$. (From [32].)

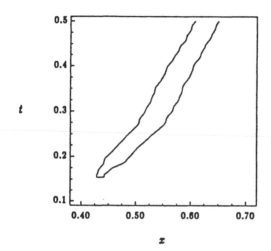

**Figure 12** Internal-layer boundary for $U^3$. (From [32].)

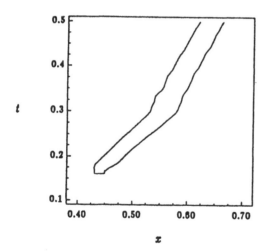

**Figure 13** Internal-layer boundary for $U^4$. (From [32].)

**Table 2** Timings

| CEs (P) | RUN TIME (seconds) | SPEEDUP (S) | EFFICIENCY (E) |
|---|---|---|---|
| 1 | 90.4 | 1.00 | 1.00 |
| 2 | 47.5 | 1.90 | 0.95 |
| 3 | 34.5 | 2.61 | 0.87 |
| 4 | 27.9 | 3.23 | 0.81 |
| 5 | 24.3 | 3.71 | 0.74 |
| 6 | 21.7 | 4.16 | 0.69 |
| 7 | 19.8 | 4.57 | 0.65 |
| 8 | 18.5 | 4.89 | 0.61 |

## 10  CONCLUSION

In this paper, asymptotics and numerics have been blended to form a new computational method. The method has potential to exploit a large amount of parallelism and provides high accuracy. Asymptotic analysis provided a theoretical basis for the domain decomposition, and guided in the derivation of rigorous local and global error bounds.

Two types of subdomains were identified by the asymptotic analysis of this problem: smooth outer regions, and an internal-layer subdomain with a shock. Pipelining the outer iteration provided large-grain parallelism. Smaller-grain parallelism was exploited by using a modification of the method of characteristics in the outer regions, and by blocking the computations in the shock layer.

The method was developed for an important problem in computational fluid dynamics; however, the method is suitable for a wide range of problems in physics and chemistry. Namely, this approach is suitable for problems with internal layers and boundary layers interspersed with regions where the solution is smooth. Examples of such problems other than transonic flow through a duct include the location of stagnation points (flow of zero velocity) where two opposing jets intersect, and combustion fronts.

The availability of estimates and bounds on the error is important in the design of numerical methods. Rigorous *a priori* error bounds were established for the method presented here. In addition, the particular numerical schemes used for the subproblems allowed *a posteriori* error estimation. The *a priori* error bounds were shown to be much larger than the errors observed in the computations; thus, sharper error bounds are expected.

## ACKNOWLEDGEMENTS.

The authors appreciate the support of Ahmed Sameh during this project. The development of this method would not have been possible without Gerald Hedstrom, Raymond C. Chin, and Fred Howes' work in the area and collaboration on this project. The authors wish to thank Barbara Stewart for her assistance in the document preparation, and to thank the reviewers for their useful comments.

## NOTATION

| | | |
|---|---|---|
| $D$ | (3) | Computational domain. |
| $D_{IL}$ | (14) | Internal-layer subdomain. |
| $D_{OR}$ | (15) | Outer-region subdomain. |
| $\hat{D}_{IL}$ | (25) | Computational internal-layer subdomain. |
| $\hat{D}_{OR}$ | (24) | Computational outer-region subdomain. |
| $L(t)$ | Sec. 6.2 | Location of the left boundary of $\hat{D}_I L$. |
| $L_n$ | Sec. 6.2 | Computed location of $L(t)$ at $t_n$. |
| $P$ | (2) | The full nonlinear operator. |
| $P_0$ | (8) | The reduced nonlinear operator. |
| $R(t)$ | Sec. 6.2 | Location of the right boundary of $\hat{D}_I L$. |
| $R_n$ | Sec. 6.2 | Computed location of $R(t)$ at $t_n$. |
| $S_i^k(t)$ | (30) | Integral for monitoring Jacobian. |
| $S_{i,j}^k$ | (31) | Discrete values of $S_i^k(t)$. |
| $T$ | (3) | Upper bound on time $t$ for $D$. |
| $U^k$ | (17),(20) | Iterate $k$. |
| $t^{\Gamma}$ | Sec. 4 | Time that the solution to the reduced problem becomes multivalued. |
| $t^*$ | Sec. 6.2 | Scaled and translated internal-layer coordinate. |
| $u$ | (2) | The solution to the full operator. |
| $\hat{u}$ | (8) | The solution to the reduced operator. |
| $\tilde{x}$ | Sec. 4 | Scaled and translated internal-layer coordinate. |
| $x^*$ | (33) | Spatial coordinate for internal-layer computations. |
| $\hat{x}^*$ | (35) | Spatial location used in MMC. |
| $x^k(t)$ | (18) | Characteristic coordinate. |
| $x_i^k(t)$ | Sec (6.1) | Characteristic grid line $i$ for iteration $k$. |
| $x_{i,j}^k$ | Sec (6.1) | Computed value of $x_i^k$ at time $t_n$. |

**References**

[1] Amiram Harten, Peter D. Lax, and Bram van Leer, On upstream differencing and Godunov-type schemes for hyperbolic conservation laws, *SIAM Review*, *25*: 35–61 (1983).

[2] C. C. Lin and L. A. Segel, *Mathematics Applied to Deterministic Problems in the Natural Sciences*, Macmillan, New York (1974).

[3] C. M. Bender and S. A. Orszag, *Advanced Mathematical Methods for Scientists and Engineers*, McGraw-Hill, New York (1978).

[4] J. R. Cannon, *The One-Dimensional Heat Equation*, vol. 23, Addison-Wesley Publishing Company, Reading, Massachusetts (1984).

[5] David Hoff and Joel Smoller, Error bounds for finite-difference approximations for a class of nonlinear parabolic systems, *Math. Comp.*, *45*: 35–49 (1985).

[6] W. Eckhaus, *Asymptotic Analysis of Singular Perturbations*, North-Holland, Amsterdam (1979).

[7] D. B. Gannon, *Self Adaptive Methods for Parabolic Partial Differential Equations*, Ph.D. Thesis, University of Illinois, Urbana-Champaign (1980).

[8] H. C. Yee and A. Harten, Implicit TVD schemes for hyperbolic conservation laws in curvilinear coordinates, *AIAA Journal*, *25*: 266–274 (1987).

[9] F. A. Howes, Some stability results for advection-diffusion equations. II, *Studies in Applied Mathematics*, *75*: 153–162 (1986).

[10] ——, *Asymptotic stability of viscous shock waves*, in Transactions of the Fourth Army Conference on Applied Mathematics and Computing (1987).

[11] ——, Multi-dimensional initial-boundary value problems with strong nonlinearities, *Arch. for Rat. Mech. Anal.*, *91*: 153–168 (1986).

[12] ——, Multi-dimensional reaction-convection-diffusion equations, in *Springer Lecture Notes*, 217–223 (1984), A. Dold, B. Eckmann, ed., Springer-Verlag.

[13] ——, Perturbed boundary value problems whose reduced solutions are nonsmooth, *Indiana Univ. Math. J.*, *30*: 267–280 (1981).

[14] J. J. Dongarra and D. C. Sorensen, *SCHEDULE: Tools for Developing and Analyzing Parallel Fortran Programs*, Tech. Rep. ANL/MCS-TM-86, Argonne National Lab. (1986).

[15] J. J. Dongarra, J. R. Bunch, C. B. Moler, and G. W. Stewart, *LINPACK Users' Guide*, SIAM, Philadelphia, PA (1979).

[16] J. Kevorkian and J. D. Cole, *Perturbation Methods in Applied Mathematics*, Springer-Verlag, New York (1981).

[17] K. W. Chang and F. A. Howes, *Nonlinear Singular Perturbation Phenomena: Theory and Applications*, Springer-Verlag, New York (1984).

[18] P. D. Lax, *Hyperbolic systems of conservation laws and the mathematical theory of shock waves*, in Regional Conference Series in Applied Mathematics (1973).

[19] A. H. Nayfeh, *Introduction to Perturbation Techniques*, John Wiley and Sons, New York (1981).

[20] ——, *Perturbation Methods*, Wiley-Interscience, New York (1973).

[21] F. W. J. Olver, *Asymptotics and Special Functions*, Academic Press, New York (1974).

[22] R. C. Y. Chin, G. W. Hedstrom, and F. A. Howes, *Moving Grid and Domain Decomposition Methods for Parabolic Partial Differential Equations*, informal report, Lawrence Livermore Natl. Lab., Livermore, CA (1985).

[23] R. E. Meyer and S. V. Parter, *Singular Perturbations and Asymptotics*, Academic Press, New York (1980).

[24] R. C. Y. Chin, Gerald W. Hedstrom, Jeffrey S. Scroggs, and Danny C. Sorensen, *Parallel computation of a domain decomposition method*, in IMACS 6th International Symposium on Computer Methods for Partial Differential Equations (1987).

[25] R. C. Y. Chin, Gerald W. Hedstrom, James R. McGraw, and F. A. Howes, *Parallel computation of multiple-scale problems*, in New Computing Environments: Parallel, Vector, and Systolic, 136–153 (1986), SIAM.

[26] R.C.Y. Chin and R. Krasny, A hybrid asymptotic-finite element method for stiff two-point boundary value problems, *SIAM J. Sci. Stat. Comp., 4*: 229–243 (1983).

[27] Richard E. Ewing and Thomas F. Russell, Multistep Galerkin methods along characteristics for convection-diffusion problems, in *Advances in Computer Methods for Partial Differential Equations-IV*, 28–36 (1981), R. Vichnevetsky and R. S. Stepleman, eds., IMACS, Rutgers Univeristy.

[28] Robert D. Richtmyer and K. W. Morton, *Difference Methods for Initial-Value Problems*, Interscience Publisher, John Wiley and Sons, New York, NY (1967).

[29] T. F. Russell, Galerkin time stepping along characteristics for Burgers' equation, in *Scientific Computing*, 183-192 (1983), R. Stepleman et. al., ed., North-Holland Publishing Company.

[30] ——, Time stepping along characteristics with incomplete iteration for a Galerkin approximation of miscible displacement in porous media, *SIAM J. Numer. Anal., 22*: 970-1013 (1985).

[31] S. Mas-Gallic and P. A. Raviart, A particle method for first-order symmetric systems, *Numerische Mathematik, 51*: 323-352 (1987).

[32] J. S. Scroggs, *The Solution of a Parabolic Partial Differential Equation via Domain Decomposition: The Synthesis of Asymptotic and Numerical Analysis*, Ph.D. Thesis, University of Illinois, Urbana-Champaign (1988).

[33] V. Ervin and W. Layton, On the approximation of derivatives of singularly perturbed boundary value problems, *SIAM J. Sci. Stat. Comp., 8*: 265–277 (1987).

[34] M. van Dyke, *Perturbation Methods in Fluid Mechanics*, Parabolic Press, Stanford (1975).

[35] B. van Leer, On the relation between the upwind-differencing schemes of Godunov, Engquist-Osher, and Roe, *SIAM J. Sci. Stat. Comp., 5*: 1–20 (1984).

[36] G. B. Whitham, *Linear and Nonlinear Waves*, John Wiley and Sons, New York (1974).

# The Rate of Convergence
# of the Modified Method of Characteristics
# for Linear Advection Equations in One Dimension

Clint N. Dawson[1], Todd F. Dupont[2], and Mary F. Wheeler[3]

## I.  Introduction

The finite element modified method of characteristics (MMOC) is a procedure for computing approximate solutions of advection-diffusion problems. Although it has the flavor of a Galerkin method, it has definite theoretical and experimental advantages over Galerkin methods in advection-dominated situations. For problems with nondegenerate diffusion, optimal order convergence rates for the MMOC can be proven, but for pure advection problems, the known estimates seem to fall short. This paper shows that the convergence rate of the MMOC depends on more than the approximation properties of the underlying function space. In the context of a very simple model problem, it is shown that the convergence rate is the same for two MMOC schemes, one built on piecewise linear functions and the other built on piecewise quadratic functions.

In a recent paper by Russell and two of the authors [2], it was demonstrated that the error for the MMOC applied to periodic linear advection problems in $n$ dimensions is at worst $O(h^r + \Delta t)$ in $L^2$ at each time level. Here $r = \min(s, k)$, where $k$ is the degree of the piecewise polynomial approximating space and $s$ is related to the smoothness of the solution. Thus, a suboptimal rate of convergence in $L^2$ was derived; the rate is suboptimal in the sense that the power of $h$ is one lower than is possible for a best approximation.

In this paper, we show that in general the rate of convergence is suboptimal, but that it is optimal in at least one case. We consider the simple one-dimensional linear advection equation

$$\begin{cases} u_t + u_x = f, & x \in (0,1) \equiv I, \\ & t \in (0,T], \\ u(x,0) = u^0(x), & x \in I, \end{cases} \tag{1.1}$$

---

[1] Mathematical Sciences Dept., Rice Univ., currently at Mathematics Dept., Univ. of Chicago
[2] Computer Science Dept., Univ. of Chicago
[3] Mathematical Sciences Dept., Rice Univ. and Mathematics Dept., Univ. of Houston

with periodic initial and boundary conditions, i.e., $u^0(0) = u^0(1)$ and $u(0,t) = u(1,t)$. In particular, we consider a finite element MMOC approximation to (1.1) with continuous piecewise linear and continuous piecewise quadratic approximating spaces. In the first case (piecewise linears), we show that under certain assumptions on the smoothness of $u$ and assuming $\Delta t = \mathcal{O}(h^2)$, one actually has optimality in $L^2$, thus, the $L^2$ error is $\mathcal{O}(h^2)$. For the second case (piecewise quadratics), assuming $\Delta t = o(h^2)$, we construct an example where the $L^2$ error is at best $\mathcal{O}(h^2)$, thus, a suboptimal order of convergence holds here.

The nature of these results and the tools used are similar to the results and tools in [3] where the rate of convergence of the continuous-time Galerkin method was studied. Our result for piecewise quadratics can be used to see that the rate of convergence of the continuous-time Galerkin method is suboptimal in this case; just take $\Delta t$ to zero.

The remainder of this paper is divided into five sections. In the next section, we establish notation, discuss the MMOC and its application to (1.1), and state the two major results of this paper. The first major result, an optimal order $L^2$ error estimate for the case of a continuous piecewise linear approximating space, is derived in Section 3. The second major result is derived in Section 4; we show by construction that for the case of continuous piecewise quadratic approximation the rate in $L^2$ is suboptimal in space. Technical lemmas used to derive the results of Sections 3 and 4 are proved in Section 5. Finally, in Section 6, we verify these theoretical results numerically by calculating experimental $L^2$ convergence rates for a smooth test problem.

# II.   Notation and Preliminaries

Let $L^2(I)$, $L^\infty(I)$, and $H^k(I)$ denote the standard Sobolev spaces defined on the unit interval. Let $\tilde{L}^2$ denote the set of functions defined on $\mathbf{R}$ which are one-periodic and whose restrictions to $I$ are on $L^2(I)$. Adopt a similar definition for $\tilde{H}^k$. For $u$ an element of $\tilde{L}^2$ let $\|u\| = \|u\|_{L^2(I)}$, and for $u \in \tilde{H}^k$, $\|u\|_k = \|u\|_{H^k(I)}$.

If $\mathcal{X}$ is a normed space with norm $\|\cdot\|_{\mathcal{X}}$ and $\phi : [0,T] \to \mathcal{X}$, let

$$\|\phi\|_{L^\infty(\mathcal{X})} = \sup_{0 \le t \le T} \|\phi(t)\|_{\mathcal{X}}.$$

For $M_h$ a finite-dimensional subspace of $\tilde{H}^1$, the finite element MMOC is a map $U : \{0 = t^0, t^1, \ldots, t^N = T\} \to M_h$ defined by

$$\begin{cases} \left(\frac{U^{n+1} - \check{U}^n}{\Delta t}, v\right) = (f^{n+1}, v), & v \in M_h, \\ U^0 = \tilde{U}^0. \end{cases} \tag{2.1}$$

Here $(\cdot, \cdot)$ denotes the $L^2$ inner product on $I$, $f^n(x) = f(x, t^n)$, $U^n(x) = U(x, t^n)$, where $t^n = n\Delta t$, $n = 0, 1, \ldots, N$,

$$\hat{U}^n(x) = U^n(x - \Delta t), \tag{2.2}$$

and $\tilde{U}^0$ is the $L^2$ projection of $u^0$ into $M_h$. (For application and analysis of the MMOC to more complicated problems, see [4], [5], [6], [7].)

We note that since the true solution satisfies

$$u(x, t^{n+1}) = u(x - \Delta t, t^n) + \int_0^{\Delta t} f(x - \Delta t + s, t^n + s) ds$$

we have

$$\left( \frac{u^{n+1} - \hat{u}^n}{\Delta t}, v \right) = (f^{n+1} + \rho^n, v), \quad v \in M_h, \tag{2.3}$$

where $\rho^n = \mathcal{O}(\Delta t)$ provided $f_\tau \in L^\infty(L^2)$, where $\tau$ is the characteristic direction.

Let $J$ be a positive integer, $h = 1/J$, and

$$\begin{aligned}
x_j &= jh, \\
X_j &= [x_{j-1}, x_j], \\
x_{j+1/2} &= (j + 1/2)h.
\end{aligned}$$

Denote by $\tilde{M}_0(1, h)$ the set of functions $v \in C^0(I)$ with $v$ linear on each $X_j$, and $v(0) = v(1)$. Moreover, denote by $\tilde{M}_0(2, h)$ the set of functions $v \in C^0(I)$, $v$ quadratic on each $X_j$, and $v(0) = v(1)$. We now study convergence of the MMOC with $M_h = \tilde{M}_0(1, h)$ and $M_h = \tilde{M}_0(2, h)$.

The main results of this paper can be stated as follows.

**Theorem 1** *Take $M_h = \tilde{M}_0(1, h)$. Let $\Delta t = \mathcal{O}(h^2)$, and assume $u$ is three times continuously differentiable, then*

$$\max_{0 \le n \le N} \|u^n - U^n\|_{L^2} = \mathcal{O}(h^2).$$

**Theorem 2** *Take $M_h = \tilde{M}_0(2, h)$. Let $\Delta t = \mathcal{O}(h^2)$ and assume $u$ is four times continuously differentiable, then*

$$\max_{0 \le n \le N} \|u^n - U^n\|_{L^2} = \mathcal{O}(h^2). \tag{2.4}$$

*In addition, if $u_{xxx}$ is not identically zero, then there exists a constant $C_0 > 0$ independent of $h$ and $\mu^n \in \tilde{M}_0(2, h)$ with $\max_n \|\mu^n\| \ge C_0$, and if $\Delta t = o(h^2)$*

$$\|u^n - U^n + h^2 \mu^n\|_{L^2} = o(h^2). \tag{2.5}$$

*Consequently, the estimate (2.4) is sharp.*

In the following analysis, $C$ represents a generic constant independent of $h$ and $\Delta t$.

# III.   Proof of Theorem 1

Let $M_h = \tilde{M}_0(1, h)$, and for each $t \in (0, T]$, let $\tilde{U}(t)$ be the $\tilde{L}^2$ projection of $u(t)$ into $M_h$, i.e.,

$$(\tilde{U}(t) - u(t), v) = 0, \quad v \in M_h. \tag{3.1}$$

Then, setting $\zeta = U - \tilde{U}$, $\xi = u - \tilde{U}$, we have by (2.1), (2.3), and (3.1)

$$\left( \frac{\zeta^{n+1} - \hat{\zeta}^n}{\Delta t}, v \right) = \left( \frac{\xi^n - \hat{\xi}^n}{\Delta t} - \rho^n, v \right), \quad v \in M_h. \tag{3.2}$$

Noting by periodicity that $\|\hat{\zeta}^n\| = \|\zeta^n\|$, and setting $v = \zeta^{n+1}$ in (3.2), we obtain

$$\|\zeta^{n+1}\|^2 - \|\zeta^n\|^2 + \|\zeta^{n+1} - \hat{\zeta}^n\|^2 = 2(\xi^n - \hat{\xi}^n - \Delta t \rho^n, \zeta^{n+1}). \tag{3.3}$$

A suboptimal $L^2$ estimate for $\zeta$ can be obtained as in [2] by noting that

$$(\xi^n - \hat{\xi}^n + \Delta t \rho^n, \zeta^{n+1}) \leq C \Delta t \left( \|\xi_x^n\|^2 + \|\rho^n\|^2 + \|\zeta^{n+1}\|^2 \right). \tag{3.4}$$

Moreover, by approximation theory

$$\|\xi_x^n\| \leq Ch\|u^n\|_2. \tag{3.5}$$

Thus, substituting (3.4) and (3.5) into (3.3) and summing on $n, n = 0, \ldots, N-1$, we find that

$$\|\zeta^N\|^2 \leq C_u(h^2 + (\Delta t)^2) + C \sum_{n=0}^{N-1} \|\zeta^{n+1}\|^2 \Delta t,$$

where $C_u$ depends on derivatives of $u$ through second order. By the discrete Gronwall Lemma,

$$\max_{0 \leq n \leq N} \|\zeta^n\| \leq C(h + \Delta t). \tag{3.6}$$

Combining (3.6) with the well-known estimate

$$\|\xi\|_{L^\infty(L^2)} \leq Ch^2 \|u\|_{L^\infty(H^2)}, \tag{3.7}$$

we have by the triangle inequality

$$\max_{0 \leq n \leq N} \|u^n - U^n\| \leq C(h + \Delta t). \tag{3.8}$$

We can see by the above argument that obtaining a "better" $L^2$ estimate requires more careful treatment of the term

$$(\xi^n - \hat{\xi}^n, \zeta^{n+1}). \tag{3.9}$$

Thus, we examine this term more closely.

We now state a lemma concerning the term (3.9), whose proof we defer to Section 5.

**Lemma 1** *Assume the hypotheses of Theorem 1 hold. Then*

$$\left| \left( \frac{\xi^n - \hat{\xi}^n}{\Delta t}, \zeta^{n+1} \right) \right| \le C(h^4 + \|\zeta^{n+1}\|^2). \tag{3.10}$$

Substituting (3.10) into (3.3), assuming $\Delta t = \mathcal{O}(h^2)$, we find

$$\frac{1}{\Delta t}\|\zeta^{n+1}\|^2 \le \frac{1}{\Delta t}\|\zeta^n\|^2 + Ch^4 + \|\zeta^{n+1}\|^2.$$

Summing on $n$ and applying the discrete Gronwall Lemma proves Theorem 1.

# IV. Proof of Theorem 2.

As in the proof of Theorem 1 with $\xi = u - \tilde{U}$ and $\zeta = U - \tilde{U}$, we obtain

$$\left( \frac{\zeta^{n+1} - \hat{\zeta}^n}{\Delta t}, \chi \right) = \left( \frac{\xi^n - \hat{\xi}^n}{\Delta t} - \rho^n, \chi \right), \quad \chi \in \tilde{M}_0(2, h).$$

Setting $\chi = \zeta^{n+1}$, we have

$$\frac{1}{\Delta t}\|\zeta^{n+1}\|^2 \le \frac{1}{\Delta t}\|\zeta^n\|^2 + \left( \frac{\xi^n - \hat{\xi}^n}{\Delta t} - \rho^n, \zeta^{n+1} \right) \tag{4.1}$$

Just as in the derivation of the optimal rate in the proof of Theorem 1 we get that

$$\frac{1}{\Delta t}\|\zeta^{n+1}\|^2 \le \frac{1}{\Delta t}\|\zeta^n\|^2 + C(h^4 + \|\zeta^{n+1}\|^2). \tag{4.2}$$

Here we have also used the fact that $\Delta t = \mathcal{O}(h^2)$. Summing (4.2) on $n$, and applying Gronwall's Lemma, we obtain (2.4).

In Lemma 4 below we construct $\mu^n \in \tilde{M}_0(2, h)$ satisfying $\max_n \|\mu^n\| \ge C_0 > 0$ and

$$\|\zeta^n + h^2\mu^n\| = o(h^2).$$

Inequality (2.5) follows immediately using approximation theory.

# V. Lemmas

## V.1 Proof of Lemma 1

Let $\mathcal{V} = \{v_0(x), v_1(x), \ldots, v_{J-1}(x)\}$ be the set of 1-periodic "hat" functions defined over the mesh $\{x_j\}$. That is, $v_j \in \tilde{M}_0(1, h)$ and

$$v_j(x_k) = \begin{cases} 1, & j = k, \\ 0, & 0 \le k < J, \ k \ne j. \end{cases} \tag{5.1}$$

Then $\mathcal{V}$ is a basis for $\tilde{M}_0(1,h)$ and

$$\zeta^n(x) = \sum_{j=0}^{J-1} \zeta_j^n v_j(x), \tag{5.2}$$

where

$$\zeta_j^n = \zeta^n(x_j).$$

It is easily seen that if

$$\chi(x) = \sum_{j=0}^{J-1} \chi_j v_j(x)$$

and

$$\|\chi\|_h^2 = \sum_{j=0}^{J-1} \chi_j^2 h, \tag{5.3}$$

then

$$\frac{1}{3}\|\chi\|_h^2 \le \|\chi\|^2 \le \|\chi\|_h^2. \tag{5.4}$$

Thus $\|\cdot\|_h$ and $\|\cdot\|$ define equivalent norms on $\tilde{M}_0(1,h)$.

Suppressing time-dependence momentarily, we express $\xi$ as

$$\xi(x) = \psi(x) + g(x), \tag{5.5}$$

where on each $X_{j+1}$

$$\psi(x) = \frac{1}{2} u_{xx}(x_{j+\frac{1}{2}})(x - y_{j1})(x - y_{j2}), \tag{5.6}$$

with $y_{j1}$ and $y_{j2}$ being the two Gauss points on $X_{j+1}$. It then follows [3] that on each $X_{j+1}$

$$|g(x)| + h|g'(x)| = \mathcal{O}(h^3); \tag{5.7}$$

i.e., $\psi$ is the leading term in the error in the $L^2$ projection.

This expression is used to show that $(\xi_x, v_j)$ is smaller than is apparent at first glance. Specifically, integration by parts shows that

$$\begin{aligned}
(\psi', v_j) &= \frac{1}{2}(x_j - y_{j1})(x_j - y_{j2})(u_{xx}(x_{j-\frac{1}{2}}) - u_{xx}(x_{j+\frac{1}{2}})) \\
&\quad + \left(\psi, v_j'\right).
\end{aligned} \tag{5.8}$$

The last integral is zero since $\psi$ vanishes at the Gauss points. Combining this with (5.7) we see that

$$(\xi_x, v_j) = \mathcal{O}(h^3). \tag{5.9}$$

Next, let

$$\xi(x) - \xi(x - \Delta t) = \Delta t \xi_x(x) + \eta(x). \tag{5.10}$$

Then for each $j$

$$\eta(x) = \begin{cases} \mathcal{O}(\Delta t^2), & \text{on } (x_j + \Delta t, x_{j+1}), \\ \mathcal{O}(h\Delta t), & \text{on } (x_j, x_j + \Delta t). \end{cases} \tag{5.11}$$

The first bound holds because $\xi_{xx} = u_{xx}$ on $(x_j, x_{j+1})$; the second holds since $\xi_x = \mathcal{O}(h)$ and

$$\xi(x) - \xi(x - \Delta t) = \int_{x-\Delta t}^{x} \xi_x(s)ds.$$

From (5.11) we see that

$$(\eta, v_j) = \mathcal{O}(h\Delta t^2). \tag{5.12}$$

Thus from (5.9), (5.10) and (5.12) we see that

$$\begin{aligned} \left(\xi - \hat{\xi}, v_j\right) &= \mathcal{O}(h^3\Delta t + h\Delta t^2) \\ &= \mathcal{O}(h^3\Delta t). \end{aligned}$$

Hence,

$$\begin{aligned} \left(\xi^n - \hat{\xi}^n, \zeta^{n+1}\right) &= \sum_j \left(\xi^n - \hat{\xi}^n, v_j\right)\zeta_j^{n+1} \\ &\leq C\Delta t \sum_j h^2|\zeta_j^{n+1}|h \\ &\leq C\Delta t(h^4 + \|\zeta^{n+1}\|^2). \end{aligned} \tag{5.13}$$

Here we used (5.4). This completes the proof of Lemma 1.

## V.2   Quadratic Case Truncation Error

Let $v_j$, $j = 0, \ldots, J-1$, be the "hat" functions defined by (5.1) and let $b_j$, $j = 0, \ldots, J-1$, be the "bump" functions given by

$$b_j(x) = \begin{cases} \frac{(x-x_j)(x_{j+1}-x)}{h^2}, & x_j \leq x \leq x_{j+1}, \\ 0, & \text{otherwise.} \end{cases} \tag{5.14}$$

Then, $\{v_j\} \cup \{b_j\}$ is a basis for $\tilde{M}_0(2, h)$. We now show that

$$\left(\frac{\xi^n - \hat{\xi}^n}{\Delta t}, v_j\right) = \mathcal{O}(h^4) \tag{5.15}$$

and

$$\left(\frac{\xi^n - \hat{\xi}^n}{\Delta t}, b_j\right) = -\frac{1}{360}u_{xxx}(x_j, t^n)h^3 + \text{higher order terms.} \tag{5.16}$$

Again suppressing time-dependence momentarily, we have

$$\frac{\xi(x) - \hat{\xi}(x)}{\Delta t} = \xi_x(x) + e(x),$$

where $e(x) = \bar{\xi}_x(x) - \xi_x(x)$ and $\bar{\xi}_x$ is the average value of $\xi_x$ over the interval $[x - \Delta t, x]$. Using Lemmas 2 and 3, the results (5.15) and (5.16) follow.

**Lemma 2** *Let $u$ be $C^4$ and 1-periodic. Then*

$$(\xi_x, v_j) = \mathcal{O}(h^4)$$

*and*

$$(\xi_x, b_j) = -\frac{1}{360} u_{xxx}(x_j) h^3 + \mathcal{O}(h^4),$$

*where $v_j$ and $b_j$ are defined by (5.1) and (5.14).*

*Proof:* Let $Iu$ denote the $C^0$ piecewise quadratic interpolant of $u$ on $X_{j+1}$ defined by interpolation of $u$ at $x_j$, $x_{j+\frac{1}{2}}$ and $x_{j+1}$. Using a Hermite-Birkhoff cubic interpolant which in addition to the conditions for $Iu$ matches $u_{xxx}$ at $x_{j+\frac{1}{2}}$, and applying the Bramble-Hilbert Lemma [1], we have on $X_{j+1}$

$$
\begin{aligned}
(u - Iu)(x) &= \frac{1}{6} u_{xxx}(x_{j+\frac{1}{2}})(x - x_j)(x - x_{j+\frac{1}{2}})(x - x_{j+1}) + g(x) \\
&\equiv \psi(x) + g(x),
\end{aligned}
\tag{5.17}
$$

where $|g(x)| + h|g'(x)| = \mathcal{O}(h^4)$. Note that

$$(\psi, v_j) = \mathcal{O}(h^5).$$

and

$$(\psi, b_j) = 0.$$

Now $\xi(x) = (u - Iu) + (Iu - \tilde{U})$. Thus, if follows easily that

$$\xi = \psi + \mathcal{O}(h^4) \tag{5.18}$$

and

$$\xi_x = \psi' + \mathcal{O}(h^3).$$

Thus, the leading order term in the expression for $\xi_x(x)$ is $\psi'(x)$. The lemma now follows since

$$(\psi', b_j) = -\frac{1}{360} u_{xxx}(x_{j+\frac{1}{2}}) h^3.$$

**Lemma 3** *Let $\bar{\xi}_x(x)$ denote the average value of $\xi_x$ over the interval $[x - \Delta t, x]$. Assume $\Delta t = \mathcal{O}(h^2)$. Then $e = \bar{\xi}_x - \xi_x$ is $\mathcal{O}(h^3)$.*

*Proof.* In the proof of Lemma 2 we showed that

$$\xi(x) = \psi(x) + \text{higher order terms},$$

where $\psi$ is given by (5.17). Now

$$|\psi'(x_j + 0) - \psi'(x_j - 0)| \le Ch^3,$$

which implies that

$$|\xi_x(x_j + 0) - \xi_x(x_j - 0)| \le Ch^3. \tag{5.19}$$

Moreover

$$\bar{\xi}_x(x) = \frac{1}{\Delta t}\int_{x-\Delta t}^{x}\xi_x(s)ds = \xi_x(x) + \frac{1}{\Delta t}\int_{x-\Delta t}^{x}(\xi_x(s) - \xi_x(x))ds. \tag{5.20}$$

By Taylor expansion and (5.19)

$$\begin{aligned} |\xi_x(s) - \xi_x(x)| &\le Ch|x - s| + Ch^3 \\ &\le Ch\Delta t + Ch^3. \end{aligned} \tag{5.21}$$

Hence, by (5.20), (5.21), and the assumption on $\Delta t$

$$\begin{aligned} |e(x)| &= |\bar{\xi}_x(x) - \xi_x(x)| \\ &\le Ch\Delta t + Ch^3 \\ &\le Ch^3. \end{aligned}$$

## V.3   The Leading Error Term

**Lemma 4** *Assume that $u_{xxx}$ is not identically zero. Let*

$$\mu^n(x) = \sum_j r(x_j, t^n)w_j(x)$$

*where $w_j = -\frac{1}{3}v_j + b_j + b_{j-1}$. Assume $\Delta t = o(h^2)$ and that the coefficients $r(x,t)$ satisfy the first order equation*

$$r_t - 5r_x = -\frac{1}{4}u_{xxx},$$

*with $r(x,0) \equiv 0$. Then, there exists $C_0 > 0$, independent of $h$, such that*

$$\max_{0 \le n \le N}\|\mu^n\| \ge C_0$$

*and*

$$\|\zeta^n + h^2\mu^n\| = o(h^2).$$

*Proof:* It is clear that $r$ is a function that is not identically zero and does not depend on $h$. The size of $\mu$ is determined by the size of $r$. In fact,

$$\|\mu^n\|^2 \ge \frac{1}{45}\sum_j |r(x_j, t^n)|^2 h.$$

Thus, the existence of $C_0 > 0$ independent of $h$ is established.

Let $\lambda = \frac{\Delta t}{h}$. A calculation yields

$$\frac{6}{h}\left(\mu^{n+1} - \hat{\mu}^n, v_j\right)$$

$$= -\frac{1}{3}[r_{j-1}^{n+1} + 4r_j^{n+1} + r_{j+1}^{n+1} - ((1 + 3\lambda)r_{j-1}^n + 4r_j^n + (1 - 3\lambda)r_{j+1}^n)]$$

$$+ \frac{1}{2}[r_{j-1}^{n+1} + 2r_j^{n+1} + r_{j+1}^{n+1} - ((1 + 2\lambda)(r_{j-1}^n + r_j^n) + (1 - 2\lambda)(r_j^n + r_{j+1}^n))]$$

$$+ \mathcal{O}(\lambda^2)$$

$$= o(\Delta t). \tag{5.22}$$

and

$$\frac{60}{h}\left(\mu^{n+1} - \hat{\mu}^n, b_j\right)$$

$$= -\frac{5}{3}[r_j^{n+1} + r_{j+1}^{n+1} - (r_j^n + r_{j+1}^n) + 2\lambda(r_{j+1}^n - r_j^n)]$$

$$+ 2[r_j^{n+1} + r_{j+1}^{n+1} - (r_j^n + r_{j+1}^n)] + o(\Delta t)$$

$$= \Delta t[\frac{2}{3}r_t(x_j, t^n) - \frac{10}{3}r_x(x_j, t^n)] + \mathcal{O}(h\Delta t) + o(\Delta t).$$

Thus,

$$\left(\frac{\mu^{n+1} - \hat{\mu}^n}{\Delta t}, b_j\right) = -\frac{h}{360}u_{xxx}(x_j, t^n) + o(h). \tag{5.23}$$

Letting $\phi^n = \zeta^n + h^2\mu^n$ and using (4.1), Lemmas 2 and 3, (5.22) and (5.23), we note that

$$\left(\frac{\phi^{n+1} - \hat{\phi}^n}{\Delta t}, \phi^{n+1}\right) \leq (o(h^2) + \|\rho^n\|)\|\phi^{n+1}\|.$$

Thus, by Gronwall's Lemma

$$\|\phi^n\| = o(h^2)$$

and the proof is complete.

# VI.  Experimental Results

In this section, we verify experimentally the theoretical results of the previous sections.

Experimental rates of convergence for the MMOC with piecewise linear and piecewise quadratic approximating spaces were computed for the test problem

$$\begin{cases} u_t + u_x = 0, & x \in (0,1) \equiv I, \\ u^0(x) = \sin(2\pi x), & x \in I, \end{cases} \tag{6.1}$$

Approximating space  Rate

| $\tilde{M}_0(1,h)$ | 2.00 |
|---|---|
| $\tilde{M}_0(2,h)$ | 1.94 |

Table 1: Experimental rates

which has solution $u(x,t) = \sin(2\pi(x-t))$. Discrete $L^2$ errors were calculated at $t = 1/16$ for $h^{-1} = 2^k$, $k = 3,\ldots,6$. For the case $M_h = \tilde{M}_0(1,h)$, $\Delta t = h^2$ and the error was calculated by the trapezoidal rule of integration. For the case $M_h = \tilde{M}_0(2,h)$, $\Delta t = h^3$ and Simpson's rule was used in the error computation. A least squares fit was used to calculated the experimental rate of convergence. Table 1 below verifies that in both the linear and quadratic cases, $h^2$ convergence was observed.

We would also like to report that, in the linear case with $f \equiv 0$, one obtains superconvergence of $h^4$ at the nodes, provided that the interpolant is used as the initial condition. This result can be verified theoretically, however, since it is very specialized, the details are not included here.

Moreover, it is clear from the proof of Theorem 2 that in the $\tilde{M}_0(2,h)$ case the convergence at the Gauss points $y_{j1}$ and $y_{j2}$ is $\mathcal{O}(h^3)$ since $\mu$ vanishes to the next higher order at these points. This was observed in our experiments.

# References

[1] J. H. Bramble and S. Hilbert, *Estimation of linear functionals on Sobolev spaces with applications to Fourier transforms and spline interpolation*, SIAM J. Numer. Anal. 7, 113-124 (1970).

[2] C. N. Dawson, T. F. Russell, and M. F. Wheeler, *Some improved error estimates for the modified method of characteristics*, to appear in SIAM J. Numer. Anal.

[3] T. Dupont, *Galerkin methods for first-order hyperbolic equations: an example*, SIAM J. Numer. Anal. 10, 890-899 (1970).

[4] J. Douglas, Jr. and T. F. Russell, *Numerical methods for convection-dominated diffusion problems based on combining the method of characteristics with finite element or finite difference procedures*, SIAM J. Numer. Anal. 19, 871-885 (1982).

[5] R. E. Ewing, T. F. Russell, and M. F. Wheeler, *Convergence analysis of an approximation of miscible displacement in porous media by mixed finite elements*

and a modified method of characteristics, Computer Methods in Applied Mechanics and Engineering 47, 73-92 (1984).

[6] K. W. Morton, A. Priestley, and E. Süli, *Stability analysis of the Lagrange-Galerkin method with non-exact integration*, Report # 86/14, Oxford University Computing Laboratory (1986).

[7] T. F. Russell, *Time-stepping along characteristics with incomplete iteration for a Galerkin approximation of miscible displacement in porous media*, SIAM J. Numer. Anal. 22, 970-1013 (1985).

# A Time-Discretization Procedure for a Mixed Finite Element Approximation of Contamination by Incompressible Nuclear Waste in Porous Media

RICHARD E. EWING    Departments of Mathematics, Petroleum Engineering, and Chemical Engineering, University of Wyoming, Laramie, Wyoming

YIRANG YUAN    Department of Mathematics, Shandong, Jinan, Shandong, China, and Department of Mathematics, University of Wyoming, Laramie, Wyoming

GANG LI    Department of Mathematics, University of Wyoming, Laramie, Wyoming

Two efficient time-stepping procedures are introduced to treat the continuous-time method proposed by the authors which employs a mixed finite element method to approximate the pressure and the fluid velocity and a standard Galerkin method to approximate the transport of heat, brine and radionuclides for the system describing the contamination by nuclear waste in porous media. Optimal order asymptotic error estimates are derived under certain constraints between the discretization parameters.

## 1. INTRODUCTION

This paper describes finite element and time-stepping methods for a model system for the simulation of incompressible flow and transport processes of nuclear contaminants in porous media. The model was developed for use by the Nuclear

Regulatory Commission for the analysis of deep geologic nuclear waste disposal facilities. The mathematical model is presented with detailed physical assumptions in [23]. A continuous time semi-discrete numerical procedure for this problem was presented and analyazed by the authors in [17]. Similar analyses for the compressible nuclear waste-disposal problem will appear in [18,19].

The mathematical model is a fully transient incompressible, two-dimensional model which solves a set of coupled equations for fluid transport and heat flow in geologic media. The processes considered are: (1) fluid flow, (2) heat transport, (3) dominant species miscible displacement (brine), and (4) trace-species miscible displacement (radionuclide).

The transport equations used here are obtained by combining appropriate continuity and constitutive relations and have been derived by several authors [2,3,5,11]. The resulting relations for the total fluid, heat, brine, and the $i^{\text{th}}$ component of radionuclides may be stated as follows:

$$-\nabla \cdot \underline{u} - q + R_s' = 0, \tag{1.1}$$

$$-\nabla \cdot (H\underline{u}) + \nabla \cdot (\mathbf{E}_H \nabla T) - q_L - qH - q_H = [\phi c_p + (1-\phi)\bar{p}_R c_{pR}]\frac{\partial T}{\partial t}, \tag{1.2}$$

$$-\nabla \cdot (\hat{c}\underline{u}) + \nabla \cdot (\mathbf{E}_c \nabla \hat{c}) - q\hat{c} - q_c + R_s = \phi\frac{\partial \hat{c}}{\partial t}, \tag{1.3}$$

$$-\nabla \cdot (c_i\underline{u}) + \nabla \cdot (\mathbf{E}_c \nabla c_i) - qc_i - q_{c_i} + q_{0i} + \sum_{j=1}^{N} k_{ij}\lambda_j K_j \phi c_j \tag{1.4}$$

$$-\lambda_i K_i \phi c_i = \phi K_i \frac{\partial c_i}{\partial t}, \quad i = 1, 2, \cdots, N.$$

Several quantities in Equations (1.1)–(1.4) require further definition in terms of the basic parameters. The tensors in Equations (1.2), (1.3), and (1.4) are defined as sums of dispersion and molecular terms:

$$\mathbf{E}_c = \mathbf{D} + D_m \mathbf{I} \tag{1.5}$$

and

$$\mathbf{E}_H = \mathbf{D}c_{pw} + K_m \mathbf{I}, \tag{1.6}$$

where

$$D_{ij} = \alpha_T |\underline{u}|\delta_{ij} + (\alpha_L - \alpha_T)u_i u_j / |\underline{u}| \tag{1.7}$$

in a cartesian system. Also, absorption of radionuclides is included via an assumption of a linear equilibrium isotherm. This yields the distribution coefficient $k_{di}$ and the retardation factor for each component

$$K_i = 1 + \rho_R k_{di}(1 - \phi)/\phi. \tag{1.8}$$

Equations (1.1)–(1.4) are coupled by a mixing rule for viscosity, $\mu = \mu(\hat{c})$, and three auxiliary relations for Darcy flux, fluid enthalpy and fluid internal energy given, respectively, by

$$\underline{u} = -(\mathbf{k}/\mu)\left(\nabla p - \rho_0 \frac{g}{g_c}\nabla z\right), \tag{1.9}$$

$$H = U_0 + U + p/\rho_0, \tag{1.10}$$

$$U = c_p(T - T_0), \tag{1.11}$$

where $U_0 = U_0(x)$ and $T_0 = T(x,0)$. The quantity $q = q(x,t)$ is a production term, $R'_s = R'_s(\hat{c}) = \frac{c_s \phi k_s f_s}{1+c_s}(1 - \hat{c})$ is a salt-dissolution term for fluid equation. $q_L = q_L(T)$ is a heat loss to under/overburden term, $q_H$ is an injected enthalpy term, $q_H = q_H(T)$ is a produced enthalpy term. $q\hat{c}$ is an injected brine term, $q_c = q_c(\hat{c})$ is a produced brine term, $R_s = R_s(\hat{c}) = \phi k_s f_s(1 - \hat{c})$ is a salt-dissolution term for brine equation. $qc_i$ are injected components terms, $q_{c_i} = q_{c_i}(c_i)$ are produced component terms, $q_{0i} = q_{0i}(c_i)$ are waste leach terms, $\sum_{j=1}^{N} k_{ij}\lambda_j K_j \phi c_j$ describe generation of component $i$ by decay of $j$, and $\lambda_i K_i \phi c_i$ is the decay for component $i$.

We shall assume that no fluid flow occurs accross the boundary and shall assume a zero Dirichlet boundary condition for $T$. In addition, the initial conditions must be given. We need the compatibility condition:

$$(q - R'_s, 1) = \int_\Omega [q(x,t) - R'_s(\hat{c})]\,dx = 0. \tag{1.12}$$

The analysis is given under a number of restrictions. The most important one is that the solution is smooth, i.e., $q, q_H, \cdots$ are smoothly distributed, the coefficients are smooth (cf. [6,13,15]), and the domain has at least the regularity required for a standard elliptic problem to have $H^2(\Omega)$-regularity. The coefficients $\phi c_p + (1 - \phi)\bar{\rho}_R c_{pR}, \phi$, and $K_i(i = 1, 2, \cdots, N)$ will be assumed bounded below positively as well as being smooth.

The authors [17] have previously defined a continuous-time finite element method based on the use of an elliptic mixed finite element method to approximate the pressure $p$ and the velocity $\underline{u}$ and a parabolic Galerkin method to approximate concentration $\hat{c}$, $c_i(i = 1, 2, \cdots, N)$ and temperature $T$. It is particularly suitable to employ the mixed method, since the velocity appears in the concentration and temperature equations. The object of this paper is to discuss a time-stepping procedure for the finite element procedure that efficiently reflects the fact that the velocity field and the pressure vary more slowly in time than the concentration and the temperature for reasonable physical data. Thus, we shall take the pressure time step to be much larger than either the concentration or temperature time steps.

Since this paper is a continuation of the authors' paper [17], we shall use the same notation as far as possible and we shall make use of the results of that paper wherever feasible to shorten our arguments here.

An outline of this paper is as follows. The time-discretization procedure will be derived in Section 2. In Section 3 some auxiliary results will be set out. In Section 4 error estimates of Scheme I will be proved and an estimate for Scheme II will be presented. Throughout, the symbols $M$ and $\epsilon$ will denote, respectively, a generic constant and a generic small positive constant.

## 2. THE FINITE ELEMENT SCHEMES

Let $V = \{\underline{v} \in H(\text{div}; \Omega); \underline{v} \cdot \nu = 0 \text{ on } \partial\Omega\}$ and $W = L^2(\Omega)/\{\psi \equiv \text{constant on } \Omega\}$. Let $W^{k,p}(\Omega)$ be the standard Sobolev space where $k$ distributional derivatives are bounded in $L^p(\Omega)$. Denote the norms on $W^{k,p}(\Omega)$ by $\|\cdot\|_{k,p}$. Note that $\|\cdot\|_{k,2} \equiv \|\cdot\|_k$ for a norm on $H^k(\Omega)$. Then a saddle-point weak form of (1.1) and (1.9) is given by the system

$$
\begin{aligned}
(a) \quad & (\nabla \cdot \underline{u}, w) = (-q + R'_s(\hat{c}), w), \quad w \in W, \\
(b) \quad & (\alpha(\hat{c})^{-1}\underline{u}, \underline{v}) - (\nabla \cdot \underline{v}, p) = (\underline{\gamma}(\hat{c}), \underline{v}), \quad \underline{v} \in V,
\end{aligned}
\tag{2.1}
$$

where $\alpha(\hat{c}) = \mathrm{k}/\mu(\hat{c}), \underline{\gamma}(\hat{c}) = \frac{\rho_o g}{g_c}\nabla z$.

The equation for concentration $\hat{c}$ is equivalent to the finding of a differentiable map $\hat{c} : J \to H^1(\Omega)$ such that

$$
\left(\phi\frac{\partial \hat{c}}{\partial t}, z\right) + (\underline{u} \cdot \nabla\hat{c}, z) + (\mathbf{E}_c(\underline{u})\nabla\hat{c}, \nabla z) = (g(\hat{c}), z),
\tag{2.2}
$$

for $z \in H^1(\Omega)$ and such that $\hat{c}(x,0) = \hat{c}_0(x)$.

The equation for concentration $c_i$ is equivalent to the finding a differentiable map $c_i : J \to H^1(\Omega)$ such that

$$
\left(\phi K_i \frac{\partial c_i}{\partial t}, z\right) + (\underline{u} \cdot \nabla c_i, z) + (\mathbf{E}_c(\underline{u})\nabla c_i, \nabla z)
$$
$$
= (f_i(\hat{c}, c_1, c_2, \cdots, c_N), z), \quad i = 1, 2, \cdots, N,
\tag{2.3}
$$

for $z \in H^1(\Omega)$ and such that $c_i(x,0) = c_{i0}(x)$.

The equation for heat change is equivalent to the finding a differentiable map $T : J \to H^1_0(\Omega)$ such that

$$\left(\hat{\phi}\frac{\partial T}{\partial t}, z\right) + c_p(\underline{u} \cdot \nabla T, z) + (\mathbf{E}_H(\underline{u})\nabla T, \nabla z) = (Q(\underline{u}, p, T, \hat{c}), z), \qquad (2.4)$$

for $z \in H_0^1(\Omega)$ and such that $T(x, 0) = T_0(x)$.

Let $h = (h_c, h_T, h_p)$ where $h_c$, $h_T$ and $h_p$, all positive, are the grid spacings for the concentration, the fluid temperature, and pressure, respectively. Let $V_h \times W_h$ be a Raviart-Thomas [22] space of index at least $k$ associated with a quasi-regular triangulation or quadrilateralization (or a mixture of the two) of $\Omega$ such that the elements have diameters bounded by $h_p$. Since in contaminant transport problems the boundary is not known exactly, we take $\Omega$ as a polygonal domain, then impose the boundary condition $\underline{v} \cdot \nu = 0$ on $\partial\Omega$ strongly on $V_h$. The approximation properties for the velocity and pressure from the spaces $V_h \times W_h$ are given by the inequalities

$$(a) \quad \inf_{\underline{v}_h \in V_h} ||\underline{v} - \underline{v}_h||_0 = \inf_{\underline{v}_h \in V_h} ||\underline{v} - \underline{v}_h||_{L^2(\Omega)^2} \leq M||\underline{v}||_{k+1} h_p^{k+1},$$

$$(2.5)$$

$$(b) \quad \inf_{\underline{v}_h \in V_h} ||\nabla \cdot (\underline{v} - \underline{v}_h)||_0 \leq M\{||\underline{v}||_{k+1} + ||\nabla \cdot \underline{v}||_{k+1}\}h_p^{k+1},$$

for $\underline{v} \in V \cap H^{k+1}(\Omega)^2$ and, in addition for (2.5)(b), $\nabla \cdot \underline{v} \in H^{k+1}(\Omega)$, and

$$\inf_{w_h \in W_h} ||w - w_h||_0 \leq M||w||_{k+1} h_p^{k+1}, \quad w \in H^{k+1}(\Omega). \qquad (2.6)$$

Let $M_h = M_{h_c} \subset W^{1,\infty}(\Omega)$ be a standard finite element space for approximation of concentrations, and assume that it is associated with a quasi-regular polygonalization of $\Omega$ and that it is of index $\ell$:

$$\inf_{z_h \in M_h} ||z - z_h||_{1,q} \leq M||z||_{\ell+1,q} h_c^\ell, \qquad (2.7)$$

for $z \in W^{\ell+1,q}(\Omega)$ and $1 \leq q \leq \infty$.

Let $R_h = R_{h_T} \subset W^{1,\infty}(\Omega) \cap H_0^1(\Omega)$ be a piecewise-polynomial space of degree at least $r$ associated with other quasi-regular polygonalization of $\Omega$, for the approximation of the temperature. Then

$$\inf_{z_h \in R_h} ||z - z_h||_{1,q} \leq M||z||_{r+1,q} h_T^r, \qquad (2.8)$$

for $z \in W^{r+1,q}(\Omega)$ and $1 \leq q \leq \infty$.

Since the pressure, in general, changes less rapidly in time than the concentrations or the temperature we will use larger time-steps for the pressure and velocity than for the other unknowns. These techniques were applied in the petroleum literature in [13,15].

For definitions of our mixed finite element approximation, we shall use the following notations:

$$\Delta t_c \quad = \quad \text{time step for } \hat{c}, c_i (i = 1, 2, \cdots, N) \text{ and } T \text{ equations,}$$

$$\Delta t_p^0 \quad = \quad \text{first time step for the pressure equation,}$$

$$\Delta t_p \quad = \quad \text{subsequent time step for the pressure equation,}$$

$$j \quad = \quad \Delta t_p / \Delta t_c \in \mathbb{Z}^+, j^0 = \Delta t_p^0 / \Delta t_c \in \mathbb{Z}^+,$$

$$t^n \quad = \quad t_c^n = n\Delta t_c, t_m = t_p^m = \Delta t_p^0 + (m-1)\Delta t_p,$$

$$\psi^n \quad = \quad \psi(t^n), \psi_m = \psi(t_m) \text{ for function } \psi(x, t),$$

$$d_t \psi^{n-1} \quad = \quad (\psi^n - \psi^{n-1})/\Delta t_c, d_t \psi_{m-1} = (\psi_m - \psi_{m-1})/\Delta t_p, m > 1,$$

$$d_t \psi_0 \quad = \quad (\psi_1 - \psi_0)/\Delta t_p^0, \delta \psi^{n-1} = \psi^n - \psi^{n-1}, \delta \psi_{m-1} = \psi_m - \psi_{m-1},$$

$$E\psi^n \quad = \quad \begin{cases} \psi_0, \ t^n \le t_1, \\ \left(1 + \frac{\gamma}{j^0}\right)\psi_1 - \frac{\gamma}{j^0}\psi_0, & t_1 < t^n \le t_2, & t^n = t_1 + \gamma\Delta t_c, \\ \left(1 + \frac{\gamma}{j}\right)\psi_m - \frac{\gamma}{j}\psi_{m-1}, & t_m < t^n \le t_{m+1}, & t^n = t_m + \gamma\Delta t_c, \ m \ge 2, \end{cases}$$

$$\check{\psi}^n \quad = \quad \begin{cases} \psi^0, \ n = 1, \\ 2\psi^{n-1} - \psi^{n-2}, \ n \ge 2, \end{cases}$$

$$\psi^{m+\frac{1}{2}} \quad = \quad \begin{cases} \psi_0, \ m = 0, \\ \frac{3}{2}\psi_m - \frac{1}{2}\psi_{m-1}, \ m \ge 1, \end{cases}$$

$$J^n \quad = \quad (t^{n-1}, t^n), J_m = (t_{m-1}, t_m).$$

Finally approximate $\hat{c}_0, c_{i0} (i = 1, 2, \cdots, N)$ and $T_0$ by function $\hat{C}_h^0 = \hat{C}_h(t_c^0) \in M_h$, $C_{hi}^0 = C_{hi}(t_c^0) \in M_h$ and $T_h^0 = T_h(t_c^0) \in R_h$; this can be done by interpolation, by $L^2$-projection, or by projection with respect to some elliptic form.

Since simultaneous solution of all these equations for large problems would swamp even the largest of the emerging supercomputers, we will use a sequential solution technique which decouples the equations and also aids in linearization. See [12–16] for similar ideas.

Now, assume $\hat{C}_h^m = \hat{C}_h(t_p^m)$ is known. Then, the velocity-pressure pair $\{\underline{U}_h^m, P_h^m\}$ at time $t_p^m$ can be calculated as the (mixed method) solution of the system

$$(a) \quad (\nabla \cdot \underline{U}_h^m, w) = (-q + R_s'(\hat{C}_h^m), w), \quad w \in W_h,$$

$$(b) \quad (\alpha(\hat{C}_h^m)^{-1}\underline{U}_h^m, v) - (\nabla \cdot \underline{v}, P_h^m) = (\underline{\gamma}(\hat{C}_h^m), \underline{v}), \quad \underline{v} \in V_h. \tag{2.9}$$

The question at hand is to discretize the concentration $\hat{c}$ equation in time for $t_p^m < t_c^n \le t_p^{m+1}$. This will be done by deriving, through several stages, a convenient variant of a backward-differenced Galerkin procedure.

## Scheme I

(a) $\quad \left(\phi\dfrac{\hat{C}_h^n - \hat{C}_h^{n-1}}{\Delta t_c}, z\right) + (E\underline{U}_h^n \cdot \nabla\hat{C}_h^n, z) + (\mathbf{E}_c(E\underline{U}_h^n)\nabla\hat{C}_h^n, \nabla z)$

$\qquad = (g(\overset{\times}{\check{C}}_h^n), z), \quad z \in M_h,$

(b) $\quad \left(\phi K_i \dfrac{C_{hi}^n - C_{hi}^{n-1}}{\Delta t_c}, z\right) + (E\underline{U}_h^n \cdot \nabla C_{hi}^n, z) + (\mathbf{E}_c(E\underline{U}_h^n)\nabla C_{hi}^n, \nabla z)$

$\qquad = (f_i(\overset{\times}{\check{C}}_h^n, \check{C}_{h1}^n, \check{C}_{h2}^n, \cdots, \check{C}_{hN}^n), z), \quad z \in M_h, \quad i = 1, 2, \cdots, N,$

(c) $\quad \left(\hat{\phi}\dfrac{T_h^n - T_h^{n-1}}{\Delta t_c}, z\right) + c_p(E\underline{U}_h^n \cdot \nabla T_h^n, z) + (\mathbf{E}_H(E\underline{U}_h^n)\nabla T_h^n, \nabla z)$

$\qquad = (Q(E\underline{U}_h^n, EP_h^n, \check{T}_h^n, \overset{\times}{\check{C}}_h^n), z), \quad z \in R_h.$

$\hspace{10cm}$ (2.10)

For Scheme I the computation is performed in the following order: $\hat{C}_h^0, C_{hi}^0(i = 1, 2, \cdots, N), T_h^0, P_h^0, \underline{U}_h^0; \cdots, \hat{C}_h^{j^0}, C_{hi}^{j^0}$ $(i = 1, 2, \cdots, N), T_h^{j^0}, P_h^1, \underline{U}_h^1; \cdots$ .

## Scheme II

(a) $\quad \left(\phi\dfrac{\hat{C}_h^n - \hat{C}_h^{n-1}}{\Delta t_c}, z\right) + \left(\underline{U}_h^{m+\frac{1}{2}} \cdot \nabla\hat{C}_h^n, z\right) + \left(\mathbf{E}_c(\underline{U}_h^{m+\frac{1}{2}})\nabla\hat{C}_h^n, \nabla z\right)$

$\qquad = ((\underline{U}_h^{m+\frac{1}{2}} - E\underline{U}_h^n) \cdot \nabla\overset{\times}{\check{C}}_h^n, z) + ((\mathbf{E}_c(\underline{U}_h^{m+\frac{1}{2}}) - \mathbf{E}_c(E\underline{U}_h^n))\nabla\overset{\times}{\check{C}}_h^n, \nabla z)$

$\qquad + (g(\overset{\times}{\check{C}}_h^n), z), \quad z \in M_h,$

(b) $\quad \left(\phi K_i\dfrac{C_{hi}^n - C_{hi}^{n-1}}{\Delta t_c}, z\right) + (\underline{U}_H^{m+\frac{1}{2}} \cdot \nabla C_{hi}^n, z) + \left(\mathbf{E}_c(\underline{U}_h^{m+\frac{1}{2}})\nabla C_{hi}^n, \nabla z\right)$

$\qquad = \left((\underline{U}_h^{m+\frac{1}{2}} - E\underline{U}_h^n) \cdot \nabla\check{C}_{hi}^n, z\right) + \left(\left(\mathbf{E}_c(\underline{U}_h^{m+\frac{1}{2}}) - \mathbf{E}_c(E\underline{U}_h^n)\right)\nabla\check{C}_{hi}^n, \nabla z\right)$

$\qquad + \left(f_i(\overset{\times}{\check{C}}_h^n, \check{C}_{h1}^n, \check{C}_{h2}^n, \cdots, \check{C}_{hN}^n), z\right), \quad z \in M_h,$

(c) $\quad \left(\hat{\phi}\dfrac{T_h^n - T_h^{n-1}}{\Delta t_c}, z\right) + c_p(\underline{U}_H^{m+\frac{1}{2}} \cdot \nabla T_h^n, z) + \left(\mathbf{E}_H(\underline{U}_h^{m+\frac{1}{2}})\nabla T_h^n, \nabla z\right)$

$\qquad = c_p\left((\underline{U}_h^{m+\frac{1}{2}} - E\underline{U}_h^n) \cdot \nabla\check{T}_h^n, z\right) + \left(\left(\mathbf{E}_H(\underline{U}_h^{m+\frac{1}{2}}) - \mathbf{E}_H(E\underline{U}_h^n)\right)\nabla\check{T}_h^n, \nabla z\right)$

$$+ \left( Q(EU_h^n, EP_h^n, \check{T}_h^n, \overset{x}{\check{C}_h^n}), z \right), \quad z \in R_h. \tag{2.11}$$

The amount of computational work would be reduced if Scheme II is used instead of Scheme I, because Scheme II only requires a single factorization of a matrix for each pressure time step, instead of one for each concentration time step.

We shall only suppose that the coefficients and data in (2.1)–(2.4) are locally bounded and locally Lipschitz continuous [17]. This is a suitable assumption for the problem of incompressible nuclear waste-disposal contamination in porous media.

## 3. SOME AUXILIARY RESULTS

The analysis of the convergence of the schemes will be given under the assumption that the imposed flow is smoothly distributed. We introduce some projections, where the constants $\lambda$, $\lambda_i (i = 1, 2, \cdots, N)$ and $\mu$ are chosen to insure the coecivity of the bilinear forms.

Let $\hat{\mathbf{C}} = \hat{\mathbf{C}}_h : J \to M_h$ be determined by the relations

$$(\mathbf{E}_c(\underline{u}) \nabla(\hat{c} - \hat{\mathbf{C}}), \nabla z) + (\underline{u} \cdot \nabla(\hat{c} - \hat{\mathbf{C}}), z) + \lambda(\hat{c} - \hat{\mathbf{C}}, z) = 0, \tag{3.1}$$

$$z \in M_h, \quad t \in J.$$

Let $\tilde{C}_i = \tilde{C}_{hi} : J \to M_h$ be determined by the relations

$$(\mathbf{E}_c(\underline{u}) \nabla(c_i - \tilde{C}_i), \nabla z) + (\underline{u} \cdot \nabla(c_i - \tilde{C}_i), z) + \lambda_i(c_i - \tilde{C}_i, z) = 0, \tag{3.2}$$

$$z \in M_h, \quad t \in J, \quad i = 1, 2, \cdots, N.$$

Let $\tilde{T} = \tilde{T}_h : J \to R_h$ be determined by the relations

$$(\mathbf{E}_H(\underline{u}) \nabla(T - \tilde{T}), \nabla z) + c_p(\underline{u} \cdot \nabla(T - \tilde{T}), z) + \mu(T - \tilde{T}, z) = 0, \tag{3.3}$$

$$z \in R_h, \quad t \in J.$$

Let

$$\begin{aligned}
(a) \quad & \hat{\varsigma} = \hat{c} - \hat{\mathbf{C}}, \quad \hat{\xi} = \hat{\mathbf{C}} - \hat{\mathbf{C}}_h, \\
(b) \quad & \varsigma_i = c_i - \tilde{C}_i, \quad \xi_i = \tilde{C}_i - C_{hi}, \\
(c) \quad & \theta = T - \tilde{T}, \quad \omega = \tilde{T} - T_h.
\end{aligned} \tag{3.4}$$

The following results can be obtained using standard analysis [7,13,16,25] in the theory of Galerkin methods for elliptic problems.

$$(a) \quad ||\hat{\varsigma}||_0 + h_c||\hat{\varsigma}||_1 \leq M||\hat{c}||_{\ell+1} h_c^{\ell+1},$$

$$(b) \quad ||\varsigma_i||_0 + h_c||\varsigma_i||_1 \leq M||c_i||_{\ell+1} h_c^{\ell+1}, \quad i = 1, 2, \cdots, N, \qquad (3.5)$$

$$(c) \quad ||\theta||_0 + h_T||\theta||_1 \leq M||T||_{r+1} h_T^{r+1},$$

and

$$(a) \quad ||\frac{\partial \hat{\varsigma}}{\partial t}||_0 + h_c||\frac{\partial \hat{\varsigma}}{\partial t}||_1 \leq M \left\{ ||\hat{c}||_{\ell+1} + ||\frac{\partial \hat{c}}{\partial t}||_{\ell+1} \right\} h_c^{\ell+1},$$

$$(b) \quad ||\frac{\partial \varsigma_i}{\partial t}||_0 + h_c||\frac{\partial \varsigma_i}{\partial t}||_1 \leq M \left\{ ||c_i||_{\ell+1} + ||\frac{\partial c_i}{\partial t}||_{\ell+1} \right\} h_c^{\ell+1}, \quad i = 1, 2, \cdots, N, \quad (3.6)$$

$$(c) \quad ||\frac{\partial \theta}{\partial t}||_0 + h_T||\frac{\partial \theta}{\partial t}||_1 \leq M \left\{ ||T||_{r+1} + ||\frac{\partial T}{\partial t}||_{r+1} \right\} h_T^{r+1},$$

for $t \in J$.

Let $\{\tilde{\underline{U}}, \tilde{P}\} = \{\tilde{U}_h, \tilde{P}_h\}$, the projection of the Darcy Velocity and the pressure be the map: $J \to V_h \times W_h$ given by

$$(a) \quad (\nabla \cdot \tilde{\underline{U}}, w) = (-q + R'_s(\hat{c}), w), \quad w \in W_h,$$

$$(b) \quad (\alpha(\hat{c})^{-1} \tilde{\underline{U}}, \underline{v}) - (\nabla \cdot \underline{v}, \tilde{P}) = (\underline{\gamma}(\hat{c}), \underline{v}), \quad \underline{v} \in V_h, \qquad (3.7)$$

for $t \in J$. Set

$$(a) \quad \eta = p - \tilde{P}, \quad \pi = \tilde{P} - P_h,$$

$$(b) \quad \underline{\rho} = \underline{u} - \tilde{\underline{U}}, \quad \underline{\sigma} = \tilde{\underline{U}} - \underline{U}_h. \qquad (3.8)$$

By the estimate of Brezzi [4] we have

$$||\underline{\rho}||_{H(\mathrm{div};\Omega)} + ||\eta||_0 \leq M||p||_{k+3} h_p^{k+1}, \qquad (3.9)$$

for $t \in J$.


## 4. CONVERGENCE ANALYSIS

We first derive an estimate of $\underline{U}_h^m - \tilde{\underline{U}}^m$ and $P_h^m - \tilde{P}^m$. Manipulation of (2.9) and (3.7) leads to the equations

(a)   $(\nabla \cdot (\underline{U}_h^m - \underline{\tilde{U}}^m), w) = (-R_s'(\hat{c}^m) + R_s'(\hat{C}_h^m), w), \quad w \in W_h,$

(b)   $(\alpha(\hat{C}_h^m)^{-1}(\underline{U}_h^m - \underline{\tilde{U}}^m), \underline{v}) - (\nabla \cdot \underline{v}, P_h^m - \tilde{P}^m)$                    (4.1)

$= (\underline{\gamma}(\hat{C}_h^m) - \underline{\gamma}(c^m), \underline{v}) + ([\alpha(\hat{c}^m)^{-1} - \alpha(\hat{C}_h^m)^{-1}]\underline{\tilde{U}}^m, \underline{v}), \quad \underline{v} \in V_h.$

Let

$$Q^h(t^n) = \max\{ \sup_{\substack{x \in \Omega \\ 0 \le t' \le t^n}} |\hat{c}(x, t')|, \sup_{\substack{x \in \Omega \\ 0 \le t' \le t^n}} |\hat{C}(x, t')|, \sup_{\substack{x \in \Omega \\ 0 \le j \le n}} |\hat{C}_h^j(x)| \}.$$

Note that $\hat{C}_h^j = \hat{C}_h^j - \hat{C}^j + \hat{C}^j = \hat{C}^j - \hat{\xi}^j$. We have

$$\sup_{\substack{x \in \Omega \\ 0 \le j \le n}} |\hat{C}_h^j(x)| \le \sup_{\substack{x \in \Omega \\ 0 \le t' \le t^n}} |\hat{C}(x, t')| + \sup_{\substack{x \in \Omega \\ 0 \le j \le n}} |\hat{\xi}^j|.$$

Let

$$\|\hat{\xi}\|_{\tilde{L}^\infty(n,\Omega)} = \sup_{\substack{x \in \Omega \\ 0 \le j \le n}} |\hat{\xi}^j|,$$

we have

$$Q^h(t^n) \le \sup_{\substack{x \in \Omega \\ 0 \le t' \le t^n}} |\hat{c}(x, t')| + \sup_{\substack{x \in \Omega \\ 0 \le t' \le t^n}} |\hat{C}(x, t')| + \|\hat{\xi}\|_{\tilde{L}^\infty(n,\Omega)}. \tag{4.2}$$

Brezzi has shown [4] that the solution operator of (4.1) is bounded, hence

$$\|\underline{U}_h^m - \underline{\tilde{U}}^m\|_V + \|P_h^m - \tilde{P}^m\|_W$$
$$\le M(Q^h(t_p^m))\{1 + \|\underline{\tilde{U}}^m\|_{L^\infty(\Omega)}\}\|\hat{c}^m - \hat{C}_h^m\|_{L^2(\Omega)}. \tag{4.3}$$

The quasi-regularity of the grid and the bound (3.9) imply that $\underline{\tilde{U}}$ is bounded in $L^\infty(J; L^\infty(\Omega))$. Note that $\hat{\xi} = \hat{C} - \hat{C}_h$ and $\hat{\varsigma} = \hat{c} - \hat{C}$. Then (2.2), (2.10.a) and (3.1) can be used to obtain the relation

$$\left(\phi \frac{\hat{\xi}^n - \hat{\xi}^{n-1}}{\Delta t_c}, z\right) + (E\underline{U}_h^n \cdot \nabla \hat{\xi}^n, z) + (\mathbf{E}_c(E\underline{U}_h^n)\nabla \hat{\xi}^n, \nabla z)$$

$$= (g(\hat{c}^n) - g(\overset{\times}{C}_h^n), z) + \left(\phi \left\{ \frac{\hat{C}^n - \hat{C}^{n-1}}{\Delta t_c} - \frac{\partial \hat{c}^n}{\partial t} \right\}, z\right) + \lambda(\hat{\varsigma}^n, z) \tag{4.4}$$

$$+ (\{E\underline{U}_h^n - \underline{u}^n\} \cdot \nabla \hat{C}^n, z) + (\{\mathbf{E}_c(E\underline{U}_h^n) - \mathbf{E}_c(\underline{u}^n)\}\nabla \hat{C}^n, \nabla z).$$

The test function will be chosen to be $z = \hat{\xi}^n$, the terms will be treated as follows.

$$|(g(\hat{c}^n) - g(\overset{X}{\hat{C}}_h^n), \hat{\xi}^n)| \le M(Q^h(t^n))\{(\Delta t_c)^2 + h_c^{2(\ell+1)} + ||\hat{\xi}^n||^2 + ||\hat{\xi}^{n-1}||^2 + ||\hat{\xi}^{n-2}||^2\} \tag{4.5}$$

where $\hat{\xi}^{n-2} = 0$ if $n = 1$.

$$\left| \left( \phi \left\{ \frac{\hat{C}^n - \hat{C}^{n-1}}{\Delta t_c} - \frac{\partial \hat{c}^n}{\partial t} \right\}, \hat{\xi}^n \right) \right|$$
$$\le M \left\{ (\Delta t_c)^{-1} \left\| \frac{\partial \hat{\xi}}{\partial t} \right\|_{L^2(J^n, L^2)}^2 + \Delta t_c \left\| \frac{\partial^2 \hat{c}}{\partial t^2} \right\|_{L^2(J^n, L^2)}^2 + ||\hat{\xi}^n||^2 \right\}, \tag{4.6}$$

where $J^n = (t^{n-1}, t^n)$. In the following, we denote $\hat{\xi}_p^m = \hat{\xi}(t_p^m)$.

$$|\lambda(\hat{s}^n, \hat{\xi}^n)| \le M\{h_c^{2(\ell+1)} + ||\hat{\xi}^n||^2\}. \tag{4.7}$$

$$|(\{E\underline{U}_h^n - \underline{u}^n\}\nabla \hat{C}^n, \hat{\xi}^n)| = |(\{(E\underline{U}_h^n - E\underline{u}^n) + (E\underline{u}^n - \underline{u}^n)\}\nabla \hat{C}^n, \hat{\xi}^n)|$$
$$\le M\{[(\Delta t_p)^2(\text{or}\Delta t_p^0) + ||\underline{U}_h^m - \underline{u}^m|| + ||\underline{U}_h^{m-1} - \underline{u}^{m-1}||]||\hat{\xi}^n||\} \tag{4.8}$$
$$\le M\{(\Delta t_p)^4(\text{or}(\Delta t_p^0)^2) + h_c^{2(\ell+1)} + h_p^{2(k+1)} + ||\hat{\xi}_p^m||^2 + ||\hat{\xi}_p^{m-1}||^2 + ||\hat{\xi}^n||^2\}.$$

$$|(\{\mathbf{E}_c(E\underline{U}_h^n) - \mathbf{E}_c(\underline{u}^n)\}\nabla \hat{C}^n, \nabla \hat{\xi}^n)| \le \epsilon||\nabla \hat{\xi}^n||^2 + M\{(\Delta t_p)^4(\text{or}(\Delta t_p^0)^2)$$
$$+ h_c^{2(\ell+1)} + h_p^{2(k+1)} + ||\hat{\xi}_p^m||^2 + ||\hat{\xi}_p^{m-1}||^2 + ||\hat{\xi}^n||^2\}. \tag{4.9}$$

$$|(E\underline{U}_h^n \cdot \nabla \hat{\xi}^n, \hat{\xi}^n)| \le \epsilon||\nabla \hat{\xi}^n||^2 + M(Q^h(t^n))\{1 + ||\hat{\xi}_p^m||^2 + ||\hat{\xi}_p^{m-1}||^2\}||\hat{\xi}^n||^2. \tag{4.10}$$

The bounds derived above can be collected to obtain the inequality

$$\frac{1}{\Delta t_c}\{(\phi \hat{\xi}^n, \hat{\xi}^n) - (\phi \hat{\xi}^{n-1}, \hat{\xi}^{n-1})\} + ((D_m + \alpha_T|E\underline{U}_h^n|)\nabla \hat{\xi}^n, \nabla \hat{\xi}^n)$$

$$\le M(Q^h(t^n))\{(\Delta t_p)^4(\text{or}(\Delta t_p^0)^2) + (\Delta t_c)^2 + h_c^{2(\ell+1)} + h_p^{2(k+1)} + ||\hat{\xi}_p^m||^2 + ||\hat{\xi}_p^{m-1}||^2$$
$$+ [1 + ||\hat{\xi}_p^m||^2 + ||\hat{\xi}_p^{m-1}||^2][||\hat{\xi}^n||^2 + ||\hat{\xi}^{n-1}||^2 + ||\hat{\xi}^{n-2}||]\} + \epsilon||\nabla \hat{\xi}^n||^2$$

$$\le M(Q^h(t^n))\{(\Delta t_p)^4(\text{or}(\Delta t_p^0)^2) + (\Delta t_c)^2 + h_c^{2(\ell+1)} + h_p^{2(k+1)} + ||\hat{\xi}_p^m||^2 + ||\hat{\xi}_p^{m-1}||^2$$
$$+ [1 + ||\hat{\xi}_p^m||_{L^\infty}^2 + ||\hat{\xi}_p^{m-1}||_{L^\infty}^2][||\hat{\xi}^n||^2 + ||\hat{\xi}^{n-1}||^2 + ||\hat{\xi}^{n-2}||^2]\} + \epsilon||\nabla \hat{\xi}^n||^2, \tag{4.11}$$

where in $\{\cdots\}$ if $n = 1$ we take $(\Delta t_p^0)^2$, if $n \geq 2$ we take $(\Delta t_p)^4$.

We shall need an induction hypothesis. We assume that

$$\sup_n \|\hat{\xi}^n\|_{L^\infty(\Omega)} \leq C. \tag{4.12}$$

Thus $Q^h$ is bounded. We have $M(Q^h(t^n)) \leq M$, where $M$ is a positive constant independent of $\hat{C}_h$ and $\hat{C}$. Now, multiply (4.11) by $\Delta t_c$ and add on the time for $0 < t_c^k \leq t_c^n$. Then, if

$$m(k) = m \quad \text{for} \quad t_p^m < t_c^k \leq t_p^{m+1}, \tag{4.13}$$

$$(\phi \hat{\xi}^n, \hat{\xi}^n) - (\phi \hat{\xi}^0, \hat{\xi}^0) + \sum_{0 < t_c^k \leq t_c^n} ((D_m + \alpha_T |E\underline{U}^k|)\nabla \hat{\xi}^k, \nabla \hat{\xi}^k)\Delta t_c$$

$$\leq M\{h_c^{2(\ell+1)} + h_p^{2(k+1)} + (\Delta t_p)^4 + (\Delta t_p^0)^3 + (\Delta t_c)^2\} \tag{4.14}$$

$$+M \sum_{0 < t_c^k \leq t_c^n} (\|\hat{\xi}^k\|^2 + \|\hat{\xi}^{m(k)}\|^2)\Delta t_c + \epsilon \sum_{0 < t_c^k \leq t_c^n} \|\nabla \hat{\xi}^k\|^2 \Delta t_c.$$

For $\epsilon$ sufficiently small the last term is covered by the diffusion term on the left-hand side, and it follows that

$$\|\hat{\xi}^n\|^2 + \sum_{0 < t_c^k \leq t_c^n} \|\nabla \hat{\xi}^k\|^2 \Delta t_c \leq M\{h_c^{2(\ell+1)} + h_p^{2(k+1)} + (\Delta t_p)^4 + (\Delta t_p^0)^3 + (\Delta t_c)^2\}$$

$$+M \sum_{0 < t_c^k \leq t_c^n} (\|\hat{\xi}^k\|^2 + \|\hat{\xi}^{m(k)}\|^2)\Delta t_c.$$

$$\tag{4.15}$$

Let

$$\alpha^n = \max\{\|\hat{\xi}^k\|^2 : \quad 0 < t_c^k \leq t_c^n\}, \tag{4.16}$$

so that

$$\alpha^n + \sum_{0 < t_c^k \leq t_c^n} \|\nabla \hat{\xi}^k\|^2 \Delta t_c \leq M\{h_c^{2(\ell+1)} + h_p^{2(k+1)} + (\Delta t_p)^4 + (\Delta t_p^0)^3 + (\Delta t_c)^2\}$$

$$+M \sum_{0 < t_c^k \leq t_c^n} \alpha^k \Delta t_c.$$

$$\tag{4.17}$$

An application of the Gronwall lemma shows that

$$\|\hat{\xi}^n\|^2 + \sum_{0 < t^k_c \le t^n_o} \|\nabla \hat{\xi}^k\|^2 \Delta t_c \le M\{h_c^{2(\ell+1)} + h_p^{2(k+1)} + (\Delta t_p)^4 + (\Delta t_p^0)^3 + (\Delta t_c)^2\}, \quad (4.18)$$

as was to have been shown. Thus, optimal order convergence will take place, provided that the induction hypothesis can be demonstrated. We require that

$$h_c^{-1}(\Delta t_c + (\Delta t_p)^2 + (\Delta t_p^0)^{\frac{3}{2}} + h_p^{k+1}) \to 0. \quad (4.19)$$

This places a restriction on discretization parameters.

Finally, (4.12) follows from the two parts of (4.19) and (4.18). Hence (4.18) is established.

We can summarize our results by combining (4.3), (3.9), (4.18) and (3.5.a). It follows that

$$(a) \quad \|\underline{u}^m - \underline{U}_h^m\|_{\bar{L}^\infty(0,\overline{T};L^2(\Omega)^2)} + \|p^m - P_h^m\|_{\bar{L}^\infty(0,\overline{T};L^2(\Omega))}$$

$$\le M\{h_c^{\ell+1} + h_p^{k+1} + (\Delta t_p)^2 + (\Delta t_p^0)^{\frac{3}{2}} + \Delta t_c\},$$

$$\quad (4.20)$$

$$(b) \quad \|\hat{c}^n - \hat{C}_h^n\|_{\bar{L}^\infty(0,\overline{T};L^2(\Omega))} + h_c\|\hat{c}^n - \hat{C}_h^n\|_{\bar{L}^2(0,\overline{T};H^1(\Omega))}$$

$$\le M\{h_c^{\ell+1} + h_p^{k+1} + (\Delta t_p)^2 + (\Delta t_p^0)^{\frac{3}{2}} + \Delta t_c\},$$

where

$$\|\varphi^k\|_{\bar{L}^\infty(0,\overline{T};X)} = \sup_{0 < t^k \le \overline{T}} \|\varphi^k\|_X, \|\varphi^k\|_{\bar{L}^2(0,\overline{T};X)} = \left(\sum_{0 < t^k \le \overline{T}} \|\varphi^k\|_X^2 \Delta t\right)^{\frac{1}{2}},$$

$M$ depends on the norms of $p$ in $L^\infty(J;W^{1,\infty})$ and $L^\infty(J;H^{k+3})$ and those of $\hat{c}$ in $H^2(J;W^{1,\infty})$ and $W^{1,\infty}(J;H^{\ell+1})$, provided that (4.19) holds.

Second, we derive the error estimate of $c_i^n - C_{hi}^n (i = 1, 2, \cdots, N)$. Note that $\xi_i^n = \tilde{C}_i^n - C_{hi}^n$ and $\varsigma_i^n = c_i^n - \tilde{C}_i^n$. Let

$$Q_i^h(t^n) = \max\{ \sup_{\substack{x \in \Omega \\ 0 \le t' \le t^n}} |c_i(x, t')|, \sup_{\substack{x \in \Omega \\ 0 \le t' \le t^n}} |\tilde{C}^i(x, t')|, \sup_{\substack{x \in \Omega \\ 0 \le j \le n}} |C_{hi}^j(x)|\}.$$

Similarly we have

$$Q_i^h(t^n) \le \sup_{\substack{x \in \Omega \\ 0 \le t' \le t^n}} |c_i(x, t')| + \sup_{\substack{x \in \Omega \\ 0 \le t' \le t^n}} |\tilde{C}_i(x, t')| + \sup_{\substack{x \in \Omega \\ 0 \le j \le n}} |\xi_i^j|.$$

Let $\overline{Q}^h(t^n) = \max\{Q^h(t^n), Q_i^h(t^n)(i = 1, 2, \cdots, N)\}$. By (2.3), (2.10.b) and (3.2) we can obtain the relation

$$
\left(\phi K_i \frac{\xi_i^n - \xi_i^{n-1}}{\Delta t_c}, \dot{z}\right) + (E\underline{U}_h^n \cdot \nabla \xi_i^n, z) + (\mathbf{E}_c(E\underline{U}_h^n)\nabla \xi_i^n, \nabla z)
$$

$$
= (f_i(\hat{c}^n, c_1^n, c_2^n, \cdots, c_N^n) - f_i(\hat{C}_h^n, C_{h1}^n, C_{h2}^n, \cdots, C_{hN}^n), z)
$$

$$
+ \left(\phi K_i \left\{\frac{\tilde{C}_i^n - \tilde{C}_i^{n-1}}{\Delta t_c} - \frac{\partial c_i^n}{\partial t}\right\}, z\right) + \lambda_i(\varsigma_i^n, z) \tag{4.21}
$$

$$
+ (\{E\underline{U}_h^n - \underline{u}^n\} \cdot \nabla \tilde{C}_i^n, z) + (\{\mathbf{E}_c(E\underline{U}_h^n) - \mathbf{E}_c(\underline{u}^n)\}\nabla \tilde{C}_i^n, \nabla z).
$$

The test function will be chosen to be $z = \xi_i^n$, the terms will be treated in a similar manner. We can obtain

$$
\frac{1}{\Delta t_c}\{(\phi K_i \xi_i^n, \xi_i^n) - (\phi K_i \xi_i^{n-1}, \xi_i^{n-1})\} + ((D_m + \alpha_T |E\underline{U}_h^n|)\nabla \xi_i^n, \nabla \xi_i^n)
$$

$$
\leq M(\overline{Q}^h(t^n))\{h_c^{2(\ell+1)} + h_p^{2(k+1)} + (\Delta t_p)^4(\text{or}(\Delta t_p^0)^2) + (\Delta t_c)^2
$$

$$
+\|\hat{\xi}_p^m\|^2 + \|\hat{\xi}_p^{m-1}\|^2 + [1 + \|\hat{\xi}_p^m\|^2 + \|\hat{\xi}_p^{m-1}\|^2][\|\xi^n\|^2 + \|\xi^{n-1}\|^2 \tag{4.22}
$$

$$
+\|\xi^{n-2}\|^2]\} + \epsilon\|\nabla \xi_i^n\|^2,
$$

where $\|\xi^n\|^2 = \sum_{i=1}^N \|\xi_i^n\|^2$.

We shall need an induction hypothesis. We assume that

$$
\sup_n \|\xi^n\|_{L^\infty(\Omega)} \leq C, \tag{4.23}
$$

where

$$
\|\xi^n\|_{L^\infty(\Omega)} = \max_{1 \leq i \leq N} \|\xi_i^n\|_{L^\infty(\Omega)}.
$$

Thus $\overline{Q}^h$ is bounded. We have $M(\overline{Q}^h(t^n)) \leq M$, where $M$ is a positive constant independent of $\hat{C}_h$ and $C_{hi}(i = 1, 2, \cdots, N)$. Now multiply (4.22) by $\Delta t_c$, add on the time for $0 < t_c^k \leq t_c^n$, and an application of the Gronwall lemma shows that

$$
\sum_{i=1}^N \|\xi_i^n\|^2 + \sum_{i=1}^N \sum_{0 < t_c^k \leq t_c^n} \|\nabla \xi_i^k\|^2 \Delta t_c
$$

$$
\leq M \quad \{h_c^{2(\ell+1)} + h_p^{2(k+1)} + (\Delta t_p)^4 + (\Delta t_p^0)^3 + (\Delta t_c)^2\}. \tag{4.24}
$$

We can summarize our results by combining (4.20), (4.24) and (3.5.b). It follows that

$$\sum_{i=1}^{N} \|c_i^n - C_{hi}^n\|_{\bar{L}^\infty(0,\bar{T};L^2(\Omega))} + h_c \sum_{i=1}^{N} \|c_i^n - C_{hi}^n\|_{\bar{L}^2(0,\bar{T};H^1(\Omega))} \tag{4.25}$$

$$\leq M \quad \{h_c^{\ell+1} + h_p^{k+1} + (\Delta t_p)^2 + (\Delta t_p^0)^{\frac{3}{2}} + \Delta t_c\},$$

where $M$ depends on the norms of $p$ in $L^\infty(J; W^{1,\infty})$ and $L^\infty(J; H^{k+3})$ and those of $\hat{c}$ and $c_i (i = 1, 2, \cdots, N)$ in $H^2(J; W^{1,\infty})$ and $W^{1,\infty}(J; H^{\ell+1})$, provided that (4.19) holds.

At last, for heat equation, note that $\theta^n = T - \tilde{T}$ and $\omega^n = \tilde{T}^n - T_h^n$. Then (2.4), (2.10.c) and (3.3) can be used to get the relation

$$\left(\hat{\phi}\frac{T_h^n - T_h^{n-1}}{\Delta t_c}, z\right) + c_p(E\underline{U}_h^n \cdot \nabla \omega^n, z) + (\mathbf{E}_H(E\underline{U}_h^n)\nabla\omega^n, \nabla z)$$

$$= \left(Q(\underline{u}^n, p^n, T^n, \hat{c}^n) - Q(E\underline{U}_h^n, EP_h^n, \check{T}_h^n, \check{C}_h^n), z\right) \tag{4.26}$$

$$+ \left(\hat{\phi}\left\{\frac{\tilde{T}^n - \tilde{T}^{n-1}}{\Delta t_c} - \frac{\partial T^n}{\partial t}\right\}, z\right) + \mu(\theta^n, z) + c_p(\{E\underline{U}_h^n - \underline{u}^n\} \cdot \nabla\tilde{T}^n, z)$$

$$+ (\{\mathbf{E}_H(E\underline{U}_h^n) - \mathbf{E}_H(\underline{u}^n)\}\nabla\tilde{T}^n, \nabla z), \quad z \in R_h.$$

Let

$$Q_T^h(t^n) = \max\{\sup_{\substack{x\in\Omega \\ 0\leq t'\leq t^n}} |T(x,t')|, \sup_{\substack{x\in\Omega \\ 0\leq t'\leq t^n}} |\tilde{T}(x,t')|, \sup_{\substack{x\in\Omega \\ 0\leq j\leq n}} |T_h^i(x)|\},$$

$$Q_p^h(t^n) = \max\{\sup_{\substack{x\in\Omega \\ 0\leq t'\leq t^n}} |p(x,t')|, \sup_{\substack{x\in\Omega \\ 0\leq t'\leq t^n}} |\tilde{P}(x,t')|, \sup_{\substack{x\in\Omega \\ 0\leq t_p^m\leq t_c^n}} |P_h^m(x)|\},$$

$$Q_{\underline{u}}^h(t^n) = \max\{\sup_{\substack{x\in\Omega \\ 0\leq t'\leq t^n}} |\underline{u}(x,t')|, \sup_{\substack{x\in\Omega \\ 0\leq t'\leq t^n}} |\tilde{\underline{U}}(x,t')|, \sup_{\substack{x\in\Omega \\ 0\leq t_p^m\leq t_c^n}} |\underline{U}_h^m(x)|\},$$

$$Q_*^h(t^n) = \max\{Q^h(t^n), Q_T^h(t^n), Q_p^h(t^n), Q_{\underline{u}}^h(t^n)\}.$$

The test function will be chosen to be $z = \omega^n$, the terms will be treated similarly. We can obtain

$$\frac{1}{\Delta t_c}\{(\hat{\phi}\omega^n, \omega^n) - (\hat{\phi}\omega^{n-1}, \omega^{n-1})\} + ((K_m + c_{pw}\alpha_T|E\underline{U}_h^n|)\nabla\omega^n, \nabla\omega^n)$$

$$\leq M(Q_*^h(t^n))\{h_c^{2(\ell+1)} + h_p^{2(k+1)} + (\Delta t_p)^4(\text{or}(\Delta t_p^0)^2) + (\Delta t_c)^2 + h_T^{2(r+1)}$$

$$+\|\hat{\xi}_p^m\|^2 + \|\hat{\xi}_p^{m-1}\|^2 + [1 + \|\hat{\xi}_p^m\|^2 + \|\hat{\xi}_p^{m-1}\|^2][\|\omega^n\|^2 + \|\omega^{n-1}\|^2$$

$$+\|\omega^{n-2}\|^2] + \|\hat{c}^n - \overset{x}{\hat{C}}_h^n\|^2 + \|\underline{u}^n - E\underline{U}_h^n\|^2 + \|p^n - EP_h^n\|^2\} + \epsilon\|\nabla\omega^n\|^2. \tag{4.27}$$

We shall need an induction hypothesis. We assume that

$$\sup_n \|\omega^n\|_{L^\infty(\Omega)} \leq C, \tag{4.28}$$

and using (4.20) we obtain that $M(Q_*^h(t^n)) \leq M$, where $M$ is independent of $\hat{C}_h, \underline{U}_h, P_h$ and $T_h$.

Now, multiply (4.27) by $\Delta t_c$, add on the time for $0 < t_c^k \leq t_c^n$, and an application of the Gronwall lemma shows that

$$\|\omega^k\|^2 + \sum_{0 < t_c^k \leq t_c^n} \|\nabla\omega^k\|^2\Delta t_c$$

$$\leq M \ \{h_c^{2(\ell+1)} + h_p^{2(k+1)} + h_T^{2(r+1)} + (\Delta t_p)^4 + (\Delta t_p^0)^3 + (\Delta t_c)^2\}. \tag{4.29}$$

We can summarize our results by combining (4.29) and (3.5.c). It follows that

$$\|T^n - T_h^n\|_{\bar{L}^\infty(0,\overline{T};L^2(\Omega))} + h_T\|T^n - T_h^n\|_{\bar{L}^2(0,\overline{T};H^1(\Omega))}$$

$$\leq M \ \{h_c^{\ell+1} + h_p^{k+1} + h_T^{r+1} + (\Delta t_p)^2 + (\Delta t_p^0)^{\frac{3}{2}} + \Delta t_c\}, \tag{4.30}$$

where $M$ depends on the norms of $p$ in $L^\infty(J; W^{1,\infty})$ and $L^\infty(J; H^{k+3})$, those of $\hat{c}$ in $H^2(J; W^{1,\infty})$ and $W^{1,\infty}(J; H^{\ell+1})$, and those of $T$ in $H^2(J; W^{1,\infty})$ and $W^{1,\infty}(J; H^{r+1})$, provided that

$$h_T^{-1}(\Delta t_c + (\Delta t_p)^2 + (\Delta t_p^0)^{\frac{3}{2}} + h_c^{\ell+1} + h_p^{k+1}) \to 0 \tag{4.31}$$

holds.

We have proved

**THEOREM 1.** Let $p(x,t), \hat{c}(x,t), c_i(x,t)(i = 1, 2, \cdots, N)$ and $T(x,t)$ be the solution of problem (1.1)–(1.12), $P_h^n, \hat{C}_h^n, C_{hi}^n(i = 1, 2, \cdots, N)$ and $T_h^n$ be the approximate solution obtained from Scheme I. Suppose $k \geq 1$, the spatial and time discretizations satisfy the relations (4.19) and (4.31). Then the error estimates (4.20), (4.25) and (4.30) hold.

As was described earlier, Scheme II allows for only one matrix factorization per pressure time step instead of one per concentration time step, resulting in reduced computation. A similar method has been analyzed for a petroleum problem by Douglas, Ewing and Wheeler [8]. Similarly using the techniques of [6,10], a single preconditioner can be used for many time-steps due to evolution nature of the equations saving in additional computation. Analysis similar to that used for Theorem 1 can be used to obtain the following result:

**THEOREM 2.** Let $p(x,t)$, $\hat{c}(x,t)$, $c_i(x,t)(i = 1, 2, \cdots, N)$ and $T(x,t)$ be the solution of problem (1.1)–(1.12), $P_h^n$, $\hat{C}_h^n$, $C_{hi}^n(i = 1, 2, \cdots, N)$ and $T_h^n$ be the approximate solution obtained from Scheme II. Suppose $k \geq 1$, the spatial and time discretizations satisfy the relations

$$(a) \quad h_p^{-1}\{h_c^{\ell+1} + h_T^{r+1} + \Delta t_c + (\Delta t_p)^2 + (\Delta t_p^0)^{\frac{3}{2}}\} \to 0,$$

$$(b) \quad (\Delta t_p + h_p^{k+1})(\log h_c^{-1})^{\frac{1}{2}} \to 0, \tag{4.32}$$

$$(c) \quad (\Delta t_p + h_p^{k+1})(\log h_T^{-1})^{\frac{1}{2}} \to 0.$$

Then the same error estimates (4.20), (4.25) and (4.30) hold as in Theorem 1.

**Remark.** The discretization methods presented in Scheme I and Scheme II are designed to linearize and formally decouple the equations for a sequential solution process. In many cases where the nonlinearities in our partial differential equations are strong, this linearization process is not sufficiently accurate for the desired application. In those cases, a better linearization and linear solution process must be considered. See [1,6,9,10,20,21,24] for effective linearization and iterative linear solution methods for problems of this type.

## ACKNOWLEDGMENTS

This research was supported in part by U.S. Army Research Office Contract No. DAAG29–84–K–0002, by U.S. Air Force Office of Scientific Research Contract No. AFOSR–85–0117, and by the National Science Foundation Grant No. DMS–8504360.

## REFERENCES

1. J.R. Appleyard, I.M. Cheshire, and R.K. Pollard, Special techniques for fully-implicit simulators, presented at the European Symposium on Enhanced Oil Recovery, Bournemouth, U.K., September, (1981).

2. K. Aziz and A. Settari, *Petroleum Reservoir Simulation*, Applied Science Publishers, (1979).

3. J. Bear, *Hydraulics of Groundwater*, McGraw-Hill, (1979).

4. F. Brezzi, On the existence, uniqueness and approximation of saddle-point problems arising from Lagrangian multipliers, *RAIRO Anal. Numér.*, *2*, 129-151 (1974).

5. H. Cooper, The equation of ground-water flow in fixed and deforming coordinates, *J. Geophys. Res.*, *71*, 4783-4790, (1966).

6. J. Douglas, Jr., T. Dupont, and R.E. Ewing, Incomplete iteration for time-stepping a Galerkin method for a quasilinear parabolic problem, *SIAM J. Numer. Anal.*, *16*, 503-522 (1979).

7. J. Douglas, R.E. Ewing, and M.F. Wheeler, The approximation of the pressure by a mixed method in the simulation of miscible displacement, *RAIRO Anal. Numér.*, *17*, 17-33 (1983).

8. J. Douglas, Jr., R.E. Ewing and M.F. Wheeler, A time-discretization procedure for a mixed finite element approximation of miscible displacement in porous media, *RAIRO Anal. Numér.*, *17*, 249-265 (1983).

9. H.C. Elman, Preconditioned conjugate-gradient methods for nonsymmetric systems of linear equations, Yale University Department of Computer Science Research Report 203, New Haven, Connecticut, 1981.

10. R.E. Ewing, Time-stepping Galerkin methods for nonlinear Sobolev partial differential equations, *SIAM, J. Numer. Anal.*, *15*, 1125-1150 (1978).

11. R.E. Ewing, Problems arising in the modeling of processes for hydrocarbon recovery, Vol. I, *Mathematics of Reservoir Simulation*, (R.E. Ewing, ed.), Research Frontiers in Applied Mathematics, SIAM, Philadelphia, 3-34 (1984).

12. R.E. Ewing, A modified method of characteristics for transport of nuclear-waste contamination in porous media, *Proceedings International Conference on Computational Engineering and Science*, Atlanta, Georgia, April 11-14, 1988 (to appear).

13. R.E. Ewing and T.F. Russell, Efficient time-stepping methods for miscible displacement problems in porous media, *SIAM J. Numer. Anal.*, *19*, 1-66 (1982).

14. R.E. Ewing, T.F. Russell, and M.F. Wheeler, Simulation of miscible displacement using mixed methods and a modified method of characteristics,

SPE No. 12241, *Proceedings Seventh SPE Symposium on Reservoir Simulation*, San Francisco, November 15–18, 1983, 71–82, in SPE Reprint Series No. 20, Numerical Simulation II, Society of Petroleum Engineers, Dallas, and *Soc. Pet. Eng. J.* (to appear).

15. R.E. Ewing, T.F. Russell, and M.F. Wheeler, Convergence analysis of an approximation of miscible displacement in porous media by mixed finite elements and a modified method of characteristics, *Computer Meth. Appl. Mech. Eng.*, *47*, (R.E. Ewing, ed.), 73–92 (1984).

16. R.E. Ewing and M.F. Wheeler, Galerkin methods for miscible displacement problems in porous media, Math. Res. Ctr. Rept. No. 1932, University of Wisconsin-Madison, 1979; and *SIAM J. Numer. Anal., 17* 351–365 (1980).

17. R.E. Ewing, Y. Yuan and G. Li, Finite element methods for contamination by nuclear waste-disposal in porous media, *Proceedings of Dundee Numerical Analysis Conference*, Dundee Scotland, (June 23–26, 1987).

18. R.E. Ewing, Y. Yuan, and G. Li, Time stepping along characteristics for a mixed finite element approximation for compressible flow of contamination by nuclear waste in porous media, *SIAM J. Numer. Anal.* (to appear).

19. R.E. Ewing, Y. Yuan, and G. Li, Numerical method for a model for compressible flow for the nuclear waste contamination in porous media, submitted to *Math. Comp.*

20. J.A. Meijerink, Iterative methods for the solution of linear equations based on incomplete block factorizations of the matrix, SPE 12262, *Proceedings, 7th SPE Symposium on Reservoir Simulation*, San Francisco (1983).

21. U. Obeysekare, M. Allen, R.E. Ewing, and J.H. George, Application of conjugate gradient-like methods to a hyperbolic problem in porous media flow, *International Journal for Numerical Methods in Fluids*, *7*, 551–566 (1987).

22. P.A. Raviart and J.M. Thomas, A mixed finite element method for 2nd order elliptic problems, Mathematical Aspects of the Finite Element Method, *Lecture Note in Math. Vol. 606*, Springer-Verlag, Berlin and New York (1977).

23. M. Reeves and R.M. Cranwell, User's Manual for the Sandia Waste-Isolation Flow and Transport Model (SWIFT) Release 4.81, *Sandia Report Nureg/CR-2324, SAND 81-2516, GR* (November 1981).

24. P.K.W. Vinsome, Orthomin, an iterative method for solving sparse sets of simultaneous linear equations, SPE 5729, *Proceedings, 4th SPE Symposium on Numerical Simulation of Reservoir Performance*, Los Angeles, CA, February 19–20, 1976.

25. M.F. Wheeler, A priori $L_2$ error estimates for Galerkin approximations to parabolic partial differential equations, *SIAM J. Numer. Anal., 10*, 723–759 (1973).

# Implementation of Finite Element Alternating-Direction Methods for Vector Computers

S. V. KRISHNAMACHARI and L. J. HAYES
The Texas Institute of Computational Mechanics
Department of Aerospace Engineering and Engineering Mechanics
The University of Texas at Austin
Austin, Texas 78712-1085

*A finite element alternating-direction method is used to solve a time dependent parabolic equation. This scheme is compared with a standard backward difference finite element scheme frequently used to solve time dependent problems. The alternating-direction method reduces multidimensional problems to a series of uncoupled one dimensional problems, which results in very low execution times. The alternating-direction method is very well suited for vector and parallel processors; whereas the backward difference scheme gives rise to large sparse systems of linear equations that are usually expensive to solve on vector processors. A variety of two and three-dimensional problems are solved on the CRAY 1-S and a timing and storage comparison is presented. These techniques can also be used on other parallel or vector machines.*

## Introduction

Many important physical phenomena can be described by parabolic boundary value problems. It is often the case when modeling multidimensional problems that computer storage and execution time limitations govern the number of unknowns and hence the degree of accuracy in the numerical models which can be used to solve these problems. The limitations are more severe when modeling three-dimensional, nonlinear problems. It is therefore essential to develop reliable numerical techniques that can be used to efficiently solve these problems on present day supercomputers. In the past decade there have been several major advances in the area of computer architecture which have allowed the engineer to solve computationally intensive problems in a reasonable time. These architectures have primarily been vector and parallel architectures. Examples of popular vector machines are the series of CRAY computers, the

CYBER-205 and the ETA-10. A variety of parallel computer architectures are also available. This paper concentrates on vector implementation. Numerical schemes that are extremely efficient on serial machines are not necessarily the best schemes for vector machines. Very often efficient vectorization requires completely rethinking the numerical algorithm and the data management.

In 1971, Douglas and Dupont [1] formulated a finite element alternating-direction procedure to solve nonlinear parabolic problems posed on rectangular regions with rectangular grid spacing. This implicit method can solve large multidimensional problems as a series of smaller one-dimensional problems, and the matrix problem that must be solved at each step of the solution process is independent of time and requires only one decomposition per problem. The storage requirements are low since they are associated with the one-dimensional problems. Dendy and Fairweather [2] extended these methods to certain unions of rectangles. Hayes [3, 4, 5] generalized these methods to non-rectangular regions that can be isoparametrically mapped onto a rectangle. In this formulation, the Jacobian of the isoparametric map is obtained at the nodes by a "patch approximation" and the alternating-direction parameter , $\lambda$, which approximates the diffusion and the Jacobian was a single constant for the entire grid. However when the Jacobian does vary greatly throughout the grid, this method requires an unreasonably small time step to produce numerically stable results [6]. This problem was remedied in the approach taken by Hayes and Krishnamachari [7] . In this formulation the alternating-direction parameter, $\lambda$, was varied throughout the grid in such a way that the coefficient matrix would still factor into a product of one-dimensional problems.

Alternating-direction methods are very attractive for both vector and parallel processing. In this paper the implementational aspects of alternating-direction methods are addressed for the CRAY 1-S. The results and techniques outlined here can also be applied for other vector processors such as the CRAY-XMP and the CYBER 205. The interested reader is referred to the literature $[1-7]$ for theoretical details of the alternating-direction method. Timing and storage comparisons between alternating-direction methods and fully implicit backward difference finite element schemes are presented for a wide variety of problems. Two different data structures, a stacked and sequential data structure are used to solve the alternating-direction equations. The vectorized IMSL and the LINPACK library routines are used to solve these equations and are compared with specialized routines written to solve multiple tridiagonal or pentadiagonal systems of equations efficiently on a vector computer. The IMSL and LINPACK routines used were compiler optimized. The LINPACK routine was available in CRAY'S SCILIB.

## Mathematical Preliminaries

The alternating-direction method can be used to solve the boundary value problem

$$\frac{\partial u}{\partial t} - \nabla \cdot k(x,y)\nabla u \; = \; f(x,y,t) \qquad \text{on } \Omega_g \tag{1}$$

subject to either Dirichlet, Neumann or Mixed boundary conditions on $\partial\Omega_g$, the boundary of $\Omega_g$. $\Omega_g$ is a curved region which can be mapped isoparametrically onto a rectangle or to unions of certain rectangles, $k(x,y)$ is the thermal conductivity and $f$ is a heat source.

For illustration, the discussion is confined to master region $\Omega = [a,b] \times [c,d]$ with tensor product finite element basis functions. $H^1(\Omega)$ and $H^1(\Omega_g)$ are the standard Sobolev spaces on $\Omega$ and $\Omega_g$. The following $L^2$ inner products are defined on $\Omega$ and $\Omega_g$

$$(f,g) = \int_\Omega fg\,d\xi\,d\eta$$

$$(f, g)_{\Omega_z} = \int_{\Omega_z} fg\, dxdy = \int_{\Omega} fg\,|\, J(F^{-1})\,|\, d\xi d\eta = (f, \rho g)$$

where $\rho = |\, J(F^{-1})\,|$ is the determinant of the Jacobian of $F^{-1}$. The map, F, is an isoparametric map which takes rectangular elements on the master region $\Omega$ onto curved elements in the general region $\Omega_z$. For any integrable function P,

$$\int_{\Omega_z} P(x, y)dxdy = \int_{\Omega} P(\xi, \eta)\rho(\xi, \eta)d\xi d\eta$$

In this discussion, two orderings of the nodes in the master region $\Omega$ are defined. The first is a tensor product ordering of the $N$ nodes in $\Omega$. The grid lines in the $\xi$ − direction are numbered $1, 2, 3, ... N_\xi$ and the grid lines in the $\eta$ − direction are numbered $1, 2, 3, ... N_\eta$. The second is a global ordering which assigns one of the numbers $1, 2, 3, ... N = N_\xi N_\eta$ to each of the $N$ nodes of $\Omega$. A $\xi$ − grid line and a $\eta$ − grid line is associated with each global node $i$, and the tensor product index of node $i$ is the pair $(\alpha(i), \beta(i))$, where $\alpha(i)$ is the number of the $\xi$ − grid line and $\beta(i)$ is the number of the $\eta$ − grid line as shown in Figure 1. The tensor product basis can be rewritten as products of one-dimensional basis functions in the following manner

$$N_i(\xi, \eta) = \phi_{\alpha(i)}(\xi)\psi_{\beta(i)}(\eta) = \phi_\alpha(\xi)\psi_\beta(\eta) \quad \text{for } 1 \le i \le N$$

These tensor product basis functions are easy to construct on $\Omega$. The following additional notation will be used

$\quad m_{pq}$.... $q$ -th grid segment on the $p$ -th $\xi$ - grid line.

$\quad n_{pq}$.... $q$ -th grid segment on the $p$ -th $\eta$ - grid line.

$\quad h_{pq}$.... Approximate chord length of grid segment $m_{pq}$

$\quad g_{pq}$.... Approximate chord length of grid segment $n_{pq}$

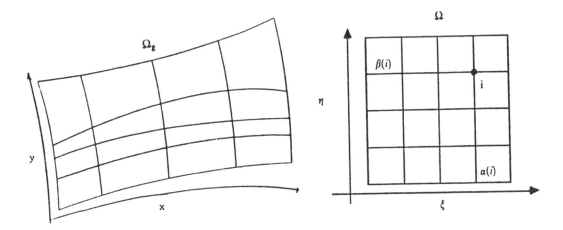

Figure 1 : The General Curved Region and the Master Region

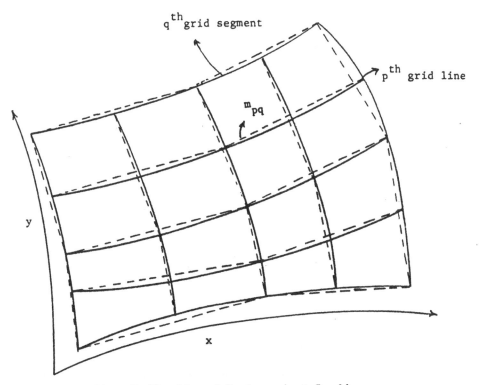

Figure 2: Chord Length for Approximate Jacobians

The approximate chord length is calculated using only the two corner nodes and ignoring any midside nodes in the mesh as shown in Figure 2.

The total time $T$ is divided into $m$ time intervals $\Delta t$, so that $T = m\Delta t$, and $t^n = n\Delta t$ for $n = 0, 1, 2 \ldots m$. Also let $u^n = u(x, y, t^n)$. The standard fully implicit backward difference finite element scheme [8] used to approximate (1) is

$$\left( \frac{u^{n+1} - u^n}{\Delta t}, v \right)_{\Omega_g} + (k\nabla u^{n+1}, \nabla v)_{\Omega_g} = (f, v)_{\Omega_g} \qquad \forall v \in H^1(\Omega_g) \qquad (2)$$

or

$$\left( \frac{u^{n+1} - u^n}{\Delta t}, v \right)_{\Omega_g} + (k\nabla(u^{n+1} - u^n), \nabla v)_{\Omega_g} = (f, v)_{\Omega_g} - (k\nabla u^n, \nabla v)_{\Omega_g} \qquad \forall v \in H^1(\Omega_g).$$

Note that the gradient, $\nabla$, in the equation is with respect to the physical $x, y$ coordinates of $\Omega_g$.

The alternating-direction method can be viewed as a perturbation of the left hand side of equation (2) and is given by

$$\left( \tilde{\rho} \frac{u^{n+1} - u^n}{\Delta t}, v \right) + (\lambda \overline{\nabla}(u^{n+1} - u^n), \overline{\nabla} v) + \Delta t \left( \lambda^2 \frac{\partial^2 (u^{n+1} - u^n)}{\partial \xi \partial \eta}, \frac{\partial^2 v}{\partial \xi \partial \eta} \right) \qquad (3)$$

$$= (f, v)_{\Omega g} - (k\nabla u^n, \nabla v)_{\Omega g} - \left( \frac{(u^n - u^{n-1})}{\Delta t}, v \right)_{\Omega g} + \left( \tilde{\rho} \frac{(u^n - u^{n-1})}{\Delta t}, v \right)$$

The gradient, $\overline{\nabla}$, is taken with respect to the master coordinates $(\xi, \eta)$ and $\lambda = \lambda(\xi, \eta)$ is a Laplace modified type parameter which approximates $k\tilde{\rho}$. The higher order cross derivative term is added so that the matrix will factor into a convenient form [3, 4, 5, 6]. Equation (3) can be written in matrix form as

$$\mathbf{K}(\mathbf{u}^{n+1} - \mathbf{u}^n) = \mathbf{F}^n$$

where

$$K_{ij} = (\tilde{\rho} N_i, N_j) + \Delta t (\lambda \overline{\nabla} N_i, \overline{\nabla} N_j) + \Delta t^2 (\lambda^2 \frac{\partial^2 N_i}{\partial \xi \partial \eta}, \frac{\partial N_j}{\partial \xi \partial \eta}) \qquad (4)$$

$$\mathbf{F}^n = \Delta t(f, v)_{\Omega_g} - \Delta t(k \nabla u^n, \nabla v)_{\Omega_g} - (\tilde{\rho}(u^n - u^{n-1}), v)_{\Omega_g} - ((u^n - u^{n-1}), v)_{\Omega_g}$$

Here $\lambda = \lambda(\xi, \eta)$ is a function of both the diffusion and the Jacobian as defined in (5). The right hand side of (4) is evaluated in the usual manner. If tensor product basis functions are used, (4) can be rewritten in the following manner

$$\mathbf{K} = \mathbf{D}^{1/2} [\mathbf{I} \otimes (\mathbf{C}_x + \Delta t \mathbf{A}_x)] \; [(\mathbf{C}_y + \Delta t \mathbf{A}_y) \otimes \mathbf{I}] \; \mathbf{D}^{1/2}$$

where $\otimes$ is the tensor product operator and

$$\mathbf{C}_x = \int_a^b \phi_l(\xi) \phi_k(\eta) |J_1| d\xi$$

$$\mathbf{C}_y = \int_c^d \psi_k(\eta) \psi_l(\eta) |J_2| d\eta$$

$$\mathbf{A}_x = \int_a^b \lambda \phi'_l(\xi) \phi'_k(\xi) d\xi$$

$$\mathbf{A}_y = \int_c^d \lambda \psi'_l(\eta) \psi'_k(\eta) d\eta$$

where the prime indicates differentiation with respect to the master coordinates. On the $q^{th}$ segment of the $p^{th}$ grid line,

$$\lambda_1^q \big|_{m_{pq}} = \frac{k \Delta \xi}{h_{pq}}, \qquad J_1^q \big|_{m_{pq}} = \frac{h_{pq}}{\Delta \xi} \qquad (5)$$

$$\lambda_2^q\big|_{n_{pq}} = \frac{k\Delta\eta}{g_{pq}}, \qquad J_2^q\big|_{n_{pq}} = \frac{g_{pq}}{\Delta\eta}$$

Here $\Delta\xi$ and $\Delta\eta$ are the width and height of the rectangular master element. $J_1$ and $J_2$ are the approximate Jacobians associated with the one dimensional mappings of the individual x and y grid lines onto the master $\xi$ and $\eta$ lines. The diagonal matrix $\mathbf{D}$ is

$$\begin{bmatrix} \tilde{\rho}_1 & & & & \\ & \tilde{\rho}_2 & & & \\ & & \tilde{\rho}_3 & & \\ & & & & \\ & & & & \\ & & & & \\ & & & & \tilde{\rho}_N \end{bmatrix}$$

where $\tilde{\rho}_i$ is obtained by a "patch approximation" [5] . The Jacobian $\rho$ is a multivalued function at each node in the mesh. The patch approximation procedure approximates the Jacobian at each node by averaging the values of the Jacobians obtained from all the elements connected to that node. The matrix problem becomes

$$\mathbf{D}^{1/2}\mathbf{K}_x\mathbf{K}_y\mathbf{D}^{1/2}(u^{n+1} - u^n) = \mathbf{F}^n \tag{6}$$

If the nodes in $\Omega_e$ are numbered in a horizontal order, then the matrix $\mathbf{K}_x$ will be block diagonal and each block corresponds to standard one-dimensional finite element problems along the horizontal lines. Each of these blocks are independent and can be solved in parallel. The matrix $\mathbf{K}_y$ will be sparse with a band width of N. If the variables are rearranged to correspond to a vertical numbering of the nodes, the structure of $\mathbf{K}_y$ is the same as the one for $\mathbf{K}_x$ and each diagonal block corresponds to one-dimensional finite element problems along the vertical grid lines. The problem (6) is solved by the following computational procedure:

1) Decompose the block diagonal forms of $\mathbf{K}_x$ and $\mathbf{K}_y$.

2) Compute $\mathbf{F}^n$ by a matrix-vector multiplication.

3) Solve $\mathbf{K}_x\phi^{n+1} = \mathbf{D}^{-1/2}\mathbf{F}^n$. (An independent series of one-dimensional problems.)

4) Rearrange $\phi^{n+1}$ in a vertical order.

5) Solve $K_y Z^{n+1} = \phi^{n+1}$. (An independent series of one-dimensional problems.)

6) $U^{n+1} = U^n + D^{-1/2} Z^{n+1}$.

7) Rearrange $U^{n+1}$ in a horizontal order and repeat (2-7) for the next time step.

This approach has several attractive properties [6, 7]. The matrices $K_x$ and $K_y$ are associated with one-dimensional problems along the grid lines and hence they are easy to create and invert. They are also independent of time, so they can be decomposed once and this decomposition can be used at each time step. They are also mutually independent so they can be decomposed in parallel. The alternating-direction method is an $O(\Delta t)$ method and since it is an implicit method it is stable under mild restrictions [3, 4, 5].

## Computational Details

Equation (1) was solved in both two and three dimensions on a unit square and unit cube respectively. Homogenous Dirichlet boundary conditions were specified on the boundary. The thermal conductivity was $k = 1$. and the heat source was $f = 1$..

It is clear from the alternating-direction algorithm that the solution process comprises of 3 distinct computational kernels:

1) The decomposition time for the matrix.

2) The forward and back subsitution phase.

3) The evaluation of the right hand side.

These kernels are examined and timed separately. This is done deliberately inorder to obtain an idea of the time spent in each kernel and also in the entire solution process.

The matrices arising from the backward difference scheme and the alternating- direction scheme are all symmetric and positive definite. These matrices are decomposed by a Cholesky factorization using the appropriate IMSL and LINPACK routines. The alternating-direction matrices are solved in two different ways. The first scheme was to sequentially append the matrices as shown in Figure 3 and solve the system as one giant system of equations by using the IMSL and LINPACK routines. The other scheme was to stack these matrices as shown in Figure 4 and perform a parallel Cholesky decomposition. The second scheme performs vector operations on vectors of length N during the decomposition and the forward and back substitution phases. This operation was done by specialized FORTRAN routines written for the CRAY. Henceforth these routines will be referred to as MULTGE (Multiple Gaussian Elimination). These routines solve multiple independent systems of tridiagonal and pentadiagonal matrices by Gauss elimination by using a Cholesky decomposition. After decomposition the forward and back substitution is done at each time step of the solution process. This substitution phase is done using both the IMSL and LINPACK routines and also by MULTGE. The matrix vector multiplications in the right hand side computations are done using a banded matrix-vector multiplication for the backward difference scheme. For the alternating-direction scheme the matrix-vector multiplication is done by a vectorized element-by-element matrix-vector multiplication developed by Hayes and Devloo [9] which proved to be extremely efficient on the CRAY.

The alternating-direction scheme is even more efficient for non-linear problems. For non-linear capacitance problems encountered in phase change, the non-linearity is included in the diagonal matrix **D** and **D** has to be formed and inverted at each time step. However this is trivial since **D** is diagonal. This scheme has been analyzed by Hayes [10] and implemented by Lewis, Morgan and Roberts [11]. If the non-linearity is due to the conductivity $,k$, the matrices $\mathbf{K}_x$ and $\mathbf{K}_y$ have to be decomposed at each time step. This can be done for the one-dimensional matrices in parallel and this is still very efficient compared to the backward-difference scheme where the large banded matrix has to be decomposed at each time step.

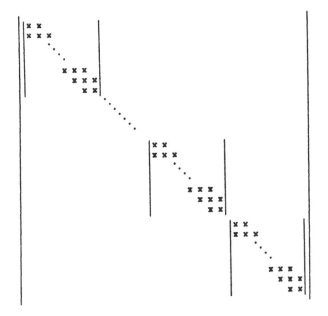

Figure 3: The Sequential Data Structure

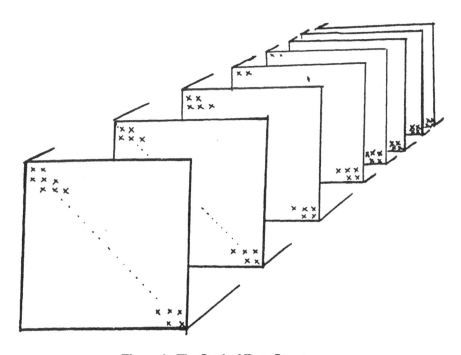

Figure 4: The Stacked Data Structure

## Computational Results

Four different tensor product elements are investigated and the three kernels are compared using both the alternating-direction method and the backward difference scheme for the temporal integration. The four element types considered are:

(i) 2-D Bilinear 4-noded elements (L-4).

(ii) 2-D Biquadratic 9-noded elements (Q-9).

(iii) 3-D Trilinear 8-noded elements (L-8).

(iv) 3-D Triquaratic 27-noded elements (Q-27).

All the matrices in the left hand side are stored in symmetric banded storage mode. However, the storage requirements are quite different. The alternating-direction matrices are tridiagonal for linear elements and pentadiagonal for quadratic elements; whereas the backward difference matrix is banded. The band width grows algebraically with the number of unknowns in the finite element mesh. Infact for two-dimensional square grids the storage requirements for the band matrix is $O(N^{3/2})$ , where $N$ is the number of unknowns in the grid. For three-dimensional problems this requirement is $O(N^{5/3})$ . However the alternating-direction scheme requires only $O(N)$ storage for both cases. Figure 5 shows the total storage versus the number of elements for linear and quadratic elements in two and three dimensions. The alternating-direction results are indicated by $\triangle$ and the the backward difference results are indicated by $\times$. From the figure it is clear that there are significant savings by using the alternating-direction scheme and these gains become more prominent as the problem size increases. In fact for 125 triquadratic elements (5 by 5 by 5 elements), the savings factor is about 20. These graphs clearly show that the storage requirements for the alternating-direction method is linear while that for standard finite element method grows superlinearly.

Figure 6 shows the time taken for the decomposition of the matrix. The alternating-direction scheme using the parallel data structure $\Lambda$-D MULTGE (The

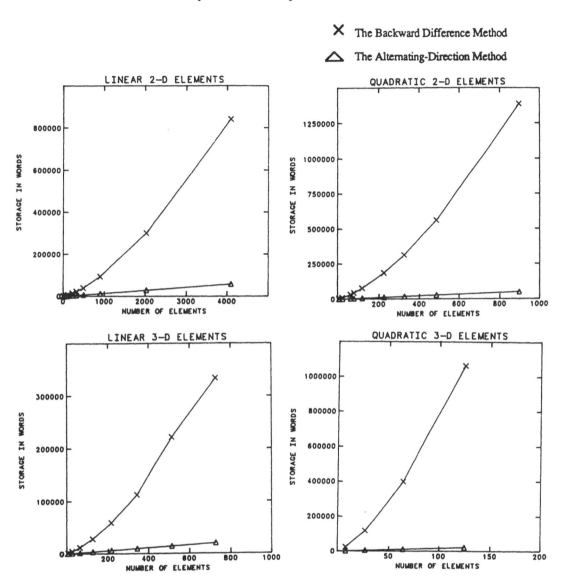

Figure 5: Storage Comparisons

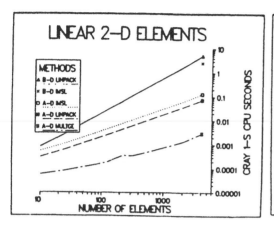

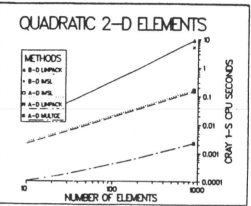

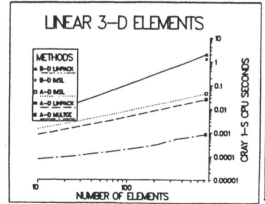

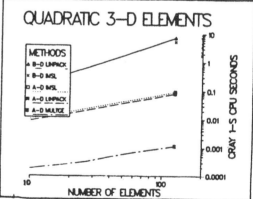

Figure 6: Comparison of the Time for Decomposition

alternating-direction method with multiple Gaussian elimination) is clearly faster than both the banded solution schemes by several orders of magnitude. This scheme exhibits a bump at 225 bilinear elements and at 343 trilinear elements. This degradation in performance is due to the occurence of memory bank conflicts in the CRAY 1-S [12]. Memory bank conflicts occur when 2 elements of an array are accessed simultaneously from the same memory bank. This occurs for these two specific cases because for 225 bilinear elements the alternating-direction matrix corresponding to a particular grid line is a tridiagonal matrix of order 16. And for 343 trilinear elements the matrix is of order 8. This problem will occur on the CRAY-XMP but not on the CYBER-205.

Figure 7 shows the time taken for forward and back substitution. One thousand time steps are taken for the backward difference scheme and two thousand time steps are taken for the alternating-direction scheme. The alternating-direction method introduces a perturbation term of $O(\Delta t^2)$ which adds to the error. In order to maintain the same order of error as the backward difference scheme, the alternating-direction scheme requires a smaller time step. The A-D MULTGE scheme is the fastest by several orders of magnitude. The A-D LINPACK (The alternating-direction scheme with a sequential data structure for the matrices and the solution is performed using a LINPACK routine) scheme performs poorly. This is because during the forward and back substitution the A-D LINPACK scheme performs SAXPY (linked triad) operations on vectors of length 1 if bilinear elements are used and this is not efficient for small vector lengths. The A-D IMSL (The alternating-direction method in conjunction with an IMSL routine for the solution of the matrix problem) scheme performs better than the B-D IMSL (The backward difference method with an IMSL routine for the matrix problem) scheme for sufficiently large problems. The B-D LINPACK (The backward difference scheme with a LINPACK routine for the matrix problem) scheme is quite fast because the SAXPY operations are on vectors whose length is equal to the band width of the matrix. The

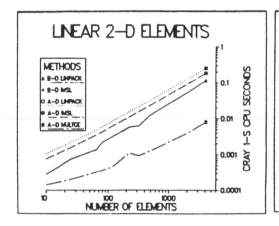

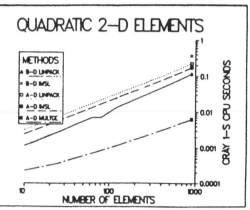

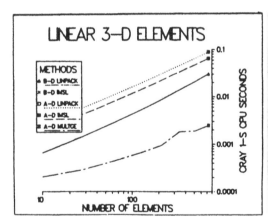

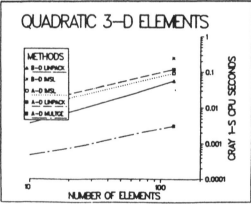

Figure 7: Comparison of the Time for Forward and Back Substitution

performance in the case of 225 bilinear elements and 343 elements is again degraded owing to memory bank conflicts.

The execution time for the right hand side computation is shown in Figure 8. This includes the matrix-vector multiplication which must be done at each step of the solution process and any other computation that is required. The element-by-element data structure is clearly superior for sufficently large problems. In the element-by-element data structure the stiffness matrix for each element is stacked and this stack is used for the matrix-vector multiplication. After the multiplication is performed the vector is assembled [10]. In the banded data structure the stiffness matrix is first assembled to form a banded matrix and a matrix-vector multiplication is done using this banded matrix. This element-by-element data structure could also be used for the backward difference scheme but this would entail maintaining two different data structures for the stiffness matrices.

Figure 9 shows the execution time for the solution of a linear parabolic problem. The A-D MULTGE scheme is compared with the B-D LINPACK scheme because the B-D LINPACK scheme was the better of the two backward difference schemes. A thousand time steps are taken for the backward difference scheme and two thousand time steps are taken for the alternating-direction scheme. For 125 triquadratic elements the alternating-direction method is about 5 times faster than the backward difference scheme.

Finally, a timing analysis is performed as if a prototypical non-linear diffusion problem is solved. An average of three inner iterations per time step is assumed and the results are shown in Figure 10. For non-linear diffusion problems a system of non-linear equations are obtained by both the backward difference and the alternating-direction method. Typically these equations are linearized by an iterative procedure which requires a few iterations in each time step. For 125 triquadratic elements, the

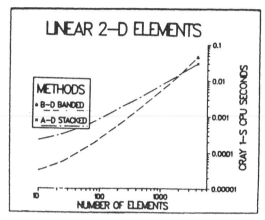
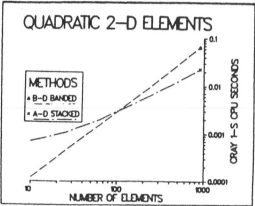

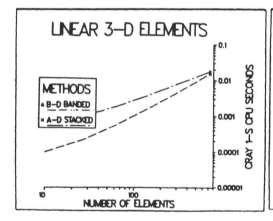
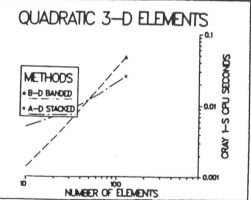

Figure 8: Comparison of the Time for the Right Hand Side Calculation

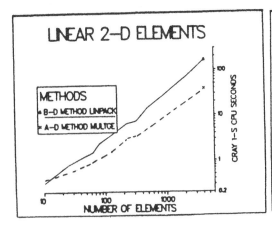

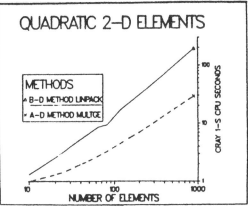

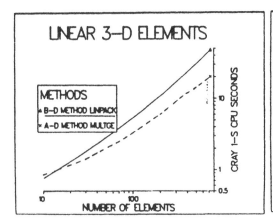

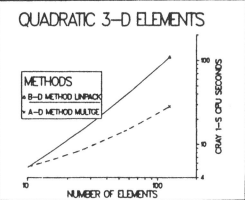

Figure 9: Comparison of the Time for the Solution of a Linear Problem

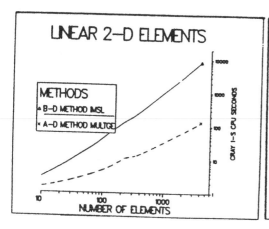

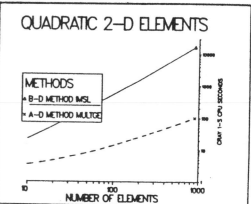

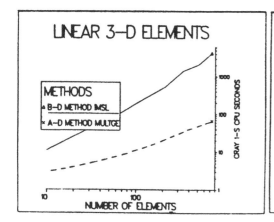

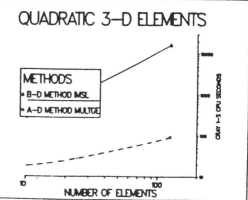

Figure 10: Comparison of the Time for a Non-Linear Problem

alternating-direction scheme is about 100 times faster than the backward difference scheme. This dramatic difference is due to the fact that the alternating-direction matrices require only $O(N)$ time for the decomposition of the matrices whereas the backward difference matrix with a band width of $m$ requires $O(m^2N)$ time for decomposition.

## Conclusions

The finite element alternating-direction scheme is superior to the backward difference scheme both in storage and execution time requirements. This is because the alternating-direction method reduces large multidimensional problems to a series of mutually independent one-dimensional problems that can be solved in parallel. On vector machines such as the CRAY the alternating-direction scheme vectorizes fully and the scheme becomes even more attractive if a stacked data structure is used. This scheme can also be used effectively on parallel machines. For non-linear diffusion problems this approach is even more efficient than conventional implicit methods. Since the scheme is implicit, it is stable with mild restrictions on the time step for curved structured grids.

## Acknowledgements

The authors wish to thank Philippe Devloo for his help. They would also like to thank Myron Ginsberg and Cleve Ashcraft of General Motors Research Laboratories for their helpful comments. This work was supported by a grant from the Army Research Office, Subcontract 8606 with the University of Houston contract no. DAAL03-86-K-0138 and a grant from the National Science Foundation grant no. DCR/8518722.

# References

1.  J. Douglas Jr. and T. Dupont, "Alternating-Direction Galerkin Methods on Rectangles", in Proc. Symp. Numerical Solutions of Partial Differential Equations - II ( Ed. B. Hubbard ) , Academic Press, New York, 1971, pp 133-214.

2.  T. Dendy and G. Fairweather, "Alternating-Direction Galerkin Schemes for Parabolic and Hyperbolic Problems on Rectangular Polygons", SIAM Jl. Numer. Anal. 2 (1975) pp 144-163.

3.  L. J. Hayes, "Galerkin Alternating-Direction Methods for Non- Rectangular Regions using Patch Approximations", SIAM Jl. of Numer. Anal. 18 (1981) pp 627-643.

4.  L. J. Hayes, "Implementation of Finite Element Alternating- Direction Methods on Non-Rectangular Regions", Intl. Jl. of Numer. Meth., 16 (1980) pp 35-49.

5.  L. J. Hayes, "A Modified Backward Time Discretization for Nonlinear Parabolic Equations Using Patch Approximations", SIAM Jl. of Numer. Anal., Vol. 18, October 1981, pp. 781-793.

6.  S. V. Krishnamachari, "An Improved Alternating-Direction Finite Element Method for Highly Nonuniform Grids", Master's Thesis, University of Texas at Austin, 1983.

7.  L. J. Hayes and S. V. Krishnamachari, "Alternating-Direction along Flow Lines in a Fluid Flow Problem", Comp. Meth. in App. Mech. and Engg., Vol. 47 (1984).

8.  J. Douglas Jr. and T. Dupont, "Galerkin Methods for Parabolic Equations", SIAM Jl. of Numer. Anal., Vol. 7, 1970, pp. 575-626.

9.  L. J. Hayes, "Generalization of Alternating-Direction Methods to Nonrectangular Regions Using Isoparametric Elements", Phd. Dissertation, The University of Texas at Austin (December 1977).

10. Ph. Devloo and L. J. Hayes, "A Fast Vector Algorithm for Matrix- Vector Multiplication with the Finite Element Method", TICOM Rept. 85-13.

11. R. W. Lewis, K. Morgan and P. M. Roberts, "Application of an Alternating-Direction Finite Element Method to Heat Transfer Problems Involving a Phase Change", (Personal Communication)

12. Boeing Computer Services, "MAINSTREAM - EKS/VSP CRAYPACK", Supplement to BCSLIB User's Manual, April 1982.

# Performance of Advanced Scientific Computers for the Efficient Solution of an Elastic Wave Code for Seismic Modeling

K. E. JORDAN, Computing and Telecommunications Division, Exxon Research and Engineering Company, Annandale, New Jersey

## 1 INTRODUCTION

The numerical analyst, in order to design and implement an efficient algorithm on modern computers, needs more than a cursory understanding of the architecture of the machine. The mathematical formulation of the physical problem and the subsequent numerical model may be better suited for one architecture than another. Careful consideration must be given not only to CPU speed but to many other factors such as number of processors or memory hierarchy. Data structuring or organization can have a dramatic effect on overall performance.

Our intent is to describe our experiences running a finite-difference elastic wave code on various high-speed computers for scientific computing. Through this comparative study, we hope to point out some of the hardware and software features of the various systems that one might take into consideration when designing numerical algorithms. This elastic wave code represents the area of seismic forward modeling. Seismic forward modeling continues to be an area which taxes the available resources of today's supercomputers. Obtaining efficient and accurate computer models continues to be of importance in the oil industry. The financial benefits from such computer modeling is significant and as such continues to make the oil industry one of the leading commercial areas for the use of advanced scientific computers.

This discussion is not meant to be a benchmark study; such a study should compare machines in a relatively short time span and simulate as much as possible the environment that will exist on the machine (for details of such a study, see [1]). Rather, the intent here is to give a general overview of factors to consider beyond computational speed alone when formulating the mathematical model with its computer implementation. Some of the factors to keep in mind include sophistication of the compilers for automatic parallelization and vectorization, memory access time, and cache size, as well as clock speed, number of arithmetic units, and number of processors.

For this comparison study, we present performance results for this elastic wave code on a variety of large-scale scientific computers: mainframe computers, mainframes with vector facilities, vector supercomputers, mini-supercomputers, and parallel-vector mini-supercomputers and supercomputers. The code that was tested is one used for computing waves in an elastic medium. It involves a finite-difference scheme that is fourth order accurate in space and second order accurate in time. It is a variant of the MacCormack scheme (Gottlieb and Turkel [2]) which is known to be effective for acoustic wave propagation (Maestrello, Bayliss, and Turkel [3]). This fourth-order scheme initially was developed for a single-processor machine [4]. On vector and shared-memory parallel computers [5,6] , the algorithm has proven to be very efficient, in part as a result of efforts to minimize memory references and to enhance chaining in the implementation. The algorithm was also among several used for a recent comparison of vector machines [1].

Two competing formulations for simulation of seismological structures exist; the homogeneous formulation and the heterogeneous formulation. The essence of the homogeneous formulation is in dividing the domain of interest up into regions of homogeneous mediums, differencing in each region and explicitly imposing interface conditions at the boundaries of the homogeneous regions. The heterogeneous formulation uses a uniform grid throughout the domain of interest. Differencing of equations is done across the interfaces. The merits and disadvantages of these two formulations are discussed in some detail below. This heterogeneous formulation is the one upon which the second order accurate in time and fourth order accurate in space elastic wave code used on the various machines discussed here is based.

Other researchers have investigated different numerical schemes for such wave propagation problems. The basic second order scheme for the coupled wave equation was introduced by Alterman and Karal [7]. Kelly et. al. [8] discussed a conservative version of this scheme appropriate for the heterogeneous formulation. Stephen ( [9], [10]) considered a variety of approximations for the problem of an acoustic-elastic interface. Numerical schemes for the first order system are considered in [11], [12] and [13]. Many authors have considered finite element approximations, for example see Lysmer and Drake [14] and Marfurt [15]. Many finite element codes are based on piecewise linear elements and are thus at most second order accurate as well (assuming that the nodal points coincide with interfaces).

Kosloff et. al. [16] have developed a spectral approximation for the elastic wave equation. The method is an extension of a spectral acoustic code described in [17]. A three-dimensional acoustic code using a Fourier pseudo-spectral method is described in [18]. The Fourier method requires a periodic extension of the domain and thus computations have to be carried out over a computational domain larger than that of interest in order to avoid spurious reflections entering from this periodic extension. In addition, the implementation of the free surface condition can not be combined with the periodicity in a natural way. The difficulties due to forcing periodicity on the problem can be avoided by using a pseudo-spectral Chebychev method and such a method is a possible extension of the fourth order difference method described in [4] and outlined below. However, the time restriction on present Chebyshev methods is prohibitive. An additional drawback of pseudo-spectral methods for heterogeneous wave propagation is that the coarse grids which are sufficient to resolve the wave propagation may not be sufficient to resolve the curvature of the interfaces.

The motivation for obtaining a numerical scheme with high performance on high-speed computer systems lies in the fact that geophysicists and seismologists typically are interested in problems whose physical domain covers 10,000 feet in the horizontal direction and 20,000 feet in the vertical direction. The physical domain is further complicated by interior interfaces across which the elastic parameters are discontinuous. The mesh size is governed by the wave speed of the faster shear waves, typically requiring mesh spacing of the order of 10 feet or less for the desired accuracy. For accuracy, the equations must be solved at many mesh points. Fast solutions are required, as the problems must be solved many times for a variety of elastic parameters.

## 2   EQUATIONS OF ELASTODYNAMICS

In this section, we give the mathematical formulation of the equations of linear elastodynamics upon which the fourth-order scheme is based. A detailed description of the numerical scheme can be found in [4]. The discussion here emphasizes only the salient points needed to understand our implementation of this numerical scheme. Before presenting the specific governing equations, we give some motivation for the choice of using the system of first order equations.

Linearized elastodynamics theory governs the response of a seismic structure to a surface or buried source. As such, the numerical computation of the equations of elastodynamics is fundamental to exploration seismology. Wave propagation in elastic media is important in other areas as well, for example, earthquake seismology and the nondestructive testing of engineering components is related to wave propagation in materials with general constitutive relations (e.g. viscoelastic materials).

The equations of linearized elastodynamics can be written as a first order system for the (infinitesimal) velocities and stresses (see Clifton[17]). These equations can also be written as a set of coupled wave equations for the displacements (or velocities) (see Alterman and Karal [7]). In either form, there are significant differences between elastodynamics and acoustics. These differences have important numerical consequences which we outline below.

The elastodynamic equations describe a much wider range of wave phenomena than the acoustic equations. In addition to compressional waves (P waves), the are shear waves (S waves), Rayleigh waves and Stonely waves. Since most of these waves have shorter wavelengths than the P waves, the resolution requirements are considerably more severe than for acoustic wave propagation in a medium whose characteristic speed is the same as the P-wave speed of the elastic medium. In addition, the large number of different waves generated at the interfaces of subsurface structures makes ray tracing considerably more expensive than for acoustic models. Due to the large number of waves moving with different speeds, adaptive grid methods which can adjust the grid to follow the waves are less useful than they are for acoustic models.

The diversity of wave phenomena also makes boundary conditions considerably more complicated. Absorbing boundary conditions must absorb many different types of waves with different dispersion relations. In addition, the free surface condition (vanishing of the normal stresses) is more difficult to implement than the pressure release boundary condition of acoustics (particularly for the coupled wave equation formulation). It is also known that some standard implementations of the free surface conditions can be unstable for sufficiently high Poisson ratio [19].

These differences suggest that the formulation as a first order system and the use of higher order differences can be advantageous in the numerical computation. The first order system, dealing directly with the stresses and the velocities, permits a more direct implementation of both the free surface condition and the absorbing boundary conditions. These conditions involve relations between the stresses and the velocities. When the stresses must be computed by one-sided differences of the displacement inaccuracies can result. This can be especially pernicious in surface seismology calculations when the solution (generally the vertical velocity) at the surface is the object of interest.

The equations of linear isotropic elastodynamics for the stresses and velocities in Cartesian coordinates are

$$\rho u_t = \tau_{11,x} + \tau_{12,y}$$

$$\rho v_t = \tau_{12,x} + \tau_{22,y}$$

$$\tau_{11,t} = (\lambda + 2\mu)u_x + \mu v_y \qquad (2.1)$$

$$\tau_{22,t} = \mu u_x + (\lambda + 2\mu)v_y$$

$$\tau_{12,t} = \mu(v_x + u_y).$$

In equation (2.1), u and v are the horizontal and vertical velocities, respectively, and $t_{ij}$ are the components of the stress tensor. The elastic parameters are the density $\rho$ and the Lamé constants $\lambda$ and $\mu$. The compressional and shear wave speeds, $C_p$ and $C_s$, are given by

$$C_p^2 = ( \lambda + \mu )/\rho$$

$$(2.2)$$

$$C_s^2 = \mu/\rho$$

and $C_p > C_s$.

Associated with these speeds, there are the spatial wave lengths $\lambda_p = C_p/f$, $\lambda_s = C_p/f$ where f is the characteristic frequency. Hence

$$\frac{\lambda_p}{\lambda_s} = \frac{C_p}{C_s} > 1.$$

The Poisson ratio is given by

$$\nu = \frac{1}{2} \left( \frac{1 - 2C_s^2/C_p^2}{1 - C_s^2/C_p^2} \right).$$

Hence, we see that

$$\frac{C_p^2}{C_s^2} = \frac{2(1-\nu)}{1-2\nu} .$$

For seismic problems, it is frequently assumed that $\nu$ is around 0.25. Therefore, one can take $C_p/C_s = \sqrt{3}$; see [4] for a detailed discussion. It follows that the shear wavelengths are about 60 percent smaller than the compressional wavelengths. The spatial resolution requirements, therefore, must be based on the shorter shear wavelengths. There may also exist interface and surface waves whose wavelengths typically are smaller than the shear wavelengths. Although in some applications these interface and surface waves are not considered important, nevertheless it is necessary to resolve these waves sufficiently to prevent a spurious transfer of energy into other wave modes. We also do not want numerical dispersion for these waves to interfere with the generation and interpretation of the waves of interest. In some cases e.g. weathering layers (near surface layers), the Poisson ratio can be considerably larger than 0.25. This further accentuates the stiffness of the problem. All these restrictions put further constraints on the spatial resolution requirements.

Equation (2.1) can be written in vector form as

$$W_t = AW_x + BW_y \qquad (2.3)$$

where $W = (u, v, \tau_{11}, \tau_{22}, \tau_{12})^T$, and the matrices A and B consist of the coefficients of equation (2.1). Equation (2.3) can be written as a symmetric system through the use of an appropriate transform matrix, $E_0$ (see Bayliss et al. 1986 for the exact form of this matrix). In symmetric form equation (2.3) is

$$E_0 W_t = A_0 W_x + B_0 W_y \qquad (2.4)$$

The matrix $E_0$ depends only on the constants $\rho$, $\lambda$, and $\mu$. Since these constants are independent of time, $E_0$ is also independent of time. The matrices $A_0$ and $B_0$ are constant matrices independent of both time and the spatial grid. Hence equation (2.4) is in divergence-free form. MacCormack schemes are usually written for equations in divergence-free form and are therefore applicable to equation (2.4). Since equation (2.4) is merely a linear manipulation of equation (2.3), the MacCormack type scheme applies equally well to equation (2.3). The numerical method is based on equation (2.3).

## 3 MODEL PROBLEMS

Next, we describe two model problems used in comparing the various machines. The first problem, referred to as the *large problem*, consists of Rayleigh wave scattering from a fluid-elastic interface. This problem was run on many different machines. The domain of definition for the large problem is

$$0 \leq x \leq L_1; \ -L_2 \leq y \leq 0 . \tag{3.1}$$

For the purposes of this example $L_1$ = 2100 and $L_2$ = 900 feet. Equation (2.1) is used in both the elastic and acoustic region. However, in the acoustic region the shear modulus is set to zero; that is, $C_S$ = 0. The physical domain is shown in Figure 1.

The large problem has one physical boundary (y = 0) and three artificial boundaries along which boundary conditions must be prescribed. Along the physical boundary, y = 0, we apply a free surface boundary condition ($\tau_{11}$ = $\tau_{12}$ = 0) over the elastic medium and a pressure release boundary condition ($\tau_{11}$ = $\tau_{22}$ = 0) over the fluid.

In addition to the physical boundary, appropriate absorbing boundary conditions are imposed on the three artificial boundaries, x = 0, x = $L_1$, and y = -$L_2$. Based on a one-dimensional analysis, characteristic equations can be derived. We impose the boundary condition that the incoming characteristic variable is zero. If the first-order system (2.1) is used, this condition becomes a Dirichlet condition. (This is somewhat of a simplification of the absorbing boundary conditions but suffices for the numerical solution of the problem. For further discussion on the appropriateness of these boundary conditions, see [4].)

A source function needs to be imposed. We take a surface line source,

$$\tau_{22} = f(t) \ \delta(t) \tag{3.2}$$

on the free surface. The source function $f$ has the form

$$f(t) = -4/\sigma^2 (e^{-2(t-t_s)^2/\sigma^2} - e^{-2(t_s/\sigma)^2}) (t - t_s) . \tag{3.3}$$

The parameters of the source function are $\sigma$ = 0.0017 sec. and the time shift $t_s$ = 0.285 sec. The delta function is approximated by a discrete function which is zero except at the location of the source, where it is 1/h ( h is the mesh size). Equations (2.1) and (3.2), together with the boundary conditions, form our model initial-boundary value problem.

The second problem, the *small problem*, consists of scattering from three layers of different materials. Each layer is a plane layer with horizontal interface given along a line y equals a constant. Each layer has a given different density and both the compressional and shear wave speeds ( $C_p$ and $C_s$ respectively) also differ from region to region (see

# FLUID-ELASTIC INTERFACE

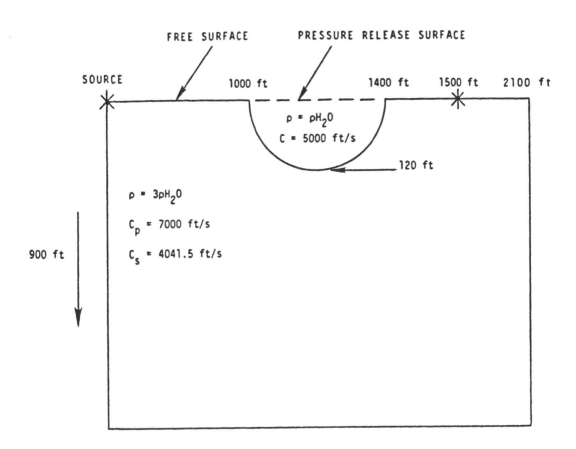

Figure 1. The physical domain for the large problem.

Figure 2 for details). The boundary conditions for the small problem are similar to those of the large problem. Again for the physical boundary (y=0), we apply a free surface boundary condition. Along x=$L_1$, and y=-$L_2$, appropriate absorbing boundary conditions are imposed. A symmetric boundary condition is imposed on x=0. In this problem, $L_1$=$L_2$=700 feet.

This problem, due to the size, is not really of physical interest but serves as a valid test problem of sufficiently size that we can use it to evaluate the effects of memory requirements. It can easily be scaled up or down as required.

One major difference between the large and the small problems is that the source and the receivers are buried 200 feet in the small problem while they are at the surface in the case of the large problem. Typically in the collection of seismic data from the field, geophones are buried. This is in part to avoid affects of the weathering layer, the near surface layer that contains many inhomogeneities.

## 4  NUMERICAL ALGORITHM

The more severe resolution requirements, due to the necessity of resolving the shear waves, clearly makes higher order accurate schemes more efficient in computing wave propagation (see [20] and [21]). It is further known that higher order schemes become more efficient as problem size increases. Thus these methods are appropriate for the very large scale problems encountered in seismology. In Bayliss et. al. [4], we showed that spatial accuracy is more important than temporal accuracy. Therefore it is  natural to consider schemes which are second order accurate in time and forth order accurate in space ( (2-4) schemes).

The numerical simulation of seismological structures involves more than just wave propagation. Typically variations in the elastic parameters are modelled by piecewise constant functions which vary across non-horizontal interfaces. The problem then is to compute reflections and transmissions from the interfaces. The solution is not smooth across the interfaces and some components of the field are not even continuous. In principle, these problems could be computed by solving the constant coefficient problem in each region using a coordinate

# Small Problem

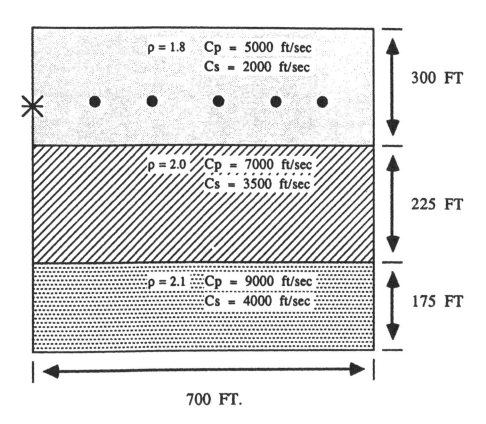

Figure 2.   The physical domain for the small problem.

system which conforms to the interface. The interface conditions are then explicitly imposed and the numerical scheme is only needed to compute wave propagation. This technique has been called the homogeneous formulation [7] and [8]. Since only wave propagation has to be computed there is very strong justification for the use of the higher order schemes([20], [21]).

The homogeneous formulation is difficult to implement with curved interfaces and requires regridding each time the model is changed. This detracts from the basic efficiency that can be obtained from the simplicity of the wave operator. This method also requires logic and data structuring that is not well suited to vector computers. A more effective approach is to use a uniform grid and difference the equations across the interfaces. This is known as the heterogeneous formulation (see [8] and [9]).

The heterogeneous formulation involves errors both in differencing functions which are not smooth (and in some cases are not even continuous) across the interfaces and in approximating the interfaces on a uniform grid. It is easy to see that even for a horizontally stratified medium the resulting approximation can be at most first order accurate due to errors in approximating the interfaces. Very little is known about the effects of curvature and of varying contrasts on the accuracy of the discrete solution. Boore [22] considered some approaches to redefining the elastic parameters in the vicinity of the interfaces, however he did not systematically study the accuracy of the resulting approximations. In [8] some differences between the homogeneous and heterogeneous formulations where noted for elastic scattering from a buried wedge, however there was no indication of which solution was more accurate. Comparisons of finite difference and pseudo spectral solutions with the analytic solution for a horizontally stratified medium were conducted in [16] and [10], however these references did not systematically consider the effects of curvature and contrast on the accuracy of the solution.

Due to the piecewise constant nature of the coefficients, it is not at all clear that higher order difference schemes will be effective in computing wave propagation in a layered medium. However, the results in [23] demonstrate that in one dimension the phase error of a solution computed with a fourth order difference scheme will be fourth order accurate. The results presented in [4] demonstrates that higher order difference schemes can give significantly more accurate solutions even in

# PARALLELISM IN CODE

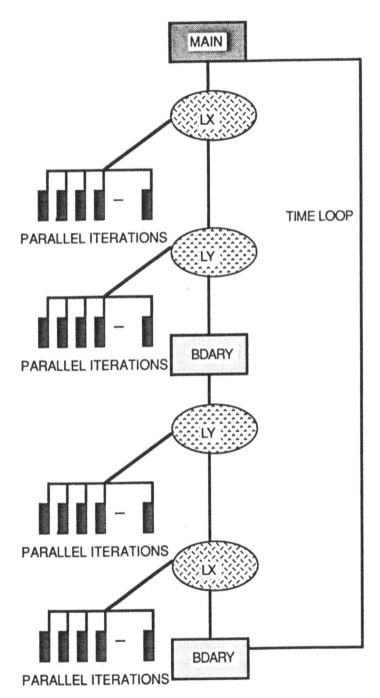

Figure 3. Schematic of Parallelism in the Elastic Wave Code.

the presence of highly curved interfaces with strong contrasts.  This is consistent with results that have been found in fluid dynamics [21].

The algorithm is based on the method of dimensional splitting (for details, see [4]) and is implemented as follows.  We update the two-dimensional problem (2.3) for one timestep by solving the equation first in the x-direction, using only terms involving the x derivatives, and then in the y-direction, using only terms involving y derivatives.  This procedure gives the solution at time level n + 1.  To update to the next time level, we repeat the procedure but reverse the order of the x and y updates, using the formula

$$W^{n+2} = L_x L_y L_y L_x W^n \tag{4.1}$$

where $L_x$, $L_y$ represent the solutions of the one-dimensional problems

$$W_t = AW_x, \quad W_t = BW_y \tag{4.2}$$

respectively.  Each of the $L_x$ and $L_y$ represents a subroutine consisting of a nested set of Fortran DO-loops two loops deep.  We solve the inner loops using a predictor-corrector MacCormack scheme that is second order in time and fourth order in space (see Gottlieb and Turkel [2]).

There are several advantages in using dimensional splitting schemes over other schemes.  One of the advantages of splitting methods is that the stability properties are governed by the one-dimensional schemes.  This generally allows allows the use of larger timesteps.  Hence our scheme is accurate with larger timesteps and spatial mesh.  We refer to [4] where the choice of timestep for our scheme is discussed in some detail.  Another advantage of our scheme, based of the first order system of equations, is that it can easily handle fluid-elastic interfaces as the first model problem is an example.  Differencing right across the fluid-elastic interface, the numerical results remain accurate and stable.  Besides these advantages, it has also been found that split schemes have smaller phase errors than a wide class of unsplit schemes (see[13]).

These inner loops vectorize on all the vector machines.  The outer loops run over the other spatial dimension not associated with the inner loops.  For example, the first equation in (4.2) is solved for each y = constant; all the x updates for each line y = constant can be done independently.  As a result of this configuration, the code is appropriate for parallel execution on parallel machines. On shared memory machines like the Alliant FX/8 and multi-processor Crays ( CRAY X-MP/416 or the CRAY Y-MP/832) the modifications to the code are minimal and can easily be implemented.

In Figure 3, we present a schematic of the parallelism in the code. The boxes from the $L_x$ or $L_y$ subroutines represent one iteration of the outer do- loops. As mentioned above, each of these iterations is independent from one another. As a consequence, these outer loops when written appropriately are amenable to parallelization such as the COVI (Concurrent Outer Vector Inner) construct the Alliant compiler recognizes or microtasking on the Crays. (For a detailed discussion of microtasking this code, see [5]. For more information on the differences between macrotasking and microtasking on the CRAY X-MP, see [24]. For a brief description of how the compiler on the Alliant FX/8 parallelizes code, see [25].)

## 5   NUMERICAL RESULTS

In this section, we present some numerical results for the elastic wave code on the model problems described in Section 3. The purpose here is to briefly demonstrate that this numerical scheme is a viable one. Figure 4 contains results for the large problem while Figure 5 shows the numerical scheme applied to the small problem.

In the case of the large problem, the acoustic fluid is treated by setting the shear modulus to zero and differencing across the interface. This model problem is designed to demonstrate that the numerical scheme is accurate and stable at a fluid-elastic interface. The dynamics of the large model problem consist of the impinging Rayleigh wave generating an interfacial wave that travels along the elastic-fluid interface and a wave that travels throughout the fluid. The slow velocity of the interfacial wave governs the resolution requirements of the problem. At the interface, body waves are generated and propagate away from the fluid.

Although we do not have an exact solution to compare with, the accuracy of the scheme can be assessed by examining time traces for different grids. By selecting a receiver location of 1500 feet, we can see the effects of the slow interfacial wave. In Figure 4, the solutions

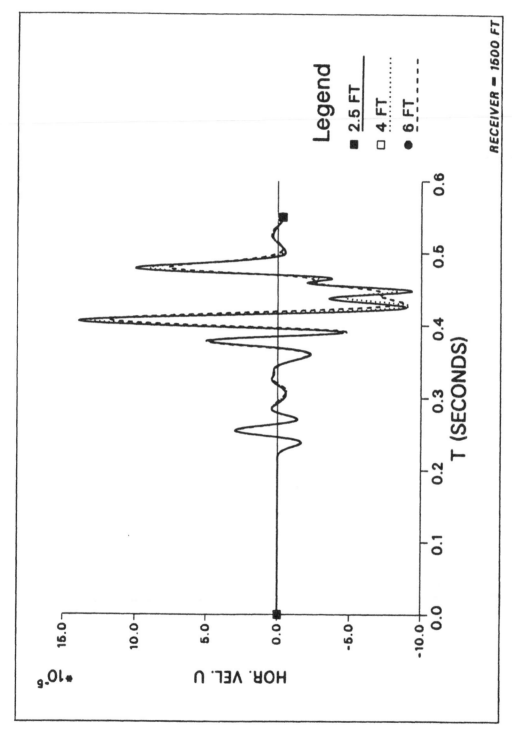

Figure 4. Comparison of horizontal velocity for the large problem on 3 different grids at a receiver location of 1500 feet.

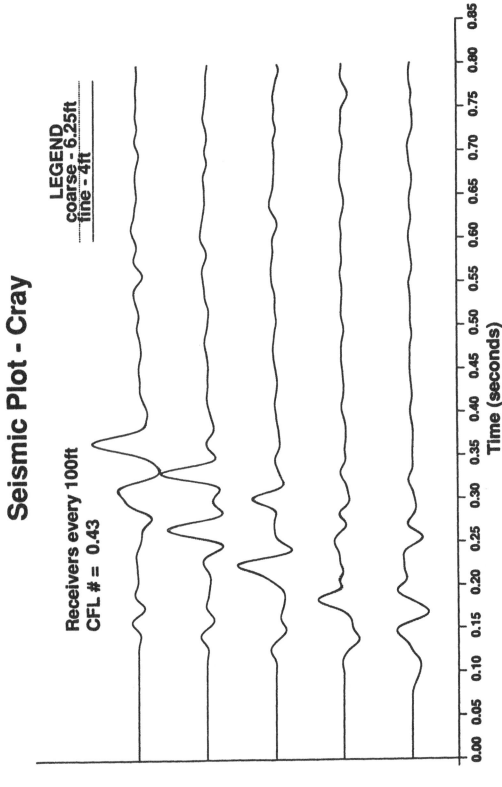

Figure 5.a. Comparison of horizontal displacement for the small problem on 2 different grids at 5 receiver locations 100 feet apart.

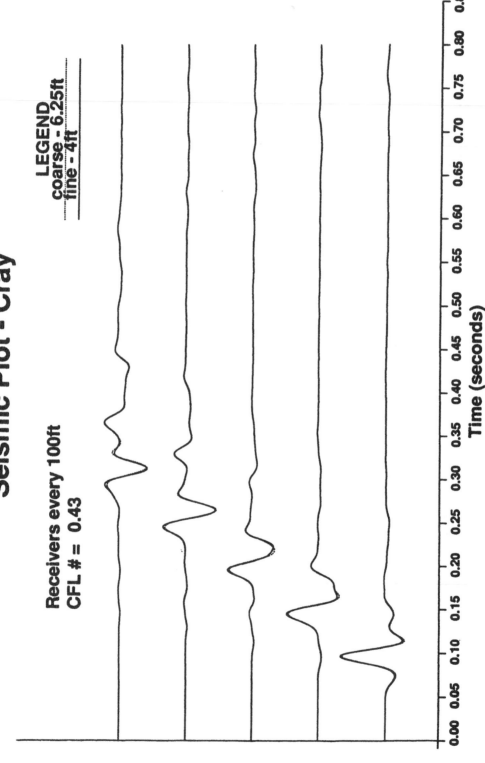

Figure 5.b. Comparison of vertical displacement for the small problem on 2 different grids at 5 receiver locations 100 feet apart.

obtained with grid sizes of 6, 4, and 2.5 feet are compared. Further refinement beyond 2.5 feet shows no appreciable difference in the solution, and we therefore conclude that the solution converges as the grid is refined. (For a more thorough discussion of the numerical accuracy and stability, see [4].)

In Figures 5.a and 5.b, the results of the numerical method applied to the small problem are given. These figures present the results from all of the receiver locations together. This is more typical of seismic plots rather than at a single receiver location. Figure 5.a gives the horizontal displacement and Figure 5.b gives the vertical displacement. The receivers are positioned every 100 feet with the first receiver coincident with the source. Here, we compare the numerical results on two spatial meshes, 6.25 ft. and 4 ft. Further refinement beyond 4 ft. produced no appreciable differences in the solution in this case. In fact, there is little discernible difference between the solution with the coarser 6.25 ft. mesh and that of the 4 ft. mesh.

## 6  PERFORMANCE RESULTS

Table 1 records the performance of the large model program with a grid of 701 x 301 points and a run of 1280 timesteps. The results range from a required 11 CPU hours on the IBM 3033 AP to 12 CPU minutes on a single-processor CRAY X-MP (9.5 ns clock), 6 CPU minutes on the Amdahl VP 1200, 3 CPU minutes on a 4-processor CRAY X-MP (8.5 ns clock), and less than a minute on an 8-processor CRAY Y-MP. The results listed in Table 1 were collected over time. Some minor changes were made to the code in order to take full advantage of the features of the computer architecture. In particular, these changes were made for multi-processing purposes. The changes involved restructuring key nested loops so compilers could recognize the parallelism or inserting directives to highlight the parallelism.

Some comments on the IBM 3090 performance are in order. The code actually ran longer on the IBM 3090 VF than on the IBM 3090 without the vector facility. Since this code has a high vector content (using one of the tools available on the Amdahl machines, we estimated the single-processor implementation to have a 99% vector content), there was some

**TABLE 1.   Comparison of CPU Time on Various Processors**

| Machines | CPU Time (sec) |
|---|---|
| Alliant FX/8 (1 proc.) | 112507.09 |
| IBM 3033AP | 39648.00 |
| IBM 3090 | 12164.00 |
| Convex C1-XP | 20906.40 |
| Alliant FX/8 (8 proc.) | 16078.62 |
| IBM 3090VF | 16025.28 |
| Cray 1-S | 1145.65 |
| Amdahl VP 500 | 778.65 |
| Cray X-MP/48 (1 proc.) | 714.90 |
| Amdahl VP 1200 | 360.00 |
| Cray X-MP/416 (4 proc.) | 187.00 |
| Cray Y-MP/832 (8 proc.) | 46.74 |

speculation that either the compiler or the cache was responsible for the performance encountered here.

The IBM vectorizing compiler has been known to generate less than optimal code through the choice of loops it vectorizes in a set of nested loops.   Careful inspection of the compiler listing, however, indicated that the compiler vectorized the expected vector loops.

Others have noted cache stumbling when the vector facility deals with long vectors.   Cache stumbling results from the fact that the vectors cannot fit in cache and, hence, data must be fetched from main memory at a slower access time than the cache.   The data structure of this code would, for large dimensions, lead to cache misses and result in the cache stumbling problem.

In the case of our elastic wave code, cache stumbling most likely contributed to the unexpected performance on the IBM 3090 VF.   The reason for this is that in a dimensional splitting type code upon which this elastic wave code is formulated, the memory access pattern of the two dimensional arrays thought of as matrices, switches from a column oriented access pattern to a row oriented access pattern,   During the column oriented part of the scheme, cache will be filled with useful data however, during the row oriented part of the scheme, little data in cache will be useful and cache misses will occur. The CPU will have to access the slower main memory to find the data it needs. The vector registers will take longer to fill and the functional units will be idle for longer periods.

One remedy for the cache stumbling problem on the elastic wave code is to carry out a matrix transpose of the data in the two dimensional arrays. However, the matrix transpose will add some extra overhead to the code. In our particular code, several two dimensional arrays are involved and each would have to be transposed. For each time step (see Figure 3), a minimum of two and possibly four depending on how the boundary updates are handled, matrix transposes would be necessary for the many two dimensional arrays required in the solution. The overhead associated with this remedy for this particular code could be significant.

For a detailed description of both the choice of vectorized loops and the cache stumbling problems, refer to Schonauer [26]. What is interesting here is that the code developer must now consider the effects that cache may have on a code and must design the data structure appropriately. Since cache misses are hard to monitor, it is difficult to know without extensive testing whether one has, in fact, obtained an optimal data structure.

The mini-supercomputers (Alliant FX/8 and Convex C1-XP) and the vector supercomputers (the one-processor CRAYs and the Amdahls) perform relatively well on this high-vector code. The mini-supercomputers are a factor of two or more faster than the IBM 3033 AP, and the vector supercomputers range from a factor of fifty to over one hundred times faster than the IBM 3033 AP. (The one-processor Alliant FX/8 run is included for comparison purposes. Its slower clock speed is evident here.)

The Convex C1-XP, like the IBM 3090 and the Alliant, has a cache. However, for vector processing, the Convex C1-XP bypasses the cache. This approach can give improved performance for vector codes, provided the vectors are unit stride (the increment of the vector loop is one) and are not offset by some constant stride greater than one. In our case, however, the elastic wave code has vectors with both unit stride and nonunit constant stride. Thus, the subroutines that have unit stride performing essentially the same work as the subroutines with nonunit constant stride ran two to three times faster.

With regard to the parallel and vector processors, some work was needed to parallelize the elastic wave code. For the parallel-vector version of the numerical scheme for the CRAY X-MP, the work involved modifying the original implementation to take advantage of microtasking. One often wonders how much effort must be expended in converting a sequential code to a multitasking code. Like vectorizing, conversion to multitasking is code dependent and problem dependent. For the wave problem, the microtasked version of the code on the CRAY X-MP required changing only 17 lines of Fortran, along with inserting the microtasking directives.

In evaluating performance on multiprocessor machines, one should consider the speedup, or ratio of uniprocessor CPU time to execution time on N processors. The efficiency factor can then be determined from the ratio of speedup to the number of processors. The efficiency factor provides an indication of the microtasking overhead associated with this implementation of the numerical scheme. Tables 2 and 3 give the speedup and efficiency factors obtained on the CRAY X-MPs tested. Because the wave code also measured MFLOPS (Millions of Floating-point Operations Per Second), these values have also been included in the tables.

In Table 2, we give the results obtained using the early version of microtasking on the older CRAY X-MP/48 with a clock cycle of 9.5 ns. These runs were done with the older microtasking library that used calls rather than directives for the microtasking preprocessor. Note that the improvement in using four processors with the microtasked code is 3.9, which gives an efficiency factor of 97.5%. Also note that a sustained rate of 83 MFLOPS was obtained on the uniprocessor version, and 329 MFLOPS when using four processors.

Table 3 contains the results of microtasking the numerical scheme on the CRAY X-MP/416, using the newer microtasking library with directives instead of subroutine calls. The Cray X-MP/416 has a faster clock (8.5 ns) than the Cray X-MP/48. Both the CRAY X-MP/416 and the CRAY X-MP/48 used for microtasking the elastic wave code have 64-way memory interleaving. However, when we ran the microtasked version of the elastic wave code on the X-MP/416 appeared as though it had only 32-way interleaving. Although extensive tests have not yet been carried out, all indications point to apparent 32-way memory interleaving as the reason

**TABLE 2. Microtasking on the CRAY X-MP/48**
**(9.5 ns cycle)**

| Number of Processors | CPU Time (sec) | MFLOPS Rate | Speedup Factor | Efficiency Factor (%) |
|---|---|---|---|---|
| 1 | 756 | 83 | 1.0 | 100.0 |
| 4 | 191 | 329 | 3.9 | 97.5 |

for the poorer performance on the newer machine. This conjecture is supported by the fact that there were twice as many I/O waits on the CRAY X-MP/416 as on the CRAY X-MP/48. We were informed after these tests by collaborators at Cray that at the time we ran this code there was a bug in the microtasking library on the Cray X-MP/416. This bug showed up as the machine having 32-way interleaved memory. This apparent 32-way interleave memory points out that performance can be affected not only by the machine speed but also by the way memory is set up; that is, a faster machine does not always imply better performance.

**TABLE 3. Microtasking on the CRAY X-MP/416**
**(8.5 ns cycle)**

| Number of Processors | CPU (sec) | MFLOPS Rate | Speedup Factor | Efficiency Factor (%) |
|---|---|---|---|---|
| 1 | 676 | 100 | 1.0 | 100 |
| 2 | 336 | 199 | 2.0 | 100 |
| 3 | 238 | 298 | 2.8 | 95 |
| 4 | 187 | 385 | 3.6 | 90 |

The Alliant FX/8, like the CRAY X-MP, is a shared-memory multiprocessor machine. The Alliant FX/8 can have up to eight processors. Table 4 contains the CPU time, speedup factors, and machine efficiency for the large model problem running on various numbers of processors on the Alliant FX/8. Note that when four processors are used on this problem, quite good machine efficiency is achieved better than on the CRAY X-MP. Even with all eight processors, the machine efficiency remains high at 87.5%. This corresponds closely with that of the CRAY X-MP.

The effort involved in multitasking the elastic wave code on the Alliant FX/8 was more than on the CRAY X-MP. This fact is surprising when one considers that the compiler generates the multitasked code. However, the compiler can make the outer loop of a set of nested loops concurrent only if the set of nested loops is relatively simple. In this case, the more complicated nested loops had to be split into two sets of simpler loops. Moreover, several one-dimensional arrays used to store some intermediate values had to be promoted to two-dimensional arrays. The addition of these one-dimensional arrays resulted in 30% more storage requirements. No attempt at conserving memory was made when promoting these arrays for intermediate values. We believe with some effort a restructuring of the data could reduce some of the additional memory.

### TABLE 4.   Multitasking on the Alliant FX/8
### (8 processors in a cluster)

| Number of Processors | CPU (sec) | Speedup Factor | Efficiency Factor (%) |
|:---:|:---:|:---:|:---:|
| 1 | 112507 | 1.00 | 100.0 |
| 2 | 56994 | 1.97 | 98.7 |
| 3 | 38652 | 2.91 | 97.0 |
| 4 | 29820 | 3.80 | 94.7 |
| 5 | 24251 | 4.64 | 92.8 |
| 6 | 20480 | 5.50 | 91.6 |
| 7 | 18073 | 6.23 | 88.9 |
| 8 | 16078 | 7.00 | 87.5 |

**TABLE 5. Comparison of the Alliant FX/8 and CRAY X-MP**

| Problem Size | FX/8-CPU sec (8 processors) | X-MP-CPU sec (1 processor) | Ratio |
|---|---|---|---|
| Small | 513.5 | 81.8 | 6.3 |
| Large | 16078.6 | 714.9 | 22.5 |

Even though the speedup factors and machine efficiency were high for the large model problem, the CPU time seemed to be larger than expected compared with the single-processor CRAY X-MP (see Table 5). Since the Alliant FX/8 has a cache, and since the memory requirements are high for the large problem, we conjectured that the comparison to the X-MP might be more favorable for a problem requiring less memory. This conjecture led to the formulation of the small model problem described in Section 3. In Table 5, one sees the improved ratio between the FX/8 and X-MP for the small problem. This implies that as the problem size increases, performance falls off. For more information on this phenomenon on the Alliant FX/8, see Abu-Sufah and Maloney [27].

In early 1988, the Alliant FX/8 at Argonne National Laboratory where the original problem was run, was upgraded in memory and cache size. The original configuration consisted of 32 megabytes of memory and 128 kilobytes of cache. The new configuration consisted of doubling both memory and cache to 64 megabytes of memory and 512 kilobytes of cache. We had the opportunity to rerun the large problem with the new configuration. We reran the code on 4 processors and 8 processors in the cluster. The improvement in both runs was less than 8 percent. Again the large problem may still be out of cache.

Finally, we note one other phenomenon observed while running this small model problem on the Alliant FX/8: when the 8-processor run was compared with the 1-processor run, the speedup factor dropped to 5. Although we have not carried out any experiments to investigate this, Hanson [28] has conducted parameter studies in which he varies the mesh sizes for a specific problem and has observed that with smaller meshes, the speedups were less than with larger meshes on the FX/8. Further work to understand this phenomenon should be carried out.

Results of combined autotasked and microtasked version of our elastic wave code running on a Cray Y-MP/832 are given in Table 6. The reason this is a combined autotasking and microtasking code is because the autotasking facility does not modify subroutines with microtasking directives in them. (For more details on the Cray Autotasking Facility see [29]). As a consequence the parts of the code that had previously been microtasked remained as before. These parts of the code were the most compute intensive. However, other parts of the code through the autotasking facility were computed in parallel where before they were sequential.

Some specific comments about the results appearing in Table 6 are needed. Most notably, the results for the Y-MP recorded on eight processors in Table 1 and here differ. The results in Table 1 are from earlier test runs. They are obtained from running on essentially a dedicated machine, that is all eight processors are available to a single user. The results in Table 6 come from test runs at a later date subsequent to some changes in hardware and software on the Y-MP. Among the changes are a slower clock cycle (6.5 nanoseconds versus 6.3 nanoseconds) and two of the eight processors are newer and differ from the other six. In addition, the results in Table 6 come from test runs on a machine that was accessible by other users as well. Running in a non-dedicated mode can have some impact on the overall performance. For one thing when requesting eight processors, one can not always be guaranteed of having all eight processors working on a single problem as processors are taken away to service networking, editing and other requests.

Other points of significance that can be derived from Table 6 are that the autotasking-microtasking version of a code efficiently takes full advantage of the available processors and that on the Cray Y-MP/832 sustained rates of over 1 gigaflops ( in this case, 1.4 gigaflops or 1.2 gigaflops) are attainable. Also, the results in Table 6 exhibit the same trends with respect to the speedups as were seen on the Alliant FX/8 (Figure 4). Of course, the speeds on the Cray Y-MP/832 are significantly faster than those on the Alliant FX/8.

**TABLE 6.** **Autotasking/Microtasking on the CRAY Y-MP/832**

| Number of Processors | CPU (sec) | MFLOPS Rate | Speedup Factor | Efficiency Factor (%) |
|---|---|---|---|---|
| 1 | 372 | 165 | 1.00 | 100 |
| 2 | 190 | 326 | 1.96 | 98 |
| 3 | 129 | 484 | 2.88 | 96 |
| 4 | 99 | 641 | 3.77 | 94 |
| 5 | 80 | 795 | 4.63 | 93 |
| 6 | 68 | 945 | 5.45 | 91 |
| 7 | 60 | 1092 | 6.24 | 89 |
| 8 | 53 | 1234 | 7.00 | 88 |

The elastic wave code because of its know properties for parallel computation was used as a test code for a package that aids in converting Fortran programs written for older, serial machines to programs that efficiently exploit the features of machines with advanced architectures. The software tools that aid in conversion are being incorporated into Toolpack. For more details on Toolpack and how it handles the data dependency analysis see [30] and [31]. In using a sequential version of the elastic wave code Toolpack correctly flagged those lines that had to be changed in order to parallelize the code. Upon making the required modifications, the elastic wave code was executed on a Sequent Balance 21000, a twenty four processor shared memory machine. Figure 6 shows the speedup curve for the elastic wave code running on 1 to 23 processors. It is interesting to note that the speedup curve is almost linear. This can be seen by comparing the actual speedup curve with the theoretical speedup curve (the dotted line) in Figure 6.

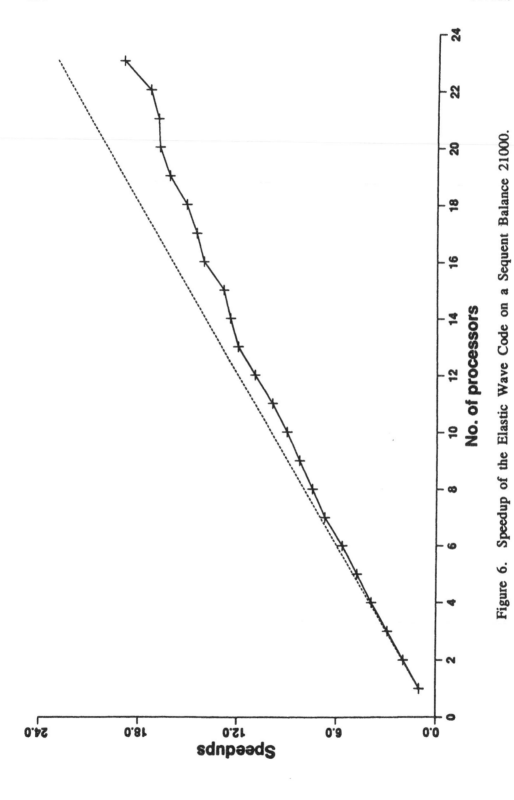

Figure 6.  Speedup of the Elastic Wave Code on a Sequent Balance 21000.

## 7 CONCLUSIONS

We have shown that the elastic wave code which is an example of dimensional splitting schemes is well suited for today's large-scale scientific computers. As we illustrated here with our code, not only are dimensional splitting schemes efficient on vector computers, they are relatively easily modified to take full advantage of shared memory parallel and parallel-vector machines such as the latest supercomputers. Efficiency on advanced scientific computers should not be the overriding issue. The mathematical formulation and subsequent implementation should also be amenable to produce accurate and stable solutions. The elastic wave code outlined here being second order accurate in time and fourth order accurate in space fits this criteria as well. The fluid-elastic model problem illustrates the robustness of this code.

Using the elastic wave code as an example, we have related several observations that can have implications for performance on many high-speed computer systems. The work presented is not intended to imply that one system is better than another. How a particular computer system handles one code is not conclusive of its overall merits. Certainly the cost/performance ratio is a factor that needs to be taken into account when judging the value of a high-speed computer system. No cost/performance ratio is given, nor is any implied.

Rather, our objective has been to show that certain hardware and software features can have an adverse effect on performance. Such features clearly warrant closer scrutiny. Code developers and numerical analysts can no longer have just a superficial acquaintance with the architecture of the high-speed computer system. They must now understand the many different components such as memory access pattern and how they work together. High performance cannot be achieved through mere use of compiler options but may require rewriting portions of code in order to fully and efficiently utilize the machine. Even with more sophisticated compilers, some "hand-coding" will be required for parallelizing codes for the foreseeable future.

One may argue that this situation is not new. What is important is that with the variety of architectures now available, different components of the architecture, hardware and software, can dramatically affect performance of a particular code. This is particularly true as one starts using multiprocessor vector machines.


## ACKNOWLEDGEMENTS

I thank M. Booth, C. Grassl, and R. Selva of Cray Research, Inc., for their assistance in microtasking this code and for showing me how to use the CRAY X-MP efficiently. Further, I thank the Benchmark Services Group of Cray Research, Inc., for providing access to the Crays at Mendota Heights.

I also acknowledge A. Bayliss for his many helpful discussions and support during the efforts to multitask the elastic wave program.

Finally, I thank my collaborators and colleagues in the Mathematics and Computer Science Division at Argonne National Laboratory for their help while I was visiting Argonne on a Department of Energy Laboratory Technology Exchange Program, and also for access to the Advanced Computing Research Facility at Argonne where all the Alliant FX/8 tests were run.


## REFERENCES

[1] K.E. Jordan, "Performance comparison of large-scale scientific computers: scalar mainframes, mainframes with vector facilities, and supercomputers," Computer, pp. 10-23, March 1987.

[2] D. Gottlieb, and E. Turkel, "Dissipative two-four method for time dependent problems," Math. Comp. vol. 30, pp. 703-723, 1976.

[3] L. A. Maestrello, A. Bayliss, and E. Turkel, "On interaction of a sound pulse with the shear layer of an axisymmetric jet," J. Sound Vib. vol. 74, pp. 281-301, 1981.

[4] A. Bayliss, K. E. Jordan, B. J. LeMesurier, and E. Turkel, "A fourth-order accurate finite-difference scheme for the computation of elastic waves," Bul. Seism. Soc. of Am., vol. 76, pp. 1115-1132, August 1986.

[5] K. E. Jordan, "On Multitasking of an Elastic Wave Code," in Advances in Computer Methods for Partial Differential Equations VI, R. Vichnevetsky and R. S. Stepleman (eds.), IMACS, New Brunswick, N.J., pp. 220-224, June 1987.

[6] K. E. Jordan, "The Good, the Bad, and the Ugly: Comparing High-Speed Computer Systems," Proc. 33rd IEEE Computer Society International Conference, COMPCON 88, San Francisco, pp 90-96, 1988.

[7] Z. S. Alterman, and F. C. Karal, Jr., "Propagation of Elastic Waves in Layered Media by Finite Difference Methods," Bull. Seism. Soc. Am., v. 58, pp. 367-398, 1968.

[8] K. R. Kelly et. al. , "Synthetic Seismograms: A Finite Difference Approach," Geophysics, v. 41, pp. 2-27, 1976.

[9] R. A. Stephen, "A Comparison of Finite Difference and Reflectivity Seismograms for Marine Models," Geophys. J. R. Astr. Soc. v. 72, pp. 39-58, 1983.

[10] R. A. Stephen and M. Burton, " Finite Difference Solution to the Elastic Wave Equation with Variable Coefficients," Preprint, 1984.

[11] R. J. Clifton, "A Difference Method for Plane Problems in Elasticity," Quart. Appl. Math., v. 25, pp 97-116, 1967.

[12] S. H. Emerman, et. al., "An Implicit Finite Difference Formulation of the Elastic Wave Equation," Geophysics, v. 47, pp. 1521-1526, 1982.

[13] E. Turkel, "Phase Error and Stability of Second Order Methods for Hyperbolic Problems," I. J. Comp. Phys., v. 15, pp. 226-250, 1974.

[14] J. Lysmer and L. A. Drake, "A Finite Element for Seismology," In Methods in Computational Physics, Academic Press, v. 11, pp. 181-216, 1972.

[15] K. J. Marfurt, "Accuracy of Finite Difference and Finite-Element Modeling of the Scalar and Elastic Wave Equation," Geophysics, v. 49, pp. 533-549, 1984.

[16] D. D. Kosloff et. al., "Elastic Wave Calculations by the Fourier Method," Bull. Seism. Soc. Am., v. 74, pp. 875-891, 1984.

[17] D. D. Kosloff and E. Baysal, "Forward Modeling by a Fourier Method," Geophysics, v. 47, pp. 1402-1412, 1982.

[18] O. G. Johnson, "Three-Dimensional Wave Equation Computations on Vector Computers," Proceedings of the IEEE, v. 72, pp. 90-95, 1984.

[19] A. Ilan and D. Loewenthal, "Instability of Finite Difference Schemes Due to Boundary Conditions in Elastic Media," Geophysical Prospecting, v. 24, pp. 431-453, 1976.

[20] H. O. Kreiss and J. Oliger, "Comparison of Accurate Methods for the Integration of Hyperbolic Systems," Tellus, v. 24, pp. 199-215, 1972.

[21] E. Turkel, "On the Practical Use of High-Order Methods for Hyperbolic Systems," J. Comp. Phys., v. 35, pp. 319-340, 1980.

[22] D. M. Boore, "Finite Difference Methods for Seismic Waves," In Methods in Computational Physics, Academic Press, v. 11, pp. 1-37, 1972.

[23] D. L. Brown, "A Note on the Numerical Solution of the Wave Equation with Piecewise Smooth Coefficients," Math. Comp., v. 42, pp. 369-391, 1984.

[24] Multitasking User's Guide, Cray Computer Systems Technical Note, Publication No. SN-0222, Cray Research Inc., March 1986.

[25] Alliant FX/Series Product Summary, Alliant Computer Systems Corporation, Acton MA, June 1985.

[26] W. Schonauer, **Scientific Computing on Vector Computers**, Private communication, 1987.

[27] W. Abu-Sufah, and A. D. Maloney, "Vector Processing on the Alliant FX/8 Multiprocessor," in Proceedings of 1986 Int. Conf. Par. Processing, K. Hwang et al., eds., IEEE Computer Society Press, Piscataway, N.J., pp. 559-566, 1986.

[28] F. B. Hanson, "Computational Dynamic Programming for Stochastic Optimal Control on a Vector Multiprocessor," Private communication, 1987.

[29] Autotasking User's Guide, Cray Computer Systems Technical Note, Publication No. SN-2088, Cray Research Inc., October 1988.

[30] W. R. Cowell and C. P. Thompson, "Tools to Aid in Discovering Parallelism and Localizing Arithmetic in Fortran Programs," Argonne National Laboratory preprint MCS-P6-0988, 1988.

[31] W. R. Cowell, "Users' Guide to Toolpack/1 Tools for Data Dependency Analysis and Program Transformation," Argonne National Laboratory report ANL-88-17, 1988.

# Generalized Gray Codes and Their Properties

LINDA S. BARASCH, S. LAKSHMIVARAHAN AND SUDARSHAN K. DHALL
Parallel Processing Institute, School of Electrical Engineering and Computer Science, University of Oklahoma, Norman, Oklahoma.

## 1 INTRODUCTION

With the widespread availability of distributed memory architectures based on the binary hypercube topology [1] [9], there is a growing interest in the portability of algorithms developed for architectures based on other topologies such as linear arrays, rings, multidimensional grids, (binary) trees, etc. to (binary) hypercube based architectures. Clearly, the question of portability of algorithms across architectures reduces to the question of embedding certain graphs in a target graph which in this case is a binary hypercube. It has been known for a long time that the general graph embedding problem is NP-complete [11]. Recently, it was shown [6] that the embedding of general graphs in binary hypercubes is also NP-complete. Further strengthening of this result indicates that the problem of embedding even arbitrary trees in a binary hypercube is NP-complete [7]. These results point to the fact that for easy embeddability, one must look for highly structured graphs. Indeed, embeddings of linear arrays, rings, multidimensional grids, binary (both complete and arbitrary) trees in a binary hypercube are well known [1] [2] [3] [5] [12]. A class of binary codes called binary reflected Gray codes [10] provides easy algorithms for embedding linear arrays, rings, and multidimensional grids in a binary hypercube [1] [2]. For a general account of the various facets of parallel processing using the Hypercube, refer to [14].

Recently, a generalized hypercube topology called the base-$b$ hypercube was introduced [3] [8]. Let $N = b^k$ for some $b \geq 2$ and $k \geq 1$. Let $<Z> = \{ 0, 1, 2, \cdots, Z-1 \}$ for some integer $Z \geq 1$. An $N$-node, base-$b$ hypercube of dimension $k$, called a $(N, b, k)$ cube for short, is a graph $G = (V, E)$, where

This research was supported in part by the Amoco Research Center in Tulsa, Oklahoma, the Energy Research Institute, University of Oklahoma and the Oklahoma Governor's Council on Science and Technology.

$$V = \{x \mid x = x_k x_{k-1} \cdots x_2 x_1 \ where \ x_i \in \ <b>, \ i = 1 \ to \ k\}$$

and

$$E = \{(x, y) \mid x, y \in V, \ x = x_k x_{k-1} \cdots x_2 x_1, \ y = y_k y_{k-1} \cdots y_2 y_1 \ and \ there$$

$$exists \ j, \ 1 \le j \le k, \ such \ that \ x_j \ne y_j \ and \ x_i = y_i \ for \ i \ne j\}$$

Examples of a $(4, 4, 1)$ cube, $(9, 3, 2)$ cube and $(8, 2, 3)$ cube are given in Figure 1.

The usefulness and importance of base-$b$ cubes [4] motivates the study of embedding standard topologies in a base-$b$ cube. To this end, a class of generalized Gray codes called base-$b$ reflected Gray codes was introduced in [3]. However, much of the analysis and developments in [3] are confined to base-$b$ reflected Gray codes where $b = 2^k$ for some $k \ge 1$. It turns out that Gray codes for general b have curious and interesting properties. For example, properties of Gray codes for odd and even b are quite different. While these generalized Gray codes have immediate applications to the embedding of graphs in a base-$b$ cube, the analysis of their properties holds independent interest as well. This paper is devoted to the analysis of the properties of generalized Gray codes. A thorough discussion of binary Gray codes is given in [10].

Section 2 of this paper provides various definitions of generalized Gray codes and brings out the intrinsic difference between the Gray codes with odd and even base. A discussion of the encoding and decoding of generalized Gray codes is included in Section 3. Section 4 summarizes a number of open problems related to generalized Gray codes.

## 2  GENERALIZED BASE-B GRAY CODES - DEFINITIONS

DEFINITION 2.1  Let $x = x_k x_{k-1} \cdots x_1$ and $y = y_k y_{k-1} \cdots y_1$ be two k-digit, base-$b$ integers, where $x_i, y_i \in \ <b>$. Then define the *base-b Hamming distance* between $x$ and $y$ as $h_b(x,y) = m$ if there exists $i_1 < i_2 < \cdots < i_m$ such that $x_{i_j} \ne y_{i_j}, \ j = 1 \ to \ m$ and $x_p = y_p, \ p \notin \{i_1, \ldots, i_m\}$.

That is, $h_b(x, y)$ represents the number of base-$b$ digit positions in which x and y differ.

EXAMPLE 2.1  $h_2(001, 110) = 3$,  $h_3(0122, 1120) = 2$,  $h_4(3210, 0123) = 4$.

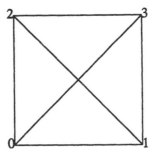

a.  Example of a (4,4,1) cube

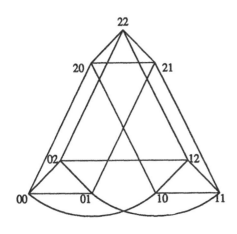

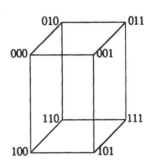

b. Example of a (9,3,2) cube

c.  Example of a (8,2,3) cube

Figure 1

DEFINITION 2.2 Let $N = b^k$. A *base-b Gray code* $G^{(b)}(k)$ is an *ordered* sequence $\{G_0^{(b)}(k), G_1^{(b)}(k), ..., G_{N-1}^{(b)}(k)\}$, of the N k-digit base-$b$ integers, where $G_i^{(b)}(k)$ is the $i^{th}$ code word and the ordering is such that

$$h_b(G_i^{(b)}(k), G_{i+1}^{(b)}(k)) = 1, \quad 0 \le i < N-1$$

EXAMPLE 2.2   Table 1 shows two examples of $G^{(2)}(4)$; Table 2 shows two examples of $G^{(3)}(3)$; Table 3 shows three examples of $G^{(4)}(3)$.

Referring to Table 1, notice that $h_b(G_0^{(b)}(k), G_{N-1}^{(b)}(k))$ equals 3 in example 1a and equals 1 in example 1b. This latter property motivates the following:

DEFINITION 2.3  Let $N = b^k$. A *base-b cyclic Gray code* $G^{(b)}(k)$ is a base-$b$ Gray code such that $h_b(G_0^{(b)}(k), G_{N-1}^{(b)}(k)) = 1$.

Of the examples in Tables 1 through 3, 1b, 2a, 3b and 3c are cyclic Gray codes.

As has been seen, for given values of $b$ and $k$, there exist many different base-$b$ Gray codes. However, there is a close relation between the cyclic Gray codes and a class of Gray codes called *reflected Gray codes*. Reflected Gray codes may be defined recursively and to this end we introduce the following definitions.

Table 1  -  Examples of $G^{(2)}(4)$

|     |         | 1a           | 1b           |
|-----|---------|--------------|--------------|
| i   | $[i]_2$ | $G_i^{(2)}(4)$ | $G_i^{(2)}(4)$ |
| 0   | 0000    | 0000         | 0000         |
| 1   | 0001    | 0010         | 0001         |
| 2   | 0010    | 1010         | 0011         |
| 3   | 0011    | 1011         | 0010         |
| 4   | 0100    | 1111         | 0110         |
| 5   | 0101    | 1101         | 0111         |
| 6   | 0110    | 1001         | 0101         |
| 7   | 0111    | 1000         | 0100         |
| 8   | 1000    | 1100         | 1100         |
| 9   | 1001    | 0100         | 1101         |
| 10  | 1010    | 0101         | 1111         |
| 11  | 1011    | 0001         | 1110         |
| 12  | 1100    | 0011         | 1010         |
| 13  | 1101    | 0111         | 1011         |
| 14  | 1110    | 0110         | 1001         |
| 15  | 1111    | 1110         | 1000         |

Table 2 - Examples of $G^{(3)}(3)$

| i | $[i]_3$ | 2a $G_i^{(3)}(3)$ | 2b $G_i^{(3)}(3)$ |
|---|---------|---------|---------|
| 0 | 000 | 000 | 000 |
| 1 | 001 | 001 | 001 |
| 2 | 002 | 002 | 002 |
| 3 | 010 | 012 | 012 |
| 4 | 011 | 010 | 011 |
| 5 | 012 | 011 | 010 |
| 6 | 020 | 021 | 020 |
| 7 | 021 | 022 | 021 |
| 8 | 022 | 020 | 022 |
| 9 | 100 | 120 | 122 |
| 10 | 101 | 122 | 121 |
| 11 | 102 | 121 | 120 |
| 12 | 110 | 111 | 110 |
| 13 | 111 | 110 | 111 |
| 14 | 112 | 112 | 112 |
| 15 | 120 | 102 | 102 |
| 16 | 121 | 100 | 101 |
| 17 | 122 | 101 | 100 |
| 18 | 200 | 201 | 200 |
| 19 | 201 | 202 | 201 |
| 20 | 202 | 212 | 202 |
| 21 | 210 | 210 | 212 |
| 22 | 211 | 211 | 211 |
| 23 | 212 | 221 | 210 |
| 24 | 220 | 222 | 220 |
| 25 | 221 | 220 | 221 |
| 26 | 222 | 200 | 222 |

Table 3 - Examples of $G^{(4)}(3)$

| | | 3a | 3b | 3c | | | 3a | 3b | 3c |
|---|---|---|---|---|---|---|---|---|---|
| $i$ | $[i]_4$ | $G_i^{(4)}(3)$ | $G_i^{(4)}(3)$ | $G_i^{(4)}(3)$ | $i$ | $[i]_4$ | $G_i^{(4)}(3)$ | $G_i^{(4)}(3)$ | $G_i^{(4)}(3)$ |
| 0 | 000 | 000 | 000 | 222 | 32 | 200 | 111 | 200 | 322 |
| 1 | 001 | 002 | 001 | 221 | 33 | 201 | 112 | 201 | 321 |
| 2 | 002 | 001 | 002 | 223 | 34 | 202 | 113 | 202 | 323 |
| 3 | 003 | 003 | 003 | 220 | 35 | 203 | 110 | 203 | 320 |
| 4 | 010 | 103 | 013 | 210 | 36 | 210 | 310 | 213 | 310 |
| 5 | 011 | 101 | 012 | 213 | 37 | 211 | 311 | 212 | 313 |
| 6 | 012 | 102 | 011 | 211 | 38 | 212 | 313 | 211 | 311 |
| 7 | 013 | 100 | 010 | 212 | 39 | 213 | 312 | 210 | 312 |
| 8 | 020 | 300 | 020 | 232 | 40 | 220 | 322 | 220 | 332 |
| 9 | 021 | 303 | 021 | 231 | 41 | 221 | 321 | 221 | 331 |
| 10 | 022 | 302 | 022 | 233 | 42 | 222 | 323 | 222 | 333 |
| 11 | 023 | 301 | 023 | 230 | 43 | 223 | 320 | 223 | 330 |
| 12 | 030 | 201 | 033 | 200 | 44 | 230 | 020 | 233 | 300 |
| 13 | 031 | 202 | 032 | 203 | 45 | 231 | 021 | 232 | 303 |
| 14 | 032 | 200 | 031 | 201 | 46 | 232 | 022 | 231 | 301 |
| 15 | 033 | 203 | 030 | 202 | 47 | 233 | 023 | 230 | 302 |
| 16 | 100 | 223 | 130 | 102 | 48 | 300 | 123 | 330 | 002 |
| 17 | 101 | 222 | 131 | 101 | 49 | 301 | 121 | 331 | 001 |
| 18 | 102 | 221 | 132 | 103 | 50 | 302 | 120 | 332 | 003 |
| 19 | 103 | 220 | 133 | 100 | 51 | 303 | 122 | 333 | 000 |
| 20 | 110 | 230 | 123 | 130 | 52 | 310 | 132 | 323 | 030 |
| 21 | 111 | 233 | 122 | 133 | 53 | 311 | 133 | 322 | 033 |
| 22 | 112 | 232 | 121 | 131 | 54 | 312 | 131 | 321 | 031 |
| 23 | 113 | 231 | 120 | 132 | 55 | 313 | 130 | 320 | 032 |
| 24 | 120 | 211 | 110 | 112 | 56 | 320 | 030 | 310 | 012 |
| 25 | 121 | 212 | 111 | 111 | 57 | 321 | 032 | 311 | 011 |
| 26 | 122 | 210 | 112 | 113 | 58 | 322 | 033 | 312 | 013 |
| 27 | 123 | 213 | 113 | 110 | 59 | 323 | 031 | 313 | 010 |
| 28 | 130 | 013 | 103 | 120 | 60 | 330 | 331 | 303 | 020 |
| 29 | 131 | 012 | 102 | 123 | 61 | 331 | 333 | 302 | 023 |
| 30 | 132 | 010 | 101 | 121 | 62 | 332 | 332 | 301 | 021 |
| 31 | 133 | 011 | 100 | 122 | 63 | 333 | 330 | 300 | 022 |

**DEFINITION 2.4** A *fundamental ordering* $\alpha$ is a permutation of $< b >$. That is, $\alpha(i) = j$, $i, j \in < b >$ and $\alpha(i) = \alpha(j)$ if and *only* if $i = j$.

The *reflected fundamental ordering* $\beta$ of $\alpha$ is a permutation of $< b >$ such that $\beta(i) = \alpha(b - i - 1)$.

In the definition of reflected Gray code that follows, for the purposes of illustration, the fundamental ordering $\iota(i) = i$ (identity permutation) will be assumed. All the results carry over to any choice of fundamental ordering $\alpha$.

**DEFINITION 2.5** A base-$b$ reflected Gray code $G^{(b)}(k) = \{G_0, \cdots, G_{b^k -1}\}$ is an

ordered sequence of length $b^k$, where the $i^{th}$ element of the sequence is the $i^{th}$ codeword.
    As the basis for recursion, let

$$G^{(b)}(1) = \{0, 1, \cdots, b-2, b-1\}.$$

Let $G_i$ denote the $i^{th}$ codeword of $G^{(b)}(k)$. Then define, for $b$ even:

$$G^{(b)}(k+1) = \{0G_0, 0G_1, 0G_2, \cdots, 0G_{b^k-1},$$

$$1G_{b^k-1}, 1G_{b^k-2}, \cdots, 1G_0,$$

$$2G_0, 2G_1, \cdots, (b-2)G_{b^k-2}, (b-2)G_{b^k-1},$$

$$(b-1)G_{b^k-1}, (b-1)G_{b^k-2}, \cdots, (b-1)G_1, (b-1)G_0\}$$

and for $b$ odd:

$$G^{(b)}(k+1) = \{0G_0, 0G_1, 0G_2, \cdots, 0G_{b^k-1},$$

$$1G_{b^k-1}, 1G_{b^k-2}, \cdots, 1G_0$$

$$2G_0, 2G_1, \cdots, (b-2)G_1, (b-2)G_0, (b-1)G_0,$$

$$(b-1)G_1, \cdots, (b-1)G_{b^k-2}, (b-1)G_{b^k-1}\}$$

Let $G^{(b)}(k)^R$ be $G^{(b)}(k)$ reflected (reverse ordering of $G^{(b)}(k)$), that is, $G_i^{(b)}(k)^R = G_{b^k-i-1}^{(b)}(k)$. Then Definition 2.5 may also be recursively written as:

$$G^{(b)}(k+1) = \{0G^{(b)}(k), 1G^{(b)}(k)^R, 2G^{(b)}(k), \cdots,$$

$$(b-2)G^{(b)}(k), (b-1)G^{(b)}(k)^R\}$$

if $b$ is even, and

$$G^{(b)}(k+1) = \{0G^{(b)}(k), 1G^{(b)}(k)^R, 2G^{(b)}(k), \cdots,$$

$$(b-2)G^{(b)}(k)^R, (b-1)G^{(b)}(k)\}$$

if $b$ is odd.
    An alternate recursive definition of these base-$b$ reflected Gray codes is possible. If

$$G^{(b)}(k) = \{G_0, G_1, G_2, \cdots, G_{b^k-1}\}$$

is the k-digit base-$b$ code, then $G^{(b)}(k+1)$ is defined:

$$G^{(b)}(k+1) = \{G_00, G_01, G_02, \cdots, G_0(b-1), G_1(b-1), G_1(b-2), \cdots, G_11,$$

$$G_10, G_20, G_21, G_22, \cdots\}$$

Notice that $G^{(b)}(1)$ is essentially $\iota$, the identity permutation of $<b>$. Although the definitions given for reflected Gray codes was based on the fundamental ordering $\iota(i) = i$, corresponding definitions exist for any fundamental ordering. In Tables 1 through 3, 1b, 2b, 3b and 3c are examples of reflected Gray codes. 1b is based on the fundamental ordering $\{0,1\}$, 2b on the ordering $\{0,1,2\}$, 3b on the ordering $\{0,1,2,3\}$, and 3c on the ordering $\{2,1,3,0\}$. Notice that 1b, 3b and 3c are cyclic Gray codes, while 2b is not. In fact, whether a reflected Gray code is a cyclic Gray code depends critically on whether b is odd or even. The following two theorems establish the relationship between cyclic Gray codes and reflected Gray codes. To make the presentation simple, the fundamental ordering $\iota(i) = i$ is assumed. This assumption, however, is inessential for the proof.

THEOREM 2.1  When b is even, a base-$b$ reflected Gray code is always a cyclic Gray code.

Proof: The proof is by induction on $k$. As a basis, observe that for $k = 1$,

$$G^{(b)}(1) = \{\,0, 1, \cdots, b-1\,\}$$

and

$$h_b(0, b-1) = 1.$$

Thus, $G^{(b)}(1)$ is cyclic.
    Now assume $G^{(b)}(k)$ is cyclic.

$$G_0^{(b)}(k+1) = 0G_0^{(b)}(k)$$

$$G_{b^{k+1}-1}^{(b)}(k+1) = (b-1)G_{b^k-1}^{(b)}(k)^R = (b-1)G_0^{(b)}(k)$$

Thus $h_b(G_0^{(b)}(k+1), G_{b^{k+1}-1}^{(b)}(k+1)) = 1.$

THEOREM 2.2  When $b$ is odd and $k \geq 2$, a base-$b$ reflected Gray code is *not* a cyclic Gray code.

Proof:  Again, the proof is by induction on $k$. As a basis, observe that for $k = 2$,

$$G^{(b)}(2) = \{\,00, 01, ..., 0(b-1), 1(b-1), ...,$$

$$10, 20, ..., 2(b-1), ..., (b-1)0, ..., (b-1)(b-1)\,\}$$

and

$$h_b(00, (b-1)(b-1)) = 2.$$

Hence $G^{(b)}(2)$ is not cyclic.
    Now, assume $G^{(b)}(k)$ is not a cyclic Gray code for some $k \geq 2$. That is,

$$h_b(G_0^{(b)}(k), G_{b^k-1}^{(b)}(k)) = r, \quad r > 1.$$

Then

$$G_0^{(b)}(k+1) = 0G_0^{(b)}(k).$$

$$G_{b^{k+1}-1}^{(b)}(k+1) = (b-1)G_{b^k-1}^{(b)}(k),$$

and

$$h_b(0G_0^{(b)}(k), (b-1)G_{b^k-1}^{(b)}(k)) = r+1 > 1.$$

So $G^{(b)}(k+1)$ is not cyclic.

Although a reflected Gray code for odd $b$ in general is not cyclic, a Gray code can be defined which is. To this end, we need the following.

**DEFINITION 2.6** Define the *shift operator,* $\sigma$, which when applied to an ordered sequence of codewords, causes a *cyclic* shift of the codewords in the ordered sequence one position to the left. Define $\sigma_i$ to be a unary operator which causes a *cyclic* shift of the rightmost $i$ codewords in the sequence one position to the left.
Note that if the length of the sequence is $N$, then $\sigma = \sigma_N$.

**EXAMPLE 2.3** Let $G^{(5)}(1) = \{0, 1, 2, 3, 4\}$. Then $\sigma [G^{(5)}(1)] = \{1, 2, 3, 4, 0\}$ and $\sigma_2 [G^{(5)}(1)] = \{0, 1, 2, 4, 3\}$.
Lemma 2.3 follows readily from the definitions:

**LEMMA 2.3** If $G^{(b)}(k)$ is a cyclic Gray code, then $\sigma[G^{(b)}(k)]$ is a cyclic Gray code.

The following theorem provides a natural definition of cyclic Gray codes for odd $b > 1$.

**THEOREM 2.4** For odd b, $b \geq 3$, the following recursive definition provides an algorithm for generating a cyclic Gray code:

$$G^{(b)}(1) = \{0, 1, 2, \cdots, (b-1)\}$$

$$G^{(b)}(k+1) = \{0G^{(b)}(k), 1G^{(b)}(k)^R, 2G^{(b)}(k), \cdots, (b-3)G^{(b)}(k),$$

$$(b-2)\sigma_2[G^{(b)}(k)^R], (b-1)\sigma[G^{(b)}(k)]\}$$

Proof: It is clear that the above definition leads to a cyclic Gray code as long as $\sigma_2[G^{(b)}(k)]$ is still a Gray code. To show this, we first show by induction on k that the first three codewords of $G^{(b)}(k)$ by this construction are $0^{k-1}0$, $0^{k-1}1$, and $0^{k-1}2$. Let $k = 1$. Then $G^{(b)}(1) = \{0, 1, 2,...\}$ and the claim is true.
Assume that the claim is true for k, then

$$G^{(b)}(k+1) = 0G^{(b)}(k)...$$

$$= 00^{k-1}0, 00^{k-1}1, 00^{k-1}2,...$$

$$= 0^k0, 0^k1, 0^k2,...$$

Thus, since $G^{(b)}(k)^R = \{ \cdots 0^{k-1}2, 0^{k-1}1, 0^{k-1}0\}$,

$$\sigma_2[G^{(b)}(k)^R] = \{ \cdots 0^{k-1}2, 0^{k-1}0, 0^{k-1}1 \}.$$

Since $G^{(b)}(k)^R$ is a Gray code, so is $\sigma_2[G^{(b)}(k)^R]$ Note, however, that it is not necessary for $\sigma_2[G^{(b)}(k)^R]$ to be cyclic. Notice that the last code word in $(b-2)\sigma_2 [G^{(b)}(k)^R]$ is $(b-2)0^{k-1}1$ and the first code word in $(b-1)\sigma[G^{(b)}(k)]$ is $(b-1)0^{k-1}1$. Combining this with Lemma 2.3, the theorem follows.

In Table 2, column 2a is a cyclic Gray code formed by the construction given in Theorem 2.4.

## 3 ENCODING AND DECODING OF BASE-B GRAY CODES

So far, definitions of reflected Gray codes have been purely recursive. This makes it difficult to determine $G_i^{(b)}(k)$ for a given $i$, or given a codeword, the index $i$ that corresponds to it. These questions clearly relate to the *encoding* and *decoding* of Gray codes. Let $E : < b > \longrightarrow G^{(b)}(k)$, where, $E(i) = G_i^{(b)}(k)$, be the encoding function, and $D : G^{(b)}(k) \longrightarrow < b >$, where $D(G_i^{(b)}(k)) = i$, be the decoding function. As may be expected, it turns out that the encoding and decoding are different for even and odd b. The following definitions of encoding and decoding functions are consequences of the properties of reflected base-$b$ Gray codes.

DEFINITION 3.1 Let $G^{(b)}(k)$ be a reflected Gray code, with $b$ even. Let

$$0 \le [i]_{10} < b^k,$$

$$[i]_{10} = [0 b_k b_{k-1} \cdots b_2 b_1]_b,$$

and

$$G_i^{(b)}(k) = [g_k g_{k-1} \cdots g_2 g_1]_b.$$

Then the *Encoding* function $E : i \longrightarrow G_i^{(b)}(k)$ is such that

$$g_i = f(b_i), \quad i = 1, 2, ..., k.$$

$$where \ f = \begin{cases} \alpha \ \text{if} \ (b_{i+1}) \bmod 2 \equiv 0 \\ \beta \ \text{if} \ (b_{i+1}) \bmod 2 \equiv 1 \end{cases}$$

The *Decoding* function $D : G_i^{(b)}(k) \longrightarrow i$ is such that

$$b_{k+1} = 0,$$

$$b_{k-i} = f^{-1}(g_{k-i}), i = 0, \cdots, k-1$$

$$where \quad f = \begin{cases} \alpha \ \text{if} \ (b_{k-i+1}) \bmod 2 \equiv 0 \\ \beta \ \text{if} \ (b_{k-i+1}) \bmod 2 \equiv 1 \end{cases}$$

Notice that while all bits of encoding might trivially be done in parallel, decoding in parallel is not obvious, since $f^{-1}(b_i)$ cannot be calculated until $b_{i+1}$ is decoded (which in turn depends on $b_{i+2}$, etc.).

EXAMPLE 3.1 Consider $G^{(4)}(3)$ given in Table 3b. Here $\alpha = (0, 1, 2, 3)$ and $\beta = (3, 2, 1, 0)$.

*Encoding:*

Let $i = 20$. Then $[20]_4 = 0110 = (0b_3b_2b_1)$.

Thus, $g_3 = \alpha(1) = 1$, $g_2 = \beta(1) = 2$, $g_1 = \beta(0) = 3$. So, $G_{20}^{(4)}(3) = 123$.

*Decoding:*

Let $G_i^{(4)}(3) = 233 = (g_3g_2g_1)$.

Thus, $b_4 = 0$, $b_3 = \alpha^{-1}(2) = 2$, $b_2 = \alpha^{-1}(3) = 3$, $b_1 = \beta^{-1}(3) = 0$. So, $[i]_4 = 0230$, $i = 44$.

EXAMPLE 3.2 Consider $G^{(4)}(3)$ given in Table 3c. Here $\alpha = (2, 1, 3, 0)$ and $\beta = (0, 3, 1, 2)$.

*Encoding:*

Let $i = 20$. $[20]_4 = 0110 = (0b_3b_2b_1)$.

Thus, $g_3 = \alpha(1) = 1$, $g_2 = \beta(1) = 3$, $g_1 = \beta(0) = 0$. So, $G_{20}^{(4)}(3) = 130$.

*Decoding:*

Let $G_i^{(4)}(3) = 233 = (g_3g_2g_1)$.

Thus, $b_4 = 0$, $b_3 = \alpha^{-1}(2) = 0$, $b_2 = \alpha^{-1}(3) = 2$, $b_1 = \beta^{-1}(3) = 2$. So, $[i]_4 = 0022$, $i = 10$.

DEFINITION 3.2 Let $G^{(b)}(k)$ be a reflected Gray code, with $b$ odd. Let

$$0 \leq [i]_{10} < b^k$$

$$[i]_{10} = [0b_kb_{k-1} \cdots b_2b_1]_b$$

and

$$G_i^{(b)}(k) = [g_kg_{k-1} \cdots g_2g_1]_b.$$

Then the *Encoding* function E: $i \rightarrow G_i^{(b)}(k)$ is such that

$$g_i = f(b_i) \quad i = 1, 2, ..., k.$$

$$where \quad f = \begin{cases} \alpha & if \left[ \sum_{j=i+1}^{k+1} b_j \right] mod\ 2 \equiv 0 \\ \\ \beta & if \left[ \sum_{j=i+1}^{k+1} b_j \right] mod\ 2 \equiv 1 \end{cases}$$

The *Decoding* function D: $G_i^{(b)}(k) \to i$ is such that

$$b_{k+1} = 0.$$

$$b_{k-i} = f^{-1}(g_{k-i}), \quad i = 0, ..., k-1.$$

$$where \quad f = \begin{cases} \alpha & if \left[ \sum_{j=k-i+1}^{k+1} b_j \right] mod\ 2 \equiv 0 \\ \\ \beta & if \left[ \sum_{j=k-i+1}^{k+1} b_j \right] mod\ 2 \equiv 1 \end{cases}$$

EXAMPLE 3.3 Consider $G^{(3)}(3)$ given in Table 2b. Here $\alpha = (0, 1, 2)$ and $\beta = (2, 1, 0)$.

*Encoding:*

Let $i = 15$. $[15]_3 = 0120 = (0b_3b_2b_1)$.

Thus, $g_3 = \alpha(1) = 1$, $g_2 = \beta(2) = 0$, $g_1 = \beta(0) = 2$. So, $G_{15}^{(3)}(3) = 102$.

*Decoding:*

Let $G_i^{(3)}(3) = 121 = (g_3g_2g_1)$.

Thus, $b_4 = 0$, $b_3 = \alpha^{-1}(1) = 1$, $b_2 = \beta^{-1}(2) = 0$, $b_1 = \beta^{-1}(1) = 1$. So, $[i]_3 = 0101$, $i = 10$.

## 4  CONCLUSION

This paper is a shortened version of a technical report [13], wherein a number of other topics including transition sequences, and properties of subsequences of base-*b* Gray codes are also described. We conclude by mentioning some open questions about base-*b* Gray codes. For example, it is not known whether the decoding of $G^{(b)}(k)$ might be parallelizable. It should be interesting to develop an alternate definition of cyclic Gray codes for odd bases that permits elegant encoding and decoding functions. Gray codes have enormous potential in helping map certain topologies onto hypercubes, as well as purely theoretical interest. The more the properties of Gray codes are understood, the greater their potential.

## ACKNOWLEDGEMENT

We thank an anonymous referee whose suggestions helped to tie up a few loose ends. Thanks are also due to Professor J.C. Diaz who invited us to submit this paper for this volume.

## REFERENCES

1. Y. Saad and M. H. Schultz, Some Topological Properties of the Hypercube Multiprocessor, RR-389, Department of Computer Science, Yale University (1984).
2. S. L. Johnsson, Odd-Even Cyclic Reduction on Ensemble Architectures and the Solution of Tri-diagonal Systems of Equations, RR-339, Department of Computer Science, Yale University (1985).
3. S. Lakshmivarahan and S.K. Dhall, A New Hierarchy of Hypercube Interconnection Scheme for Parallel Computers: Theory and Applications, OU-PPI-TR 86-02, Parallel Processing Institute, School of EECS, University of Oklahoma (August 1986). (Also to appear in the Journal of Supercomputing (1988).)
4. S. Lakshmivarahan and S.K. Dhall, A Lower Bound on the Communication Complexity in Solving Linear Tri-diagonal Systems on Cube Architectures, Second SIAM Conf. on Hypercube Multiprocessors, *Conf. Proc.*, pp 560-568 (1987).
5. S. N. Bhatt and E. C. F. Ipsen, How to Embed Trees in Hypercubes, RR-443, Department of Computer Science, Yale University (1985).
6. G. Cybenko, K. N. Venkataraman, and D. W. Krumme, Hypercube Embedding is NP-Complete, First SIAM Conference on Hypercube Multiprocessors, *Conf. Proc.*, pp 148-160, (1986).
7. A. Wagner and D. G. Corneil, Embedding Trees in a Hypercube is NP-Complete, Technical Report 197/87, Department of Computer Science, University of Toronto (1987).
8. L. N. Bhuyan and D. P. Agrawal, Generalized Hypercube and Hypercube Structures for a Computer Network, *IEEE Trans. Computers*, Vol 33, pp 323-333 (1984).
9. C. L. Seitz, The Cosmic Cube, *Communications of ACM*, Vol 28, pp 22-33 (1985).
10. E. M. Reingold, J. Nievergelt, and N. Deo, *Combinatorial Algorithms*, Prentice Hall (1977).

11. M. R. Garey and D. S. Johnson, *Computers and Intractability - A Guide to the Theory of NP-Completeness*. W.H. Freeman, San Francisco (1979).
12. J. E. Brandenberg and D. S. Scott, Embedding of Communication Trees and Grids in Hypercubes, Technical Report, Intel Scientific Computers, Beaverton, Oregon (1985).
13. L.S. Barasch, S. Lakshmivarahan and S.K. Dhall, Generalized Gray Codes and Their Properties, Technical Report, Parallel Processing Institute, School of EECS, University of Oklahoma (June 1987).
14. S. Lakshmivarahan and S.K. Dhall. *Analysis and Design of Parallel Algorithms (to appear)*.

# Nested Block Factorization Preconditioners
# for Convective-Diffusion Problems in Three Dimensions

G. K. LEAF and M. MINKOFF  Mathematics and Computer Science Division, Argonne National Laboratory, Argonne, Illinois

J. C. DÍAZ  Center for Parallel and Scientific Computing, University of Tulsa, Tulsa, Oklahoma

## 1  INTRODUCTION

The solution of the linear system that arises from the discretization of a partial differential equation (PDE) can be accomplished by direct or iterative methods. Direct methods can lead to high computational complexity and to considerable memory requirements, both of which limit the practical size of the discretization that can be used to solve the PDE. These memory requirements arise from fill-in in forming the matrix factorization and from pivoting which may be required in solving other than positive definite symmetric problems. As an alternative approach iterative methods for solving

Work supported in part by the Applied Mathematical Sciences subprogram of the Office of Energy Research, U.S. Department of Energy, under contract W-31-109-Eng-38, and National Science Foundation Grant RII-OK-8610676 (Task 10).

the linear system have been considered. Iterative methods permit the maintenance of the sparsity pattern, reducing storage requirements while decreasing computation time as well.

A variety of techniques use a combination of direct and iterative approaches to obtain the advantages of both methods. In these techniques an approximation of the factorization of the matrix or an approximation of the inverse of the matrix is used as a preconditioner. An iterative approach is then used to deal with the resulting preconditioned linear system. The primary purpose of using the preconditioner is to accelerate the convergence in the iterative phase of the computation. Thus, these combined techniques have the advantage that the preconditioner selected circumvents the storage limitations of direct methods and yet provides a system that the iterative techniques can solve in relatively few iterations (so that the iterative phase can be conducted relatively rapidly).

Our intent is to develop several types of preconditioners based on a nested factorization approach for a class of matrices that typically arise from finite difference approximations to general convective-diffusion PDEs. Our interest in this approach stems from experiences with the use of direct methods to solve linear systems associated with a method-of-lines approach to solving systems of PDEs for kinetic-diffusion models [1]. Such linear systems which arise from general convective-diffusion PDEs are nonsymmetric and can be indefinite but nonsingular. We consider a convective-diffusion problem with which we can study both nonsymmetric and indefinite linear systems. In constructing this problem we intend to provide a class of matrices that typically arise in scientific and engineering applications so that it can be used to evaluate the methods developed in this paper.

We consider a time-dependent evolution equation of convective-diffusion type in three spatial dimensions. If an implicit time-differencing scheme is used, then at each time step a PDE of the following type can be generated [2]:

$$- \nabla^2\phi + \vec{V}(x,y,z) \cdot \nabla\phi = F(x,y,z) \; . \tag{1.1a}$$

Consider this equation over the unit cube subject to the boundary conditions

$$\frac{\partial \phi}{\partial n} = 0 \quad \text{on all four vertical faces} \tag{1.1b}$$

$$\phi = 1 \quad \text{on bottom} \tag{1.1c}$$

$$\phi = 2 \quad \text{on top} . \tag{1.1d}$$

Here the velocity field $\vec{V}(x,y,z) = (V^x, V^y, V^z)(x,y,z)$ and the source field $F(x,y,z)$ are specified functions of position. For this study we use the following class of functions:

$$V^x = \alpha^x x(1 - x)y(1 - y)z \tag{1.2a}$$

$$V^y = \alpha^y xyz^2 \tag{1.2b}$$

$$V^z = \alpha^z x(1 - x)y(1 - y)z \tag{1.2c}$$

$$F = x^2 yz , \tag{1.2d}$$

where $\alpha^x$, $\alpha^y$, and $\alpha^z$ are specified constants. This problem describes a flow field. Setting these parameters to zero produces a positive-definite symmetric linear system. Using positive values for these parameters produces a more skewed and potentially non-symmetric indefinite system.

As an example of the typical discretization procedure, we have employed the usual donor cell finite-difference approximation (see [3]). This approximation generates a seven-point finite-difference approximation that leads to a linear system having a non-symmetric block tridiagonal coefficient matrix (to be discussed later). We use a uniform discretization in each of the three coordinate directions to obtain an associated linear system.

In the area of preconditioners we study the incomplete *LU* factorization, the modified ILU factorization, row-oriented and column-oriented nested incomplete factorizations, and forms of block factorizations with incomplete approximate inverses. The iterative method we use is normal-form conjugate gradient. Normal form was selected because it is guaranteed to handle indefinite systems. It should be noted that the normal equations are never formed.

We provide results of combining these preconditioners and the iterative method by comparing the preconditioned spectrum, computational time, and iteration histories for specific cases. Finally, we comment about possible further work with respect to vector and parallel computation.

## 2 OVERVIEW OF THE APPROACH

As described above, we shall consider approaches that involve the use of preconditioners for the linear system combined with the iterative method for solving the resulting preconditioned problem. The objective of the preconditioner is to approximate the inverse of the system in some appropriate sense.

We denote the linear system that arises from the PDE as

$$Ax = h \, , \tag{2.1}$$

where $A$ is the matrix of coefficients, $h$ is known, and $x$ is to be determined. With lexicographic ordering, the form of the matrix $A$ is a seven-stripe matrix $(e,d,c,a,b,f,g)$ where each stripe is parallel to the main diagonal — $a$:

$$A = \begin{bmatrix} a_1 & b_1 & 0 & f_1 & 0 & g_1 & 0 & 0 \\ c_2 & \cdot & \cdot & 0 & \cdot & 0 & \cdot & 0 \\ 0 & \cdot & \cdot & \cdot & 0 & \cdot & 0 & g_{\bar{k}} \\ d_{\underline{i}} & 0 & \cdot & \cdot & \cdot & 0 & \cdot & 0 \\ 0 & \cdot & 0 & \cdot & \cdot & \cdot & 0 & f_{\bar{j}} \\ e_{\underline{k}} & 0 & \cdot & 0 & \cdot & \cdot & \cdot & 0 \\ 0 & \cdot & 0 & \cdot & 0 & \cdot & \cdot & b_{n-1} \\ 0 & 0 & e_n & 0 & d_n & 0 & c_n & a_n \end{bmatrix} , \tag{2.2}$$

where $0$ denotes blocks of stripes of zeros, $\cdot$ denotes continuation of the elements of the vector along that diagonal, $\underline{i}$ denotes the beginning index for the $d$ vector, $\bar{j}$ denotes ending index for the $f$ vector, $\underline{k}$ denotes the beginning index for the $e$ vector, and $\bar{k}$ denotes ending index for the $g$ vector. The $c$ and $b$ stripes are the sub- and superdiagonals immediately adjacent to $a$, thus forming a tridiagonal system. The $c$ and $b$ vectors also have few zero elements at places corresponding to discretizations near the boundary. For efficiency we store those zero entries. The $d$ and $f$ vectors are sub- and superdiagonals arising from the discretization associated with the second spatial variable. The $e$ and $g$ vectors are further away from $a$ and arise from the discretization of the third spatial variable.

An alternative description of the form of the matrix (which will be useful later) is

$$A = E + D + C + V + B + F + G \tag{2.3}$$

where $V$ is a diagonal matrix, the matrices $E$, $D$, and $C$ are strictly lower triangular matrices, and the matrices $B$, $F$, and $G$, are strictly upper triangular matrices. In particular, the main diagonal of $V$ is $a$. The appropriate subdiagonals of $E$, $D$, and $C$ are $e$, $d$, and $c$, respectively. The appropriate superdiagonals of $G$, $F$, and $B$ are $g$, $f$, and $b$, respectively. The remaining elements of $E$, $D$, $C$, $G$, $F$, and $B$ are zero.

Our primary concern is with the development of preconditioners of the block nested factorization type. We shall evaluate the performance of these preconditioners by using them in a particular iterative scheme. This scheme will be a variant of a preconditioned normal-form conjugate gradient algorithm described in [4]. This variant can be described in general terms as follows. Given a linear system as in Eq. (2.1), we seek a preconditioner P with at least the following two general properties. First, the linear system $Px = f$ is easy to solve in the sense that it is fast, stable, and does not require excessive storage. Second, the matrix P is a good approximate inverse for A in the sense that $P^{-1}A \approx I$ and $A^T P^{-T} \approx I$. Here we have chosen to consider left approximate inverses and measure the quality of the approximation of the identity in terms of the eigenvalue spectrum of the above matrix products. If we set $Z = P^{-1}A$ and $g = P^{-1}h$, Eq. (2.1) can be rewritten in the form

$$Z^T Z x = Z^T g . \tag{2.4}$$

The left preconditioned normal form conjugate-gradient algorithm [4] is then

Step 1: Initialization

Begin with an initial solution estimate $x_0$ .

Specify a required solution accuracy $\varepsilon$ .

Calculate the initial residual $R_0 = h - A x_0$ .

Form the scaled residual $r_0 = Z^T P^{-1} R_0$ .

Set the initial search direction $p_0 = r_0$ .

Step 2: Iterate until a norm of the residual of the original system $\|R_i\| = \|A x_i - h\| < \varepsilon$ .

$$a_i = \frac{r_i^T r_i}{p_i^T Z^T Z p_i}$$

$$x_{i+1} = x_i + a_i p_i$$

$$r_{i+1} = r_i - a_i Z^T Z p_i$$

$$b_i = \frac{r_{i+1}^T r_{i+1}}{r_i^T r_i}$$

$$p_{i+1} = r_{i+1} + b_i p_i$$

The quality of $\mathbf{P}^{-1}$ as an approximate inverse to $\mathbf{A}$ determines the rate of convergence of this procedure, whereas the ease of solving a linear system manifests itself in the formation of the initial scaled residual

$$r_0 = \mathbf{Z}^T \mathbf{P}^{-1} R_0 \tag{2.5a}$$

and the vectors

$$s_i = \mathbf{Z}^T \mathbf{Z} p_i \qquad u_i = \mathbf{Z} p_i . \tag{2.5b}$$

For example, $u_i$ is determined from $\mathbf{P} u_i = \mathbf{A} p_i$ and $s_i$ is determined from $\mathbf{P}^T v_i = u_i$ and $s_i = \mathbf{A}^T v_i$. These steps are, in fact, the major computational costs in the above algorithm. In Section 6 we shall elaborate on the procedure involved in solving these equations from the viewpoint of the preconditioners developed in this paper.

Note that we monitor the true residual ($\|R_i\|$) rather than the norm of the residual of the preconditioned system ($\|r_i\|$). We do this so that we can compare results of the iteration process for several preconditioners. By using the true residual, the error norm being used is not dependent on the choice of preconditioner. The above left preconditioned normal-form conjugate-gradient algorithm is the vehicle used to evaluate the preconditioners in this study. This approach has been found to be effective for our class of model problems [4]. By using the normal form of the preconditioned linear system, the standard conjugate gradient method can be applied. While this approach does lead to increasing the condition number and the operation count per iteration, it provides a generally robust technique for indefinite systems. In [2] we studied the use of ORTHO-MIN(7) and ORTHODIR(7) as alternative iterative methods. While both approaches treat nonsymmetric systems, the truncated forms cannot be assured to converge for general indefinite systems. We have therefore restricted our attention to the normal-form conjugate-gradient approach.

In the following sections we focus on the choice of preconditioners used in our study. We shall consider three groups of preconditioners:

1. incomplete LU factorizations,
2. nested block incomplete LU factorizations, and
3. nested block incomplete LU factorizations with approximate inverses.

## 3   INCOMPLETE LU FACTORIZATIONS

A straightforward approach to solving the linear system (2.1) would be to use a direct solver. In particular, we would perform Gaussian elimination to produce an *LU* factorization of the matrix A. This approach, however, is well known to be computationally expensive and produces fill-in which leads to large storage requirements. Thus an obvious initial choice for a preconditioner would be to take the same approach but not allow the factorization to produce any fill-in. We shall refer to this preconditioner as incomplete factorization (ILU). The general form of this preconditioner can be written as

$$P = (\bar{L} + \bar{V})(I + \bar{U}) \tag{3.1}$$

where $\bar{L}$ is a strictly lower triangular matrix, $\bar{V}$ is a diagonal matrix, and $\bar{U}$ is a strictly upper triangular matrix.

There are, however, pitfalls in this simple approach. It is possible that in conducting this modified Gaussian elimination we may produce a singular system. It is therefore necessary to allow pivot candidates to be perturbed to generate the approximate factorization (see, for example, [5]). Another disadvantage of this approach is that the rowsums of the approximate factorization are not equal to the rowsums of the original matrix. Having the rowsums preserved is known to lead to a stable preconditioner [5]. We shall use this approach in generating a preconditioner and refer to it as modified incomplete *LU* factorization (MILU).

There are many extensions of the (ILU) approach that allow fill-in at specified locations or in cases when the candidate elements to be dropped are larger than a specified threshold [5]. Based on our experience in [4], we shall limit ourselves to the case with no fill-in. Our approach is based on software developed in [4] for general sparse systems and extended to stripe data structures by Kaufman [6]. This preconditioner is characterized by the fact that its factors have similar sparsity patterns to A. It has the property that if none of the diagonals needed to be perturbed, then

$$sparse(\bar{L}) = sparse(C + D + E) \tag{3.2}$$

$$sparse(\bar{U}) = sparse(B + F + G)$$

where *sparse* indicates the sparsity pattern of a matrix. It has been shown in [12] that the ILU factorization of an M-matrix is as stable as the complete factorization. Unfortunately, for a general positive matrix the ILU factorization may be unstable [5].

A modified approach involves adding to the diagonal the values of elements in off-diagonal locations that would be dropped in the ILU decomposition. It has the additional property that

$$rowsum(\mathbf{P} - \mathbf{A}) = 0 \qquad\qquad (3.3)$$

where *rowsum* produces a diagonal matrix whose elements are the rowsums of the matrix it operates on. This preconditioner was analyzed by Gustafsson [5] and shall be referred to as the MILU approach. That is, the sums of elements in each row of the matrix and the preconditioner are the same. Gustafsson [5] proves that for a weakly diagonally dominant matrix the MILU approach produces a stable preconditioner.

In summary we shall consider two incomplete *LU* preconditioners. The first (ILU) is a straightforward application of approximate factorization. The second one (MILU) is a modified procedure that preserves rowsums.

## 4   NESTED BLOCK INCOMPLETE LU FACTORIZATIONS

The second group of preconditioners we discuss takes advantage of the block structure inherent in the linear system that arises in discretizing the partial differential equation. In fact, since we are considering a three-dimensional system, there are three levels of structure in the system—there are blocks within blocks. By exploiting this structure we hope to produce not only a better preconditioner but also (as shall be seen later) a faster procedure to generate the preconditioner.

Our approach is based on a nested block factorization approach and is similar to one used by Appleyard and Cheshire [7]. As will be seen, however, the crucial steps in the process of generating the factorizations involve the selection of a principle for the approximation and the nature of approximations that are used in the method. In our approach we recognize that the **A** matrix is a block-structured tridiagonal system and use a recursive *ILU* approximation. Here, we do not mean that the factors are lower and upper triangular matrices. Rather, they are matrix products that will be recursively defined. Appleyard and Cheshire also sought a nested factorization that preserved columnsums of the original matrix. Their motivation was to maintain a material balance in a reservoir simulation.

Alternatively, one can develop an approach in which the approximation preserves the rowsums of the original matrix as in the MILU preconditioner. This alternative is motivated by the desire to reduce the condition number of the preconditioned matrix (see, for example, [7]). We shall develop preconditioners based upon each approach.

Unlike the ILU factorization (3.1) discussed in the previous section, we consider nested incomplete factorizations where the Gaussian elimination steps are not performed on the strictly lower triangular elements of the original matrix.

The general form of the nested incomplete *LU* factorization can be expressed as

$$P = (E + Q)(I + S) \tag{4.1a}$$

$$Q = (D + T)(I + W) \tag{4.1b}$$

$$T = (C + M)(I + U) \tag{4.1c}$$

where S, W, and U are appropriately defined. The nested factorization introduced in [7] requires $S = Q^{-1}G$, $W = T^{-1}F$, and $U = M^{-1}B$. Thus, the columnsum condition enforced in [7] takes the form

$$M = V - CU - colsum(DW + ES) . \tag{4.1d}$$

Upon expanding these equations we obtain

$$P = A + (DW + ES) - colsum(DW + ES) . \tag{4.2}$$

It will be shown this is not the only possible nested factorization algorithm. Our intent is to develop and investigate other algorithms based on nested factorization. The notation used in expressing Eq. (4.1) does reflect the nested aspect of the procedure. However, it does not reflect the block structure which is inherent in the matrix A. We shall utilize this block structure, and for this reason we shall use an alternative notation to describe our matrices and the algorithms which we develop. We consider a three-dimensional problem with the axes labeled $x,y,z$ and number the variables associated with the $x$ axis on the inside, then the $y$ axis, and the $z$ axis on the outside. We define the matrix as

$$A = (E^k, V^k, G^k) , \tag{4.3}$$

where $k$ is an index on the blocks in the $z$ variable with $1 \le k \le n_3$, $E^k$ is a diagonal matrix made up of the diagonal stripe $e$ in Eq. (2.2), $G^k$ is a diagonal matrix made up of the diagonal stripe $g$ in (2.2), and $V^k$ is a block tridiagonal matrix. $V^k$ is defined as

$$V^k = (D_j^k, V_j^k, F_j^k) , \tag{4.4}$$

where j is an index on the blocks in the y variable with $1 \le j \le n_2$, $D_j^k$ is a diagonal matrix made up of the diagonal stripe $d$ in (2.2), $F_j^k$ is a diagonal matrix made up of the diagonal stripe $f$ in (2.2), and $V_j^k$ is a tridiagonal matrix. In fact,

$$V_j^k = (c_{i,j}^k, a_{i,j}^k, b_{i,j}^k) \tag{4.5}$$

where the elements are made up of $c$, $a$, and $b$, respectively with $1 \le i \le n_1$. The inherent recursive nature of the block structure is clear from these definitions.

To discuss our factorizations in detail, we use a basic property of tridiagonal factorization which we call the tridiagonal identification. Consider a tridiagonal matrix

$$T = (c_i , a_i , b_i) , \tag{4.6}$$

where $a$ is the diagonal, $c$ is the subdiagonal, and $b$ is the superdiagonal stripe. Using this notation, we can write an *LU* factorization as

$$(c_i , a_i , b_i) = (c_i , m_i , 0)(0 , 1 , q_i) . \tag{4.7}$$

The factorization of this tridiagonal system follows from the standard recurrence relations

$$b_{i-1} = m_{i-1}q_{i-1} \tag{4.8a}$$

$$a_i = c_iq_{i-1} + m_i . \tag{4.8b}$$

If we wish to preserve columnsums, we find that we must have

$$b_{i-1} + a_i + c_{i+1} = m_{i-1}q_{i-1} + (c_iq_{i-1} + m_i) + c_{i+1} . \tag{4.9}$$

This relationship is satisfied by Eq. (4.8) since an exact factorization must, of course, preserve columnsums. Alternatively, we can seek to preserve rowsums. In this case we obtain

$$b_i + a_i + c_i = m_iq_i + (c_iq_{i-1} + m_i) + c_i . \tag{4.10}$$

Again, this relationship is satisfied by Eq. (4.8), since the exact factorization also preserves rowsums. The tridiagonal identification (TI) is the process by which quantities satisfying equations similar to (4.9) or (4.10) are approximated by requiring these quantities to satisfy equations similar to Eq. (4.8).

This situation does not occur, however, when we consider nested incomplete factorization preconditioners. That is, a nested incomplete factorization may preserve rowsums or columnsums or neither. We illustrate this point in the two-dimensional case. Consider (4.4) for a specific $k$. Dropping the superscript $k$ in Eqs. (4.4) and (4.5), we note that $A \equiv V$ and seek an approximation to the matrix $A$ where

$$A = (D_j , V_j , F_j) \tag{4.11}$$

with

$$V_j = (c_{i,j} , a_{i,j} , b_{i,j}) . \tag{4.12}$$

In parallel to (4.7), we consider a nested incomplete factorization of $A$ denoted by $Q$ as

$$\mathbf{Q} = (D_j , T_j , 0)(0 , I_j , W_j) \tag{4.13a}$$

where $T_j$ is

$$T_j = (c_{i,j} , m_{i,j} , 0)(0 , 1 , q_{i,j}) = (c_{i,j} , \tilde{a}_{i,j} , \tilde{q}_{i,j}) \tag{4.13b}$$

with

$$\tilde{q}_{i-1,j} = m_{i-1,j} q_{i-1,j} , \tag{4.13c}$$

$$\tilde{a}_{i,j} = m_{i,j} + c_{i,j} q_{i-1,j} , \tag{4.13d}$$

and $W_j$ is an unknown matrix to be determined. Expanding Eq. (4.13), we obtain

$$\mathbf{Q} = (D_j , D_j W_{j-1} + T_j , T_j W_j) \tag{4.14a}$$

$$T_j = (c_{i,j} , c_{i,j} q_{i-1,j} + m_{i,j} , m_i q_i) = (c_{i,j} , \tilde{a}_{i,j} , \tilde{q}_{i,j}) . \tag{4.14b}$$

The remaining issues are the determination of $m_{i,j}$, $q_{i,j}$, the structure of $W_j$ and its numerical value. If we wish to maintain columnsums, we note that

$$columnsum(\mathbf{A}) = \underline{\mathbf{K}^T}\mathbf{A} = \underline{K_{j-1}^T}F_{j-1} + \underline{K_j^T}V_j + \underline{K_{j+1}^T}D_{j+1} \tag{4.15a}$$

$$columnsum(\mathbf{Q}) = \underline{K_{j-1}^T}T_{j-1}W_{j-1} + (\underline{K_j^T}D_j W_{j-1} + \underline{K_j^T}T_j) + \underline{K_{j+1}^T}D_{j+1} . \tag{4.15b}$$

In Eq. (4.15b) we have introduced $\mathbf{K}^T$ as a "summing" block matrix. Each block $K_j^T$ is a diagonal matrix with ones on the diagonal. The underbar in Eq. (4.15b) denotes treating a diagonal matrix as a vector. Thus $\underline{K_j^T}$ is a row vector with all elements equal to one. If we want to preserve columnsums, we require that the two expressions (4.15a) and (4.15b) be equal. One way to achieve this is by the identifications we have referred to as TI, namely:

$$\underline{K_j^T}T_j = \underline{K_j^T}(V_j - D_j W_{j-1}) \tag{4.16a}$$

$$\underline{K_j^T}T_j W_j = \underline{K_j^T}F_j . \tag{4.16b}$$

The matrices $T_j$ and $W_j$ are as yet not defined. We first restrict $W_j$ to be a diagonal matrix which will be determined subsequently. First we consider Eq. (4.16a). Since

$$V_j - D_j W_{j-1} = (c_{i,j} , a_{i,j} - d_{i,j} w_{i,j-1} , b_{i,j}) \tag{4.17}$$

the right-hand side has the form

$$\underline{K_j^T}(V_j - D_j W_{j-1}) = [b_{i-1,j} + a_{i,j} - d_{i,j} w_{i,j-1} + c_{i+1,j}] . \tag{4.18}$$

From Eq. (4.14b) we find

$$K_j^T T_j = [\tilde{q}_{i-1,j} + \tilde{a}_{i,j} + c_{i+1,j}] \,, \tag{4.19}$$

which gives

$$\tilde{q}_{i-1,j} + \tilde{a}_{i,j} = b_{i-1,j} + a_{i,j} - d_{i,j} w_{i,j-1} \tag{4.20a}$$

or

$$m_{i-1,j} q_{i-1,j} + m_{i,j} + c_{i,j} q_{i-1,j} = b_{i-1,j} + a_{i,j} - d_{i,j} w_{i,j-1} \,. \tag{4.20b}$$

While there is no unique choice, we shall make the following identification (based upon TI)

$$m_{i,j} + c_{i,j} q_{i-1,j} = a_{i,j} - d_{i,j} w_{i,j-1} \tag{4.21a}$$

$$m_{i-1,j} q_{i-1,j} = b_{i-1,j} \,. \tag{4.21b}$$

Now return to Eq. (4.16b). Using Eq. (4.14b) and the assumption that $W_j$ is diagonal, we see that Eq. (4.16b) when written in detail gives

$$(\tilde{q}_{i-1,j} + \tilde{a}_{i,j} + c_{i+1,j}) w_{i,j} = f_{i,j} \tag{4.22a}$$

or

$$(m_{i-1,j} q_{i-1,j} + m_{i,j} + c_{i,j} q_{i-1,j} + c_{i+1,j}) w_{i,j} = f_{i,j} \,. \tag{4.22b}$$

We can summarize the two-dimensional columnsum algorithm as

$$m_{i,j} = a_{i,j} - c_{i,j} q_{i-1,j} - d_{i,j} w_{i,j-1} \tag{4.23a}$$

$$m_{i,j} q_{i,j} = b_{i,j} \tag{4.23b}$$

$$(m_{i-1,j} q_{i-1,j} + m_{i,j} + c_{i,j} q_{i-1,j} + c_{i+1,j}) w_{i,j} = f_{i,j} \,. \tag{4.23c}$$

Note that this definition of $W_j$ is merely a scaling of the $F_j$.

Similarly, we consider a rowsum approach based on the use of TI. Using the block "summing" matrix K introduced above, we can write the rowsum expression as

$$rowsum(\mathbf{A}) = \mathbf{A}\underline{K} = D_j \underline{K}_j + V_j \underline{K}_j + F_j \underline{K}_j \tag{4.24a}$$

$$rowsum(\mathbf{Q}) = \mathbf{Q}\underline{K} = D_j \underline{K}_j + (D_j W_{j-1} + T_j) \underline{K}_j + T_j W_j \underline{K}_j \,. \tag{4.24b}$$

There are many ways to achieve this condition. Using TI, we obtain the rowsum condition by setting

$$V_j \underline{K}_j = D_j W_{j-1} \underline{K}_j + T_j \underline{K}_j \tag{4.25a}$$

$$T_j W_j \underline{K}_j = F_j \underline{K}_j .$$ (4.25b)

With $W_j$ a diagonal matrix, we have

$$T_j W_j = (c_{i,j} w_{i-1,j} , \; c_{i,j} q_{i-1,j} w_{i,j} + m_{i,j} w_{i,j} , \; m_{i,j} q_{i,j} w_{i+1,j}) ,$$ (4.26)

so that

$$T_j W_j \underline{K}_j = [c_{i,j} w_{i-1,j} + (c_{i,j} q_{i-1,j} + m_{i,j}) w_{i,j} + m_{i,j} q_{i,j} w_{i+1,j}]^T .$$ (4.27)

That is, $T_j W_j \underline{K}_j = T_j \underline{W}_j$, and since $F_j \underline{K}_j = \underline{F}_j$, we see that Eq. (4.25b) reduces to the tridiagonal linear system:

$$c_{i,j} w_{i-1,j} + (c_{i,j} q_{i-1,j} + m_{i,j}) w_{i,j} + m_{i,j} q_{i,j} w_{i+1,j} = f_{i,j} .$$ (4.28)

Note that unlike the columnsum case, Eq. (4.28) leads to a tridiagonal system for each fixed $j$. In the columnsum case Eq. (4.23c) a scalar equation was obtained for each fixed $i$ and $j$.

Now we consider Eq. (4.25a) in detail. Since

$$V_j \underline{K}_j = [c_{i,j} + a_{i,j} + b_{i,j}]^T , \quad \text{and} \quad D_j W_{j-1} \underline{K}_j = [d_{i,j} w_{i,j-1}]^T ,$$ (4.29)

we find

$$(V_j - D_j W_{j-1}) \underline{K}_j = [c_{i,j} + a_{i,j} + b_{i,j} - d_{i,j} w_{i,j-1}]^T .$$ (4.30)

Thus Eq. (4.25a) implies

$$c_{i,j} + (c_{i,j} q_{i-1,j} + m_{i,j}) + m_{i,j} q_{i,j} = c_{i,j} + a_{i,j} + b_{i,j} - d_{i,j} w_{i,j-1} .$$ (4.31)

This equation can be satisfied by using TI with

$$c_{i,j} q_{i-1,j} + m_{i,j} = a_{i,j} - d_{i,j} w_{i,j-1}$$ (4.32a)

$$m_{i,j} q_{i,j} = b_{i,j} .$$ (4.32b)

The rowsum preconditioner method thus consists of the application of (4.32) and the solution of the tridiagonal system (4.28) to obtain the $\{w_{i,j}\}$ vector. Note that (4.28) arises from the linkage of the elements across rows and is not a scaling of the $\{f_{i,j}\}$ vector. Rather, it requires the solution of a tridiagonal system for each $j$. One would expect that the columnsum approach would be faster to compute, since it lumps the tridiagonal systems for each $j$ into a scalar equation.

Before we extend these algorithms to three dimensions, we observe that the nested block factorization algorithm based on columnsum preservation at each level does not lead to the nested incomplete *LU* factorization that is described in [7]. To see this, we observe that in two dimensions, the quantity $\mathbf{W}$ in Eq. (4.1b) corresponds to the quantity $\{W_j\}$ as generated in Eq. (4.23c). The nested factorization introduced in [7] requires that $\mathbf{W} = \mathbf{T}^{-1}\mathbf{F}$. In [7] $\mathbf{W}$ results from the solution of tridiagonal systems, whereas $\{W_j\}$ in Eq. (4.23c) results from a scaling of $\{f_{i,j}\}$. Thus the two procedures do not generate the same algorithms.

We further extend these two block preconditioners to the three-dimensional case. First consider the columnsum algorithm. Returning to the notation of Eqs. (4.3)-(4.5), we start with a block tridiagonal matrix

$$\mathbf{A} \equiv (\mathbf{E}^k, \mathbf{V}^k, \mathbf{G}^k), \tag{4.33}$$

and we seek an approximate block nested factorization

$$\mathbf{P} \equiv (\mathbf{E}^k, \mathbf{Q}^k, 0)(0, \mathbf{I}^k, \mathbf{S}^k) \tag{4.34}$$

with

$$Q^k = (D_j^k, T_j^k, 0)(0, I_j^k, W_j^k) \tag{4.35}$$

and

$$T_j^k = (c_{i,j}^k, m_{i,j}^k, 0)(0, 1, q_{i,j}^k) \tag{4.36a}$$

with

$$S^k = diag(S_j^k), \qquad S_j^k = diag(s_{i,j}^k) \tag{4.36b}$$

$$W_j^k = diag(w_{i,j}^k). \tag{4.36c}$$

This nested factorization is based on a block incomplete factorization without pivoting of a block tridiagonal matrix. We would like to factor $\mathbf{A}$:

$$\mathbf{A} \equiv (\mathbf{E}^k, \mathbf{V}^k, \mathbf{G}^k) \approx (\mathbf{E}^k, \mathbf{H}^k, 0)(0, \mathbf{I}^k, \mathbf{P}^k), \tag{4.37}$$

where

$$\mathbf{H}^k = \mathbf{V}^k - \mathbf{E}^k\mathbf{P}^{k-1} \tag{4.38a}$$

$$\mathbf{H}^k\mathbf{P}^k = \mathbf{G}^k. \tag{4.38b}$$

The primary tasks involve the approximation of $\mathbf{H}^k$ by $\mathbf{Q}^k$ and $\mathbf{P}^k$ by $\mathbf{S}^k$. We now describe the basic approach used in these approximations. Suppose we have defined $\mathbf{Q}^{k-1}$. We now define $\mathbf{S}^{k-1}$ and $\mathbf{Q}^k$. In Eq. (4.38b) with $k-1$ in place of $k$, we replace $\mathbf{H}^{k-1}$ by $\mathbf{Q}^{k-1}$ and $\mathbf{P}^{k-1}$ by the diagonal matrix $\mathbf{S}^{k-1}$. Hence we can generate the diagonal matrix $\mathbf{S}^{k-1}$ by solving

$$Q^{k-1}S^{k-1} = G^{k-1} . \tag{4.39}$$

The next task is to define the matrix $Q^k$. We start by combining Eqs. (4.38a) and (4.38b) yielding

$$H^k = V^k - E^k(H^{k-1})^{-1}G^{k-1} \approx V^k - E^k(Q^{k-1})^{-1}G^{k-1} . \tag{4.40}$$

Note that $E^k(Q^{k-1})^{-1}G^{k-1}$ is a full matrix. For this class of preconditioners we will approximate this matrix by a diagonal matrix. We shall consider two natural choices. First, the columnsum algorithm uses

$$E^k(Q^{k-1})^{-1}G^{k-1} \approx diag(colsum(E^k(Q^{k-1})^{-1}G^{k-1})) \tag{4.41}$$

while the rowsum algorithm uses

$$E^k(Q^{k-1})^{-1}G^{k-1} \approx diag(rowsum(E^k(Q^{k-1})^{-1}G^{k-1})) . \tag{4.42}$$

With this general approach in mind we now consider the details of each algorithm. Using the notation of Eqs. (4.3)-(4.5), we extend the block "summing" matrix to three dimensions as $K = diag(K^k)$, $K^k = diag(K^k_j)$, $K^k_j = diag(k^k_{i,j})$ where $k^k_{i,j} = 1$. The columnsums can thus be expressed as

$$columnsum(A) = \underline{K}^T A = (\underline{K}^{k-1})^T G^{k-1} + (\underline{K}^k)^T V^k + (\underline{K}^{k+1})^T E^{k+1} \tag{4.43a}$$

$$columnsum(P) = (\underline{K}^{k-1})^T Q^{k-1}S^{k-1} + (\underline{K}^k)^T [E^k S^{k-1} + Q^k] + (\underline{K}^{k+1})^T E^{k+1} . \tag{4.43b}$$

Application of TI at the $k$th level leads to

$$(\underline{K}^k)^T [E^k S^{k-1} + Q^k] = (\underline{K}^k)^T V^k \tag{4.44a}$$

$$(\underline{K}^{k-1})^T Q^{k-1}S^{k-1} = (\underline{K}^{k-1})^T G^{k-1} . \tag{4.44b}$$

We write Eq. (4.44a) at the $j$th level using the definition of $Q^k$ in Eq. (4.35), the definition of $S^{k-1}$ in Eq. (4.36b) and the definition of $V^k$ in Eq. (4.4).

$$(\underline{K}^k_{j-1})^T T^T_{j-1} W^k_{j-1} + (\underline{K}^k_j)^T D^k_j W^k_{j-1} + (\underline{K}^k_j)^T T^k_j + (\underline{K}^k_{j+1})^T D^k_{j+1}$$

$$= (\underline{K}^k_{j-1})^T F^k_{j-1} + (\underline{K}^k_j)^T V^k_j + (\underline{K}^k_{j+1})^T D^k_{j+1} - (\underline{K}^k_j)^T E^k_j S^{k-1}_j . \tag{4.45}$$

With our experience in two dimensions, it would be natural to make the following identifications in Eq. (4.45):

$$(\underline{K}^k_{j-1})^T T^k_{j-1} W^k_{j-1} = (\underline{K}^k_{j-1})^T F^k_{j-1} \tag{4.46a}$$

$$(\underline{K}^k_j)^T T^k_j = (\underline{K}^k_j)^T [V^k_j - D^k_j W^k_{j-1} - E^k_j S^{k-1}_j] . \tag{4.46b}$$

We next write Eq. (4.44b) at the $j$th level, but at the $k$th level rather than the $k$-1st level as indicated in Eq. (4.44b).

$$[(K_{j-1}^k)^T T_{j-1}^T W_{j-1}^k + (K_j^k)^T D_j^k W_{j-1}^k + (K_j^k)^T T_j^k + (K_{j+1}^k)^T D_{j+1}^k]S_j^k = (K_j^k)^T G_j^k . \tag{4.47}$$

Thus, at the $j$th level, we have the following three equations:

$$(K_j^k)^T T_j^k = (K_j^k)^T [V_j^k - D_j^k W_{j-1}^k - E_j^k S_j^{k-1}] \tag{4.48a}$$

$$(K_{j-1}^k)^T T_{j-1}^k W_{j-1}^k = (K_{j-1}^k)^T F_{j-1}^k \tag{4.48b}$$

$$[(K_{j-1}^k)^T F_{j-1}^k + (K_j^k)^T D_j^k W_{j-1}^k + (K_{j+1}^k)^T D_{j+1}^k]S_j^k = (K_j^k)^T G_j^k . \tag{4.48c}$$

Next we consider the implications of these three equations at the $i$th level. Eq. (4.48a) takes the form

$$m_{i-1,j}^k a_{i-1,j}^k + (c_{i,j}^k a_{i-1,j} + m_{i,j}^k) + c_{i+1,j}^k = b_{i-1,j}^k + (a_{i,j}^k - d_{i,j}^k w_{i,j-1}^k - e_{i,j}^k s_{i,j}^{k-1}) + c_{i+1,j}^k . \tag{4.49}$$

With the TI identification we have

$$m_{i-1,j}^k a_{i-1,j}^k = b_{i-1,j}^k \tag{4.50a}$$

$$m_{i,j}^k = a_{i,j}^k - c_{i,j}^k a_{i-1,j}^k - d_{i,j}^k w_{i,j-1}^k - e_{i,j}^k s_{i,j}^{k-1} , \tag{4.50b}$$

which generates the factored form of

$$T_j^k = (c_{i,j}^k , m_{i,j}^k , 0)(0 , 1 , q_{i,j}^k) . \tag{4.51}$$

With $T_j^k$ known we now write out Eq. (4.48b) with $j$ in place of $j-1$:

$$[m_{i-1,j}^k a_{i-1,j}^k + (c_{i,j}^k a_{i-1,j}^k + m_{i,j}^k) + c_{i+1,j}^k]w_{i,j}^k = f_{i,j}^k . \tag{4.52}$$

This defines $\{w_{i,j}^k\}$ at a given $k$ and $j$. Note that again the columnsum algorithm yields $w_{i,j}^k$ as a scaling of $f_{i,j}$. Finally, Eq. (4.48c) when written in detail yields the equation for $\{s_{i,j}^k\}$

$$[f_{i,j-1}^k + d_{i,j}^k w_{i,j-1}^k + m_{i-1,j}^k a_{i-1,j}^k + (c_{i,j}^k a_{i-1,j}^k + m_{i,j}^k) + c_{i+1,j}^k + d_{i,j+1}^k]s_{i,j}^k = g_{i,j}^k , \tag{4.53}$$

and $\{s_{i,j}^k\}$ is just a scaling of $\{g_{i,j}^k\}$. We summarize this algorithm (which we call Method C) as follows:

For $k = 1, n_3$

    For $j = 1, n_2$

        $a_{i,j}^k = a_{i,j}^k - d_{i,j}^k w_{i,j-1}^k - e_{i,j}^k s_{i,j}^{k-1} .$

        Factor $T_j^k = (c_{i,j}^k, a_{i,j}^k, b_{i,j}^k) = (c_{i,j}^k, m_{i,j}^k, 0)(0,1,q_{i,j}^k) .$

        Generate the $w_{i,j}^k$

$$[(m_{i-1,j}^k + c_{i,j}^k)q_{i-1,j}^k + m_{i,j}^k + c_{i+1,j}^k]w_{i,j}^k = f_{i,j}^k .$$

Generate $s_{i,j}^k$

$$[f_{i,j-1}^k + d_{i,j}^k w_{i,j-1}^k + (m_{i-1,j}^k + c_{i,j}^k)q_{i-1,j}^k + m_{i,j}^k + c_{i,j+1}^k + d_{i,j+1}^k]s_{i,j}^k = g_{i,j}^k .$$

end $j$

    end $k$

The rowsum method also extends to three dimensions. The rowsum requirement can be expressed as

$$rowsum(\mathbf{A}) = \mathbf{A}\underline{\mathbf{K}} = \mathbf{E}^k\underline{\mathbf{K}}^k + \mathbf{V}^k\underline{\mathbf{K}}^k + \mathbf{G}^k\underline{\mathbf{K}}^k \tag{4.54a}$$

$$rowsum(\mathbf{P}) = \mathbf{P}\underline{\mathbf{K}} = \mathbf{E}^k\underline{\mathbf{K}}^k + (\mathbf{E}^k\mathbf{S}^{k-1} + \mathbf{Q}^k)\underline{\mathbf{K}}^k + \mathbf{Q}^k\mathbf{S}^k\underline{\mathbf{K}}^k . \tag{4.54b}$$

The application of TI at the $k$th level leads to

$$\mathbf{Q}^k\underline{\mathbf{K}}^k = \mathbf{V}^k\underline{\mathbf{K}}^k - \mathbf{E}^k\mathbf{S}^{k-1}\underline{\mathbf{K}}^k \tag{4.55a}$$

$$\mathbf{Q}^k\mathbf{S}^k\underline{\mathbf{K}}^k = \mathbf{G}^k\underline{\mathbf{K}}^k . \tag{4.55b}$$

At the $j$th level, Eq. (4.55b) has the form

$$D_j^k S_{j-1}^k \underline{K}_j^k + D_j^k W_{j-1}^k S_j^k \underline{K}_j^k + T_j^k S_j^k \underline{K}_j^k + T_j^k W_j^k S_{j+1}^k \underline{K}_j^k = G_j^k \underline{K}_j^k . \tag{4.56}$$

Since $S_j^k$ is restricted to be a diagonal matrix, we observe that $S_j^k \underline{K}_j^k = \underline{S}_j^k$, i.e., a column vector. Thus Eq. (4.56) has the form of a block tridiagonal linear system for the vector $\underline{S}_j^k$,

$$D_j^k \underline{S}_{j-1}^k + (D_j^k W_{j-1}^k + T_j^k)\underline{S}_j^k + T_j^k W_j^k \underline{S}_{j+1}^k = \underline{G}_j^k ; \tag{4.57a}$$

that is, from Eq. (4.35) the above equation can be written as

$$\mathbf{Q}^k \underline{\mathbf{S}}^k = \underline{\mathbf{G}}^k . \tag{4.57b}$$

But we have $\mathbf{Q}^k$ in factored form; that is, from Eq. (4.35) this is $\mathbf{Q}^k$ in expanded form:

$$\mathbf{Q}^k = (D_j^k , T_j^k , 0)(0 , I_j^k , W_j^k) = (D_j^k , D_j^k W_{j-1}^k + T_j^k , T_j^k W_j^k) ; \tag{4.58}$$

thus we can solve Eq. (4.57a) with the following forward/backward sweeps:

$$T_j^k \underline{Z}_j = \underline{G}_j^k - D_j^k \underline{Z}_{j-1} \tag{4.59a}$$

$$\underline{S}_j^k = \underline{Z}_j - W_j^k \underline{S}_{j+1}^k . \tag{4.59b}$$

This procedure will generate the vectors $\{\underline{S}_j^k\}$.

We next consider Eq. (4.55a) at the $j$th level which will define $T_j^k$. Now,

$$(\mathbf{V}^k - \mathbf{E}^k \mathbf{S}^{k-1})\mathbf{K}^k = D_j^k \underline{K}_j^k + V_j^k \underline{K}_j^k + F_j^k \underline{K}_j^k - E_j^k S_j^{k-1} \underline{K}_j^k \tag{4.60}$$

while

$$\mathbf{Q}^k \underline{\mathbf{K}}^k = D_j^k \underline{K}_j^k + (D_j^k W_{j-1}^k \underline{K}_j^k + T_j^k \underline{K}_j^k) + T_j^k W_j^k \underline{K}_j^k . \tag{4.61}$$

Using the TI identification at this $j$th level, we have

$$T_j^k \underline{K}_j^k = V_j^k \underline{K}_j^k - E_j^k S_j^{-1} \underline{K}_j^k - D_j^k W_{j-1}^k \underline{K}_j^k \tag{4.62a}$$

$$T_j^k W_j^k \underline{K}_j^k = F_j^k \underline{K}_j^k . \tag{4.62b}$$

Since $S_j^k$ and $W_j^k$ are restricted to being diagonal, these equations take the form

$$T_j^k \underline{K}_j^k = V_j^k \underline{K}_j^k - D_j^k \underline{W}_{j-1}^k - E_j^k \underline{S}_j^k \tag{4.63a}$$

$$T_j^k \underline{W}_j^k = \underline{E}_j^k . \tag{4.63b}$$

At the $i$th level, the right-hand side of Eq. (4.63a) has the form

$$c_{i,j}^k + a_{i,j}^k + b_{i,j}^k - e_{i,j}^k s_{i,j}^{k-1} - d_{i,j}^k w_{i,j-1}^k \tag{4.64}$$

and with

$$T_j^k = (c_{i,j}^k , m_{i,j}^k , 0)(0 , 1 , q_{i,j}^k) \tag{4.65}$$

we have

$$T_j^k \underline{K}_j^k = c_{i,j}^k + (c_{i,j}^k q_{i-1,j}^k + m_{i,j}^k) + m_{i,j}^k q_{i,j}^k . \tag{4.66}$$

Making the TI identification, we have

$$m_{i,j}^k + c_{i,j}^k q_{i-1,j}^k = a_{i,j}^k - e_{i,j}^k s_{i,j}^{k-1} - d_{i,j}^k w_{i,j-1}^k \tag{4.67a}$$

$$m_{i,j}^k q_{i,j}^k = b_{i,j}^k \tag{4.67b}$$

which defines $T_j^k$. With $T_j^k$ defined in factored form, Eq. (4.63b) can be solved for the vector $\underline{W}_j^k$. This rowsum algorithm will be referred to as Method R and summarized below.

For $k = 1, n_3$

    For $j = 1, n_2$

        $\tilde{a}_{i,j}^k = a_{i,j}^k - d_{i,j}^k w_{i,j-1}^k - e_{i,j}^k s_{i,j}^{k-1} .$

        Factor $T_j^k = (c_{i,j}^k , \tilde{a}_{i,j}^k , b_{i,j}^k) = (c_{i,j}^k , m_{i,j}^k , 0)(0 , 1 , q_{i,j}^k) .$

        Generate the $w_{i,j}^k$ from the tridiagonal system with forward/backward sweeps

$$T_j^k \underline{W}_j^k = E_j^k \ .$$

Generate $s_{i,j}^k$ from solving the tridiagonal system

$$T_j^k \underline{Z}_j = \underline{G}_j^k - D_j^k \underline{Z}_{j-1}$$

end $j$

For $j = n_2, 1$

Set

$$\underline{S}_j^k = \underline{Z}_j - W_j^k \underline{S}_{j+1}^k$$

end $j$

end $k$

Note that Method R involves the solution to tridiagonal linear systems whereas Method C involves scalar equations.

## 5 BLOCK ILU WITH APPROXIMATE INVERSES

The third class of preconditioners we study is based on the use of approximate inverses. In this approach, rather than attempting to find an approximate factorization for A based entirely on nested factorization at every level, we seek an approximate factorization for A based on a nested factorization at the higher levels but using approximate inverses at lower levels. The advantage of this approach is that the operations in the iterative method with the preconditioner can be conducted as matrix-vector products rather than doing forward and backsolve steps involving the factorization. The resulting matrix-vector products are thus well suited to parallel and vector computation.

Approximate inverses have been proposed and studied by several authors either directly as a preconditioning or in combination with a form of nested factorization [8,14,15]. Here we use an approximation to the inverse within a nested factorization approach. That is, we use the nested factorization method, but instead of factoring a tridiagonal matrix, we use an approximation of the inverse of this matrix. As will be seen below, the derivation of the appropriate nested factorization based on approximate inverses is significantly different from the methods derived in Section 4. In [8,15] block incomplete factorizations using approximate inverses are developed and analyzed for symmetric positive definite systems. The approximate inverses presented in [8] use known properties of the true inverses of symmetric tridiagonal matrices. The approximate inverses presented in [15] use a minimization of a matrix-norm such as the Frobenius norm. These inverses are further modified by enforcing symmetry on the approximate inverse.

Here we are interested in nonsymmetric problems. Our approximate inverses are computed using a minimization of the Frobenius norm. To motivate the approach, we start from the factorization of $A$ given by Eqs. (4.38). Just as before, the primary task involves the approximation of $H^k$ by a matrix $Q^k$ and $P^k$ by a matrix $S^k$. Thus we replace Eqs. (4.38) by

$$Q^k = V^k - E^k S^{k-1} \tag{5.1a}$$

$$Q^k S^k = G^k . \tag{5.1b}$$

As before, we assume $Q^k$ to be of the form

$$Q^k = (D_j^k , T_j^k , 0)(0 , I_j^k , W_j^k) = (D_j^k , D_j^k W_{j-1}^k + T_j^k , T_j^k W_j^k) \tag{5.2}$$

where each $T_j^k$ is a tridiagonal matrix. However, at this stage we will not restrict $W_j^k$ to be diagonal as we did in the previous section. We also assume that

$$S^k = diag(S_j^k) . \tag{5.3}$$

However, we do not restrict the matrices $S_j^k$ to be diagonal at this stage either. Now the right-hand side of Eq. (5.1a) has the form

$$V^k - E^k S^{k-1} = (D_j^k , V_j^k - E_j^k S_j^{k-1} , F_j^k) . \tag{5.4}$$

Using the definitions of $Q^k$ in Eq. (5.2) and identifying corresponding terms, we find

$$T_j^k = V_j^k - D_j^k W_{j-1}^k - E_j^k S_j^{k-1} \tag{5.5a}$$

$$T_j^k W_j^k = F_j^k \tag{5.5b}$$

and from Eq. (5.1b) we have

$$D_j^k S_{j-1}^k + (T_j^k + D_j^k W_{j-1}^k) S_j^k + F_j^k S_{j+1}^k = G_j^k . \tag{5.5c}$$

These equations form the basis for our approximate inverse approach. Equation (5.5a) defines a matrix $T_j^k$, Eq. (5.5b) defines a matrix $W_j^k$, and Eq. (5.5c) defines a matrix $S_j^k$. The general approach will be to approximate inverse matrices from a given class to approximately solve for the matrices $W_j^k$ and $S_j^k$. Consider the generation of $W_j^k$ in Eq. (5.5b), and suppose that $Z_j^k$ denotes an approximate inverse to $T_j^k$. In this paper we will restrict our attention to matrices $Z_j^k$ from the class of either diagonal or tridiagonal matrices. The details of the construction will be discussed later. Having $Z_j^k$ available enables us to generate an approximation to the matrix $W_j^k$ by

$$W_j^k = Z_j^k F_j^k . \tag{5.6}$$

Note that if $Z_j^k$ is selected from the class of tridiagonal matrices, then, since $F_j^k$ is diagonal, the matrix $W_j^k$ will be tridiagonal. This is in sharp contrast to the previous algorithms where the matrix $W_j^k$ was diagonal in both cases.

From Eq. (5.5c) we see that the generation of the matrix $S_j^k$ involves the solution of a block tridiagonal matrix

$$(D_j^k , (T_j^k + D_j^k W_{j-1}^k) , F_j^k) . \tag{5.7}$$

Again it would be natural to seek our approximate block inverse from an appropriate class of matrices. For the class of block approximate inverses, two obvious choices come to mind. First the class of block tridiagonal matrices of the form

$$\Omega_j^k = (X_j^k , R_j^k , Y_j^k) \tag{5.8}$$

where $X_j^k$ , $Y_j^k$ are diagonal and $R_j^k$ is tridiagonal. A simpler class would be the class of block diagonal matrices

$$\Omega_j^k = R_j^k \tag{5.9}$$

where $R_j^k$ is a tridiagonal matrix. If we use $\Omega_j^k$ from Eq. (5.8), then $S_j^k$ is given by

$$S_j^k = \Omega_j^k G_j^k = X_j^k G_{j-1}^k + R_j^k G_j^k + Y_j^k G_{j+1}^k \tag{5.10}$$

and if Eq. (5.9) is used, then

$$S_j^k = R_j^k G_j^k . \tag{5.11}$$

Again we see that the matrices $S_j^k$ are not restricted to diagonal matrices as in the earlier algorithms. We see then that with this approach, the matrices $T_j^k$ , $W_j^k$, and $S_j^k$ are generated by

$$T_j^k = V_j^k - D_j^k W_{j-1}^k - E_j^k S_j^{k-1} \tag{5.12a}$$

$$W_j^k = Z_j^k F_j^k \tag{5.12b}$$

$$S_j^k = \Omega_j^k G_j^k \tag{5.12c}$$

where $Z_j^k$ is an approximate inverse to generate $W_j^k$ and $\Omega_j^k$ is a block approximate inverse to generate $S_j^k$.

The remaining task is to specify how the approximate inverses are defined. For the sake of brevity we will discuss the generation of approximate inverses to tridiagonal matrices as, for example, $Z_j^k$ in Eq. (5.12). The generalization to block approximate inverses such as $\Omega_j^k$ in Eq. (5.12) is just a slight generalization of the scalar case. We thus consider the generation of an approximate inverse to the tridiagonal matrix $T_j^k$ in Eq. (5.12) by $Z_j^k$. Since the approximate inverses are at the innermost level, the $k$ and $j$ indices are just parameters and can be dropped to ease the notational complexity.

We consider any tridiagonal matrix

$$T = (c_i \, , a_i \, , b_i) \tag{5.13}$$

and seek an approximation to an inverse of this matrix that is not as dense as the actual inverse matrix (which can be a full matrix). Three considerations come to mind. First, do we seek a left or a right approximate inverse? Second, from what "sparse" class of matrices will we seek the approximation? And finally, what criterion will we use to determine the approximation? Recall that in this paper we have restricted our discussion to a class of left preconditioners. Numerical evidence and conventional wisdom suggest that left approximate inverses match with left preconditioning and vice versa, whereas mixing left and right does not perform as well. For this reason we shall restrict our discussion in this section to left approximate inverses. Considering the algorithm described in above, it would be natural to restrict our attention to the class of either diagonal or tridiagonal matrices in seeking a left approximate inverse. As far as the selection criterion is concerned, we shall use the Frobenius norm.

Thus, given the matrix $T$ in Eq. (5.13), we seek a matrix $Z$ such that

$$\|I - ZT\|_F \tag{5.14}$$

is minimized where $Z$ ranges over either the class of diagonal matrices or the class of tridiagonal matrices. Recall that for an $n \times n$ matrix $A$

$$\|A\|_F = \sum_{i=1}^{n}\sum_{j=1}^{m}|a_{i,j}|^2 \, . \tag{5.15}$$

If $Z = diag(q_i)$, then

$$\|I - ZT\|_F^2 = \sum_{i=1}^{n} q_i^2 c_i^2 + (1 - q_i a_i)^2 + q_i^2 b_i^2 \, , \tag{5.16}$$

and the optimal $Z$ is given by

$$q_i = \frac{a_i}{(c_i^2 + a_i^2 + b_i^2)} \, . \tag{5.17}$$

Next we consider the class of tridiagonal matrices $Z = (x_i \, , z_i \, , y_i)$. Setting

$$F(x_2, x_3, \cdots, x_n, z_1, \cdots, z_n, y_1, \cdots, y_{n-1}) = \|I - ZT\|_F^2 , \tag{5.18}$$

and summing by rows yields the sum

$$F = \sum_{i=1}^{n} [x_i^2 c_{i-1}^2 + (x_i a_{i-1} + z_i c_i)^2 + (1 - x_i b_{i-1} - z_i a_i - y_i c_{i+1})^2 + (z_i b_i + y_i a_{i+1})^2 + y_i^2 b_{i+1}^2] . \tag{5.19}$$

Hence the stationary values are defined by

$$\frac{\partial F}{\partial y_i} = 0 , \quad \frac{\partial F}{\partial z_i} = 0 , \quad \frac{\partial F}{\partial x_i} = 0 , \quad \text{for } i = 1, \cdots, n . \tag{5.20}$$

This leads to the system

$$b_{i+1} c_{i+1} x_i + (a_i c_{i+1} + b_i a_{i+1}) z_i + (c_{i+1}^2 + a_{i+1}^2 + b_{i+1}^2) y_i = c_{i+1}$$

$$(a_{i-1} c_i + b_{i-1} a_i) x_i + (c_i^2 + a_i^2 + b_i^2) z_i + (c_{i+1} a_i + a_{i+1} b_i) y_i = a_i \tag{5.21}$$

$$(c_{i-1}^2 + a_{i-1}^2 + b_{i-1}^2) x_i + (c_i a_{i-1} + a_i b_{i-1}) z_i + c_{i+1} b_{i-1} y_i = b_{i-1}$$

for $i = 1, \ldots, n$ with $c_i = x_i = 0$ for $i \leq 2$ and $b_i = y_i = 0$ for $i \geq n$. Thus for each $i$, we have a $3 \times 3$ system to solve for $(x_i, z_i, y_i)$, and we generate the left approximate inverse $Z$ as a tridiagonal matrix. With such a left approximate inverse we generate an approximate solution to a linear system

$$Tw = f \tag{5.22a}$$

as

$$\tilde{w} = Zf . \tag{5.22b}$$

The approximate inverse generated by Eq. (5.21) will be referred to as a Left Tridiagonal Approximate Inverse (LTAI), while that generated by Eq. (5.17) will be referred to as a Left Diagonal Approximate Inverse (LDAI).

Having discussed in detail the generation of the left approximate inverse $Z_j^k$ in Eq. (5.12b), we will comment further on the block inverses $\Omega_j^k$ in Eq. (5.12c). First they are left approximate inverses, and in this study we have restricted our attention to the class of block diagonal matrices with tridiagonal blocks in Eq. (5.11). Within this class we can, of course, restrict our class further to diagonal matrices just as in the scalar case. In fact, by restricting our class to block diagonal with tridiagonal blocks for the inverses in generating $S_j^k$, this is the same class as was used to generate $W_j^k$. Thus we have restricted our study to LTAI for both $W_j^k$ and $S_j^k$ and LDAI for both $W_j^k$ and $S_j^k$. We remark that if we had used the class of block tridiagonal approximate inverses in Eq. (5.10) to generate $S_j^k$, we would have had to solve $5 \times 5$ systems rather than $3 \times 3$ when generating $\Omega_j^k$. Since factorization is not a major cost, this may be worthwhile.

With regard to parallel and vector aspects of approximate inverses, we note that Díaz and Macedo [9, 10] have observed that if one organizes the solution of Eq. (5.21) appropriately, this portion of the construction of the preconditioner can be fully vectorized. This formulation has not been implemented here, however, the application of the approximate inverse as a matrix-vector multiply has been vectorized. Some properties of the approximate inverses have been derived and are summarized in [9, 10, 14]. Further, Díaz and Cosgrove [11] have evaluated the use of the approximate inverse alone as a preconditioner. The results in [11] point out the high potential for parallelism in the scheme. This technique has also some potential for sparse problems, as the sparsity of the preconditioner could be adjusted to satisfy some numerical criteria. These preliminary studies have not been incorporated here.

## 6 SOLVING NESTED BLOCK LINEAR SYSTEMS

In conducting the preconditioned normal-form conjugate gradient algorithm described in Section 2, the major computational expense occurs in computing $s_i$ and $p_i$ in Eq. (2.5b) which we restate here:

$$s_i = Z^T Z p_i \qquad u_i = Z p_i , \qquad (2.5b)$$

where $Z = P^{-1} A$. In particular, $u_i$ is determined from

$$P u_i = A p_i \equiv f , \qquad (6.1)$$

and $s_i$ is determined from

$$P^T v_i = u_i \qquad s_i = A^T v_i . \qquad (6.2)$$

We shall consider the computation of $u_i$ from the viewpoint of the nested block preconditioners developed in Section 4 and their variation with approximate inverses described in Section 5. The treatment for the computation of $s_i$ is similar and therefore will not be presented. Our objective is to illustrate how the block structure of the preconditioner affects the solution of Eq. (6.1) and how the use of approximate inverses modifies the process.

Whether Method R or C is used, the structure of the preconditioner is given in Eq. (4.34)-(4.36) which we restate below:

$$P \equiv (E^k , Q^k , 0)(0 , I^k , S^k) \qquad (4.34)$$

with

$$\mathbf{Q}^k = (D_j^k, T_j^k, 0)(0, I_j^k, W_j^k) \tag{4.35}$$

and

$$T_j^k = (c_{i,j}^k, m_{i,j}^k, 0)(0, 1, q_{i,j}^k) \tag{4.36a}$$

with

$$S^k = diag(S_j^k), \qquad S_j^k = diag(s_{i,j}^k) \tag{4.36b}$$

$$W_j^k = diag(w_{i,j}^k) . \tag{4.36c}$$

Thus, to solve Eq. (6.1), we need to solve

$$(\mathbf{E}^k, \mathbf{Q}^k, 0)v = f \tag{6.3a}$$

$$(0, \mathbf{I}^k, \mathbf{S}^k)u = v . \tag{6.3b}$$

Expanding the equations, we obtain

$$\mathbf{Q}^k v^k = f^k - \mathbf{E}^k v^{k-1} \tag{6.4a}$$

$$u^k = v^k - S^k u^{k+1} . \tag{6.4b}$$

Note that we must determine all of $\{v^k\}$ on the forward sweep (Eq. (6.4a)) before obtaining the $\{u^k\}$ on the backward sweep (Eq. (6.4b)).

To solve Eq. (6.4a), we use Eq. (4.35) to obtain

$$(D_j^k, T_j^k, 0)(0, I_j^k, W_j^k)v^k = f^k - \mathbf{E}^k v^{k-1} , \tag{6.5}$$

which can be written as

$$(D_j^k, T_j^k, 0)x^k = f^k - \mathbf{E}^k v^{k-1} \tag{6.6a}$$

$$(0, I_j^k, W_j^k)v^k = x^k . \tag{6.6b}$$

Expanding the equations, we obtain

$$T_j^k x_j^k = f_j^k - E_j^k v_j^{k-1} - D_j^k x_{j-1}^k \tag{6.7a}$$

$$v_j^k = x_j^k - W_j^k v_{j+1}^k . \tag{6.7b}$$

Note that the situation is similar to the case at the $k$ level. Here, at the $j$ level, we must obtain all of the $\{x_j^k\}$ for a fixed $k$ in the forward sweep (Eq. (6.7a)) before obtaining the $\{v_j^k\}$ in the backward sweep (Eq. (6.7b)).

To solve Eq. (6.7a), we use Eq. (4.36a) to obtain

$$m_{i,j}^k y_{i,j}^k = f_{i,j}^k - e_{i,j}^k v_{i,j}^{k-1} - d_{i,j}^k x_{i,j-1}^k - c_{i,j}^k y_{i-1,j}^k \tag{6.8a}$$

$$x_{i,j}^k = y_{i,j}^k - q_{i,j}^k x_{i+1,j}^k . \tag{6.8b}$$

Here, at the $i$th level, we must obtain $\{y_{i,j}^k\}$ for a given $j$ and $k$ in the forward sweep (Eq. (6.8a)) before obtaining $\{x_{i,j}^k\}$ in the backward sweep (Eq. (6.8b)).

We can summarize this approach with a nested block algorithm for solving Eq. (6.1):

For $k = 1$, $n_3$

    For $j = 1$, $n_2$

        For $i = 1$, $n_1$

$$m_{i,j}^k y_{i,j}^k = f_{i,j}^k - e_{i,j}^k v_{i,j}^{k-1} - d_{i,j}^k x_{i,j-1}^k - c_{i,j}^k y_{i-1,j}^k$$

        end $i$

        For $i = n_1$ , 1

$$x_{i,j}^k = y_{i,j}^k - q_{i,j}^k x_{i+1,j}^k$$

        end $i$

    end $j$

    For $j = n_2$, 1

        For $i = 1$, $n_1$

$$v_{i,j}^k = x_{i,j}^k - w_{i,j}^k v_{i,j+1}^k$$

        end $i$

    end $j$

end $k$

For $k = n_3$ , 1

    For $j = 1$, $n_2$

        For $i = 1$, $n_1$

$$u_{i,j}^k = v_{i,j}^k - s_{i,j}^k u_{i,j}^{k+1}$$

        end $i$

    end $j$

end $k$

We next consider the use of approximate inverses developed in Section 5. In particular, we deal with the Left Tridiagonal Approximate Inverse (LTAI). The diagonal approximation of the inverse (LDAI) is a simpler variation of this case. Note that now instead of using $T_j^k$ in Eq. (6.7a) we have an approximation

$$(T_j^k)^{-1} \approx R_j^k \equiv (\gamma_{i,j}^k, \alpha_{i,j}^k, \beta_{i,j}^k) \tag{6.9}$$

where $R_j^k$ is a tridiagonal matrix. The solution of Eq. (6.1) proceeds through the $k$ and $j$ levels as before. Now Eq. (6.7a) is approximately solved as

$$x_j^k = R_j^k (f_j^k - E_j^k v_j^{k-1} - D_j^k x_{j-1}^k) \equiv R_j^k g_j^k . \tag{6.10a}$$

That is,

$$x_j^k = \gamma_{i,j}^k g_{i-1,j}^k + \alpha_{i,j}^k g_{i,j}^k + \beta_{i,j}^k g_{i+1,j}^k . \tag{6.10b}$$

We summarize this variation for solving Eq. (6.1) in the following algorithm:

For $k = 1, n_3$
    For $j = 1, n_2$
        For $i = 1, n_1$
            $g_{i,j}^k = f_{i,j}^k - e_{i,j}^k v_{i,j}^{k-1} - d_{i,j}^k x_{i,j-1}^k$
        end $i$
        For $i = 1 , n_1$
            $x_{i,j}^k = \gamma_{i,j}^k g_{i-1,j}^k + \alpha_{i,j}^k g_{i,j}^k + \beta_{i,j}^k g_{i+1,j}^k$
        end $i$
    end $j$
    For $j = n_2, 1$
        For $i = 1, n_1$
            $v_{i,j}^k = x_{i,j}^k - w_{i,j}^k v_{i,j+1}^k$
        end $i$
    end $j$
end $k$
For $k = n_3 , 1$
    For $j = 1, n_2$
        For $i = 1, n_1$
            $u_{i,j}^k = v_{i,j}^k - s_{i,j}^k u_{i,j}^{k+1}$
        end $i$
    end $j$
end $k$

We note that there are several possibilities for vector/parallel operations in this algorithm. The loop generating $\{x_{i,j}^k\}$ can be started once the first three components for $\{g_{i,j}^k\}$

are obtained ($i = 1,2,3$). Thus the generation of $\{g^k_{i,j}\}$ and $\{x^k_{i,j}\}$ can proceed in parallel. Furthermore, the calculation of $\{x^k_{i,j}\}$ can be done in vector mode.

## 7  NUMERICAL RESULTS

We have implemented Methods C and R as well as the methods based on approximate inverses with LDAI or LTAI in a software package for testing the effectiveness of these methods. In addition, we have obtained an implementation of the ILU and MILU methods from Kaufman [6]. The problem studied is given in Eqs. (1.1)-(1.2). The velocity parameters $\alpha^x$, $\alpha^y$, and $\alpha^z$ are chosen to create linear systems of varying difficulty. The diffusion coefficients are set to one in this problem. To characterize these problems, we consider the cell Peclet number to be

$$P_c = \frac{\|Velocity\ Field\|\ \|Cell\ Size\|}{\|Diffusion\ Coefficient\|} \qquad (7.1)$$

with suitably chosen norms. In our case we used

$$P_c = \left(\frac{\sqrt{3}}{N}\right)\sqrt{\left(\frac{\alpha^x}{16}\right)^2 + \left(\frac{\alpha^z}{16}\right)^2 + (\alpha^y)^2}\ , \qquad (7.2)$$

where $N$ is the number of points in each dimension. Note that we conduct experiments in which the number of variables per axis is the same in each direction. The parameter $P_c$ gives a measure of the ratio per cell of the convective term magnitude to the diffusive term magnitude and is related to the physical dimensionless parameter—the Peclet number. Note that $P_c = 0$ is a diffusion problem with no convection, and we obtain a positive definite symmetric linear system. While we consider $P_c = 0$, we do not take advantage of either the positive-definiteness or symmetry of the matrix. Other values of $P_c$ lead to possibly indefinite and nonsymmetric systems.

Once we have selected a problem, we apply all six preconditioners to it. We wish to use the same starting point for all six preconditioners and seek six orders of magnitude reduction in the initial residual. Method C, however, has a special property (see [7]) that the sum of residuals in each plane of the three-dimensional problem is preserved. Thus, the starting point should have a residual with the property that the sum of residuals on each plane is zero. In particular, such a starting point can be generated by solving

$$Px = h\ , \qquad (7.3)$$

where $P$ is the columnsum preconditioner. In the experiments we generate this starting point and use it for all six methods.

Two types of results will be presented. We present iteration histories where we plot the log of the normalized residual versus computation time. The normalized residual is the true residual error scaled by the initial residual. The computation time in seconds includes the time required to generate the preconditioner (which is negligible to graphical resolution). These computations are done on a $20 \times 20 \times 20$ grid. We also present the spectrum of the preconditioned matrices for two Peclet values on a $9 \times 9 \times 9$ grid. For the conjugate-gradient method the relevant matrix to study is $Z^T Z$. Because of the storage required to calculate the spectrum, we treat only a $9 \times 9 \times 9$ grid. We calculate the eigenvalues of this matrix to provide some insight into the relative effectiveness of the preconditioners. The spectrum should be relatively near one and should cluster near this value. However, the conjugate-gradient method can eliminate individual "outlying" eigenvalues rapidly. In numerical applications, extreme outlying eigenvalues can be reintroduced after being eliminated because of roundoff effects [13]. All of these computations were conducted on a CRAY X-MP/14 at Argonne National Laboratory.

To begin with, we consider the diffusion problem $P_c = 0$. Figure 1 gives the iteration histories. Note that since the true residual is monitored, it can increase during the iterations and does so in three cases. Note also that the tridiagonal approximate inverse and the rowsum methods are the most effective, while the MILU method is least effective. The other three methods fall in between. Figure 2 shows the spectrum for the six methods calculated on a $9 \times 9 \times 9$ grid. Note that the spectrum for the column-sum method has a maximum value of 186. The relatively poor performance of MILU can be related to existence of clusters away from one. The good performance of the tridiagonal approximate inverse and rowsum methods can be related to their relatively clustered spectrum near one. It should be noted that while the spectrum gives some insight into the performance of the method, it cannot be related completely to the specifics of a particular iteration history.

In Figure 3 we present iteration histories for $P_c = 0$ on a $9 \times 9 \times 9$ grid so that a direct comparison with Figure 2 can be made. While the quantitative details differ in the $9 \times 9 \times 9$ case from those in the $20 \times 20 \times 20$ case, the qualitative behavior is the same; i.e., the tridiagonal approximate inverse, the rowsum method, and the ILU method give the fastest performance.

In Figure 4 we use $P_c = 1$ on the $20 \times 20 \times 20$ grid. Note that the performance is quite different from that in the pure diffusion case. In particular the MILU and column-sum methods do not converge. The tridiagonal approximate inverse, columnsum

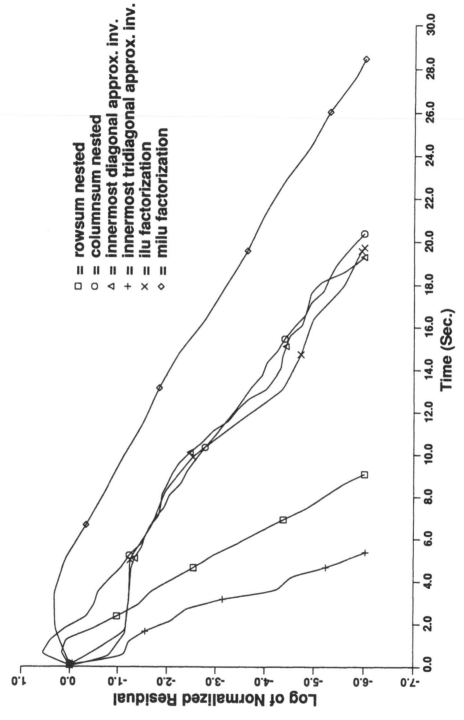

Figure 1.  Iteration History For $P_c=0$ With 20 x 20 x 20 Grid

Figure 2. Preconditioned Eigenvalues For $P_c=0$ With 9 x 9 x 9 Grid

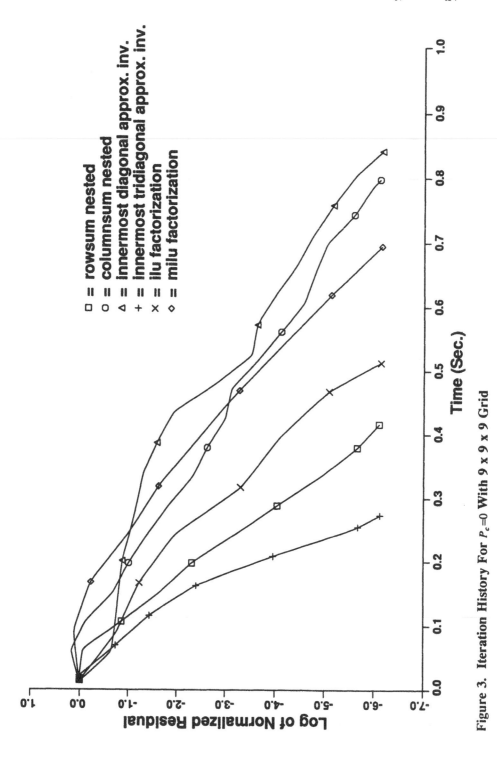

Figure 3.   Iteration History For $P_c = 0$ With 9 x 9 x 9 Grid

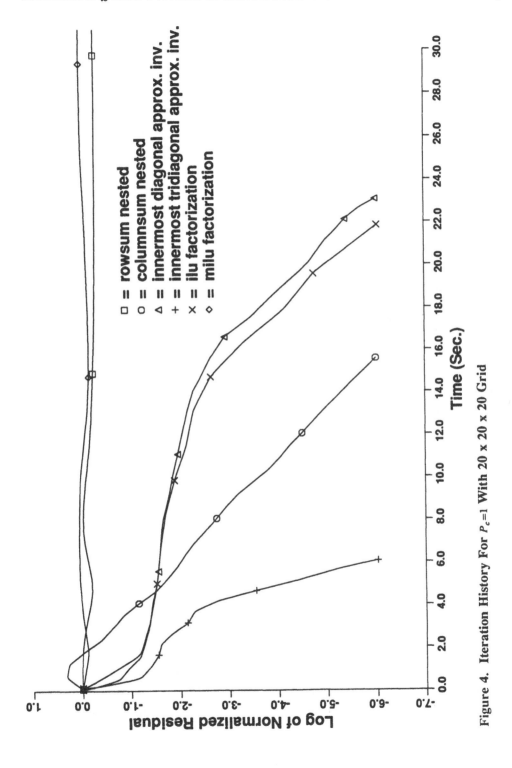

Figure 4. Iteration History For $P_c = 1$ With 20 x 20 x 20 Grid

method, and the ILU method are most efficient. Figures 5, 6, and 7 provide similar results for $P_c$ = 2, 3, and 4. In Figure 8 the spectra for the $9 \times 9 \times 9$ case associated with $P_c$ = 4 is given. The corresponding iteration history for the $9 \times 9 \times 9$ case is given in Figure 9. Note that the spectrum of the rowsum preconditioner ranges up to $10^5$ and the iterations do not converge even though only ten eigenvalues are beyond the scale of the figure. We feel that the lack of convergence is due to rounding errors reintroducing the outlying eigenvalues. While the MILU spectrum is poorly clustered, the method does eventually converge.

Finally in Figure 10 we present the iteration history for $P_c$ = 12 on a $20 \times 20 \times 20$ grid. This cell Peclet number is beyond the range where physically meaningful results would be expected. However, we present the iteration history to see the effect of larger $P_c$. In fact the results are similar to those in Figure 7 ($P_c$ = 4).

While the focus in this paper is on the selection of preconditioners, there is the question of whether the selection of the normal-form conjugate gradient method affects the results in our computations. We would hope that the use of the normal-form conjugate gradient method does not lead to the selection of a preconditioner that performs well only with that iterative method. In Figures 11-13 we studied the four preconditioners developed in this paper with ORTHOMIN(1) and $P_c$ = 0, 1, and 4. As expected, when convergence occurs with ORTHOMIN, much less computation time is required than with normal-form conjugate gradient iterations. However, the use of ORTHOMIN(1) is not as robust as the normal-form conjugate gradient method. In particular, at $P_c$ = 4 only the columnsum method converges. This situation can be improved by using more history in the ORTHOMIN procedure. In these figures the tridiagonal approximate inverse and columnsum methods perform best (when they converge). This conclusion is consistent with our results with the normal-form conjugate gradient method.

All of these results were obtained with the use of the CFT 1.15 (version 3) vectorizing compiler. While there was an effort to study the key routines to improve their performance, there remains the question of how the relative performance of the preconditioners is affected by the use of a vectorizing architecture and how well the methods utilize this architecture. To answer these question, we ran the cases $P_c$ = 0 and 4 without the use of vectorization. These results are presented in Figures 14 and 15. In terms of relative performance we find that the results are similar to those obtained with vectorization (compare Figure 1 with 14 and Figure 7 with 15). The principal result is that without vectorization the ILU and MILU methods perform relatively faster. Stated another way, the use of vectorization reduced computation time only by 10% for the

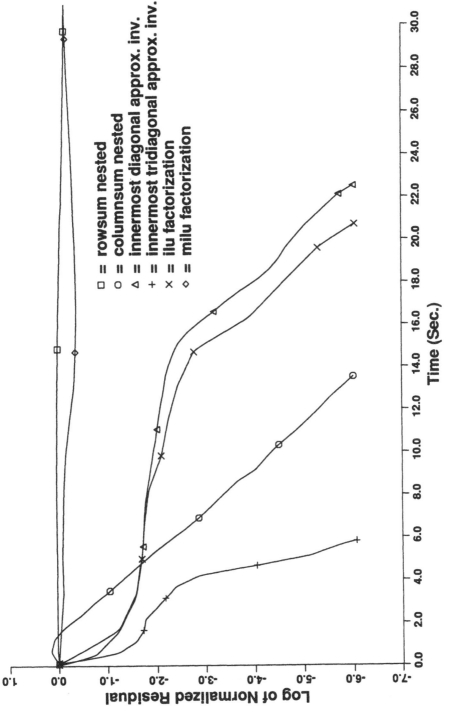

Figure 5.  Iteration History For $P_c=2$ With 20 x 20 x 20 Grid

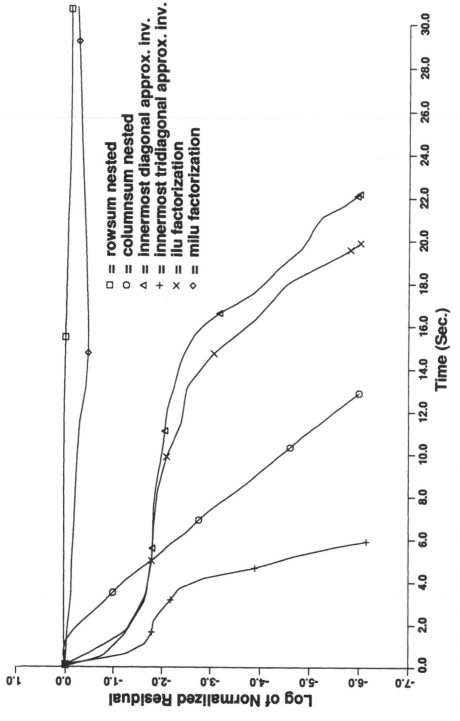

Figure 6.  Iteration History For $P_c=3$ With 20 x 20 x 20 Grid

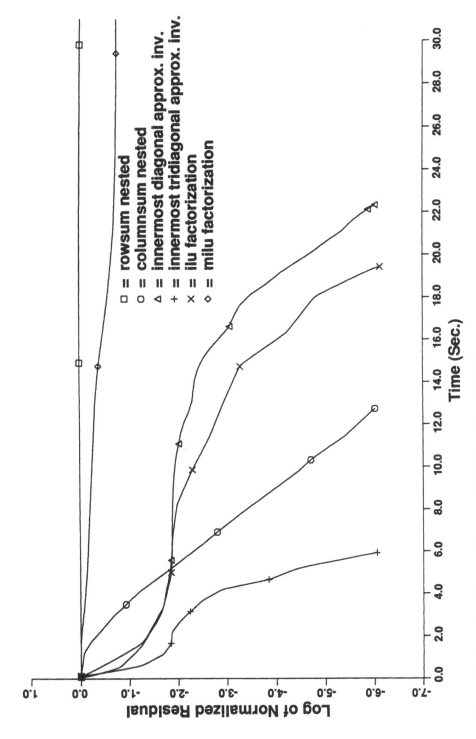

Figure 7. Iteration History For $P_c=4$ With 20 x 20 x 20 Grid

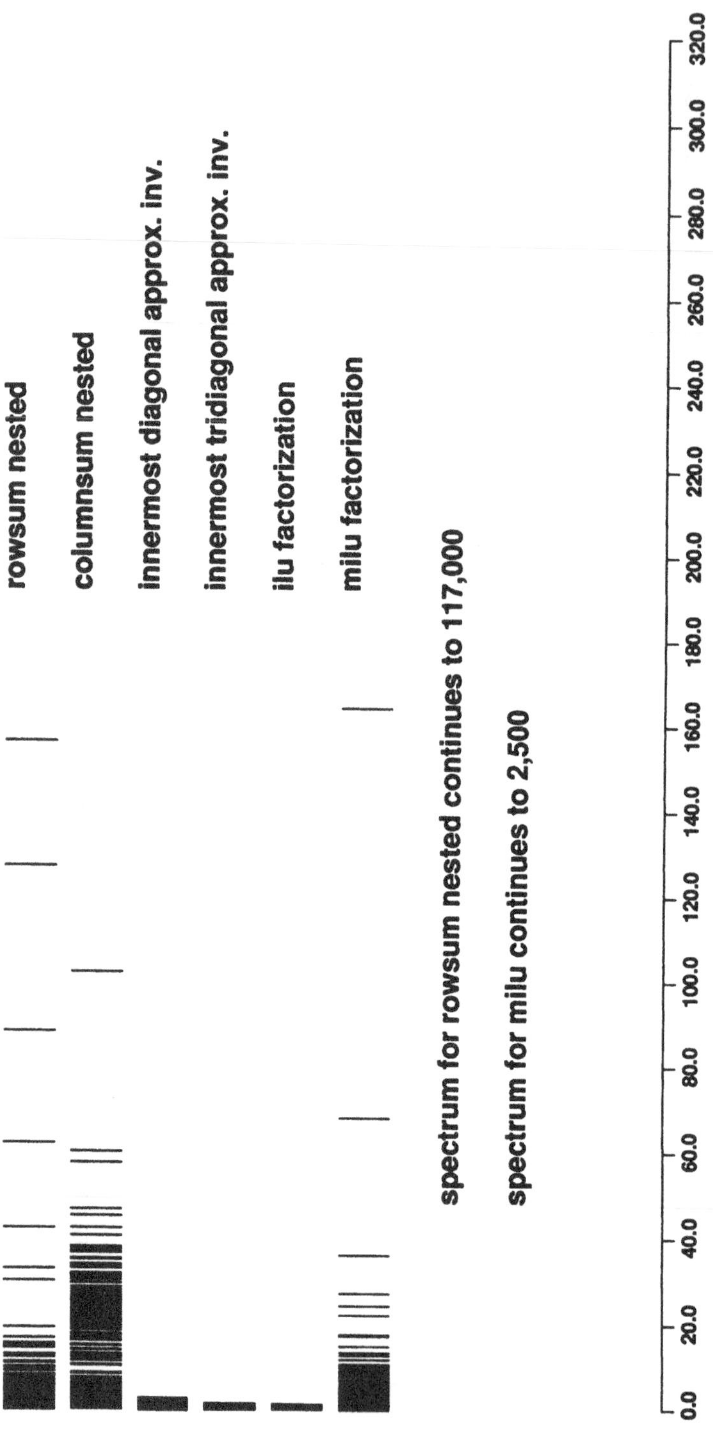

Figure 8.  Preconditioned Eigenvalues For $P_c = 4$ With 9 x 9 x 9 Grid

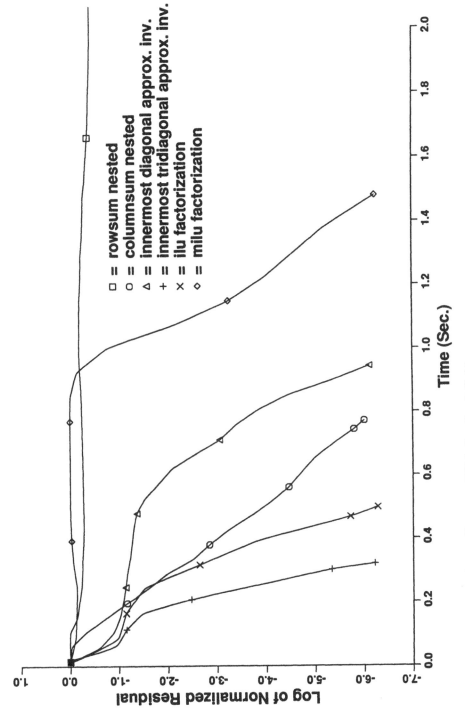

Figure 9.  Iteration History For $P_c=4$ With 9 x 9 x 9 Grid

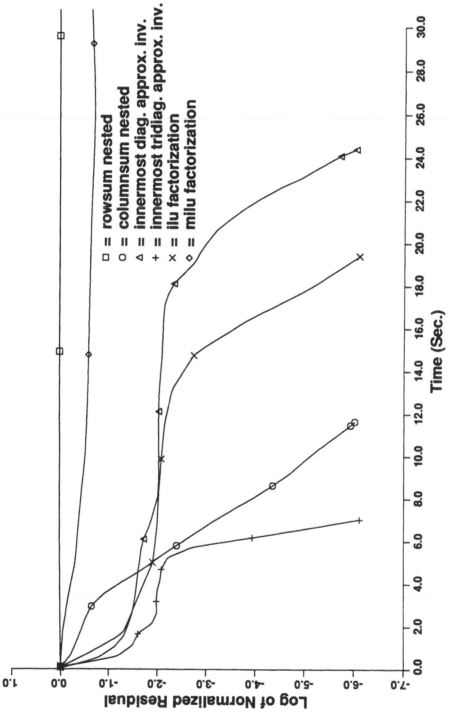

Figure 10.  Iteration History For $P_c$ =12 With 20 x 20 x 20 Grid

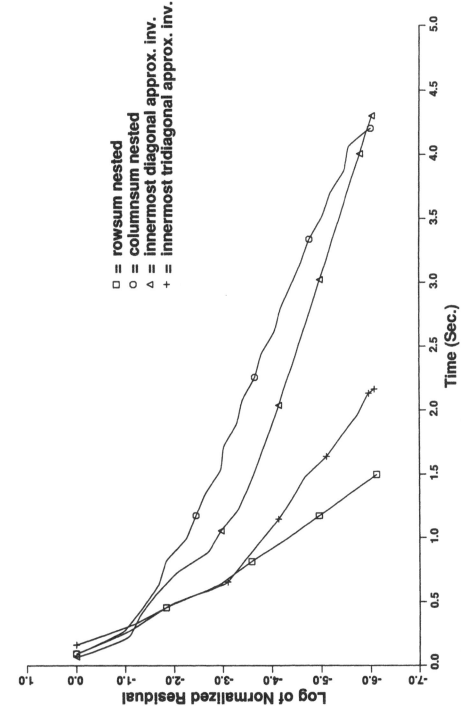

Figure 11. Iteration History Using ORTHOMIN(1) For $P_c = 0$ With 20 x 20 x 20 Grid

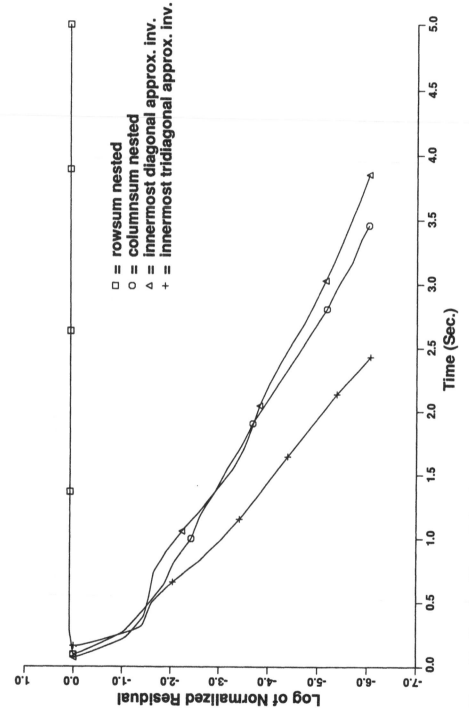

Figure 12.  Iteration History Using ORTHOMIN(1) For $P_c=1$ With 20 x 20 x 20 Grid

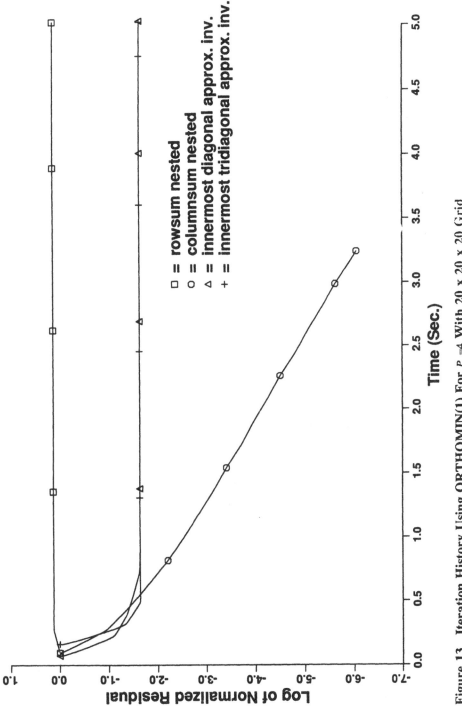

Figure 13. Iteration History Using ORTHOMIN(1) For $P_c = 4$ With 20 x 20 x 20 Grid

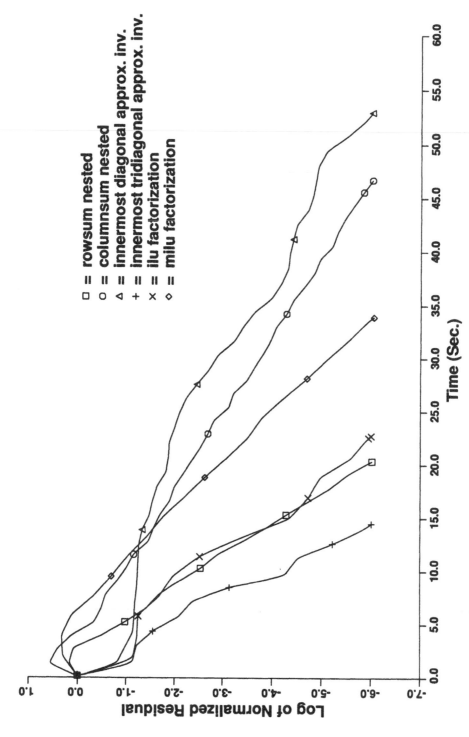

Figure 14. Iteration History Using No Vectorization For $P_c=0$ With 20 x 20 x 20 Grid

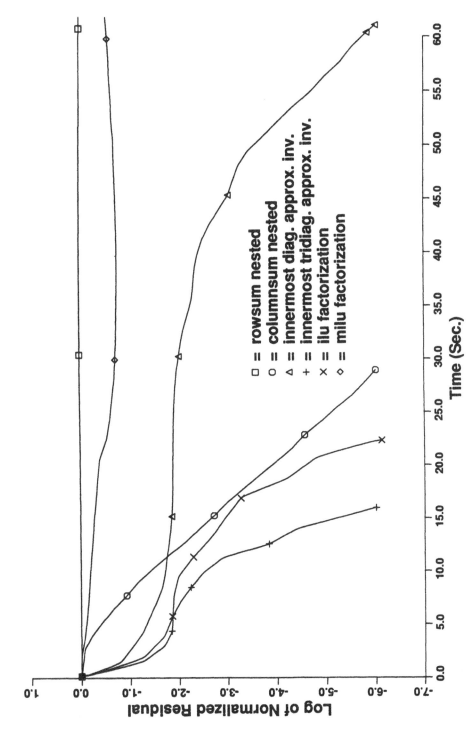

Figure 15. Iteration History Using No Vectorization For $P_c=4$ With 20 x 20 x 20 Grid

ILU and MILU methods but by a factor of two for the other methods. We feel this result occurred because the innermost level of the nested block solve steps are executed frequently and vectorize well.

## 8 CONCLUSIONS

The major conclusion of this study is that the block incomplete decomposition based on tridiagonal approximate inverses seems to be the most robust preconditioning technique among those studied. The rowsum preconditioner by itself does not converge for nonzero cell Peclet numbers. The results for the columnsum nested incomplete factorization preconditioner are very encouraging for moderately large cell Peclet numbers. The ILU preconditioner generally leads to convergence but is not the most efficient method. The Modified ILU preconditioner does not lead to convergence for nonzero cell Peclet values.

The results for the approximate inverse preconditioners lead us to study their extension to block approximate inverse methods in future work.

## REFERENCES

[1] Leaf, G. K., and M. Minkoff, "DISPL1: A Software Package for One and Two Spatially Dimensioned Kinetics-Diffusion Problems," Argonne National Laboratory Report ANL-84-56, 1984.

[2] Leaf, G. K., M. Minkoff, and J. C. Díaz, "Preconditioned Iterative Methods for Differential Equations," Proceedings of the Sixth IMACS International Symposium on Computer Methods for Partial Differential Equations, ed. Vichnevetsky and Stepleman, Bethlehem, Pennsylvania, June 1987, pp. 551-555.

[3] Roache, P. J. "Computation Fluid Dynamics," Hermosa Publications, Albuquerque, New Mexico, 1976.

[4] Dongarra, J. J., G. K. Leaf, and M. Minkoff, "A Preconditioned Conjugate Gradient Method for Solving a Class of Non-Symmetric Linear Systems," Argonne National Laboratory Report ANL-81-71, 1981.

[5] Gustafsson, I., "Modified Incomplete Cholesky (MIC) Methods)," in

*Preconditioning Methods: Theory and Applications,* ed. D. J. Evans, Gordon and Breach, New York, 1983, pp. 265-293.

[6] Kaufman, L. Private communication, 1988.

[7] Appleyard, J. R., and I. M. Cheshire, "Nested Factorization," Proceedings of the Res. Simulation Symposium, San Francisco, California, November 1983, pp. 315-324.

[8] Concus, P., G. H. Golub, and G. Meurant, "Block Preconditioning for the Conjugate Gradient Method," SIAM J. Sci. Stat. Comp., 6 (1985), 220-252.

[9] Díaz, J. C., and C. G. Macedo, Jr., "Vectorizable Nested-Factorization Preconditionings for Nonsymmetric Systems of Equations," Proceedings of the Second Workshop on Applied Computing, Tulsa, Oklahoma, March 1988, pp. 22-28.

[10] Díaz, J. C., and C. G. Macedo, Jr., "Vectorizable Block Preconditionings for Nonsymmetric Systems of Equations," Computational and Applied Mathematics, to appear.

[11] Cosgrove, J. D. F., and J. C. Díaz, "Fully Parallel Preconditioning for Sparse Systems of Equations," Proceedings of the Second Workshop on Applied Computing, Tulsa, Oklahoma, March 1988, pp. 29-34.

[12] Meijirink, J. A., and H. A. van der Vorst, "An Iterative Solution Method for Linear Systems of Which the Coefficient Matrix Is a Symmetric M-Matrix," Math. Comp., 31 (1977) 148-162.

[13] Van Der Vorst, H. A., Private Communication, 1988.

[14] Benson, M. W., and P. O. Frederickson, "Iterative Solution of Large Sparse Linear Systems Arising in Certain Multidimensional Approximation Problems," Utilitas Math., 22 (1982) 127-140.

[15] Kolotilina, L. Y., and A. Y. Yeremin, "On a Family of Two-Level Preconditionings of the Incomplete Block Factorization Type," Sov. J. Nemer. Anal. Math. Modelling, 1 (1986) 293-320.

# Performance of the Chbyshev Iterative Method, GMRES and ORTHOMIN on a Set of Oil-Reservoir Simulation Problems

SUSANA GÓMES, Departamento de Análisis Numérico, IIMAS, UNAM. Apdo. Postal 20-726, México D. F., 01000, México

JOSÉ LUIS MORALES, Departamento de Simulación. Instituto de Investigaciones Eléctricas. Interior Internado Palmira, Cuernavaca, Mor., México.

## 1. INTRODUCTION

It is well known that the numerical solution of the equations modeling complex phenomena usually require large amounts of computational resources. Frequently, most of the work is done when solving large sparse systems of algebraic linear equations.

In this paper we present some evidence on the behaviour of three iterative methods for solving the nonsymmetric linear system of equations

$$A\ x = b \tag{1.1}$$

where $A$ in an $N \times N$ nonsingular seven banded matrix and $x$ and $b$ are $N$-dimensional vectors. These systems of equations arise when a seven point centered finite differences scheme is applied to solve the set of partial differential equations that model oil reservoir problems. We assume here that apart from knowing the general method used to simulate these problems, almost nothing is known about the mathematical properties of system (1.1) *a priori*. We do know however that these systems are large, highly structured and in general, nonsymmetric.

We have selected three different methods and preconditioners. One of these methods has not been extensively applied to these problems because it is parameter dependent, (some knowledge of the spectrum of the matrix is needed). This is the Chebyshev Iterative Method, developed by Manteuffel [13]. It is based on the theory of Chebyshev polynomials that possess certain optimality and recurrence properties. Manteuffel has also developed an adaptive procedure that dynamically finds the parameters, based on the residuals information [14]. We will use this procedure in this paper.

The second method has been recently developed called Generalized Minimal Residual Method, GMRES, due to Saad and Schultz [21]. It is a variational method, because it minimizes a norm of the residual at each step over a subspace. It is cheaper per iteration than other variational methods and is a parameter free method. Only limited numerical experiments have been done with it so far particularly with real problems modelled with $PDE$'S. It is thus a good candidate for use in this paper.

The third method has been widely and successfully used for large scale simulation on a large variety of problems. This method is ORTHOMIN and was first developed by Vinsome [22]. It is a minimum residual norm technique, See Elman [8]. These three methods present the common features of low computational cost per iteration and low storage.

The preconditioners that we have selected, $DKR, DD$ and $AB$, were developed and have been frequently used for oil reservoir research. See Dupont, Kendall and Rachford [6], and Behie and Vinsome [3] for further details. These preconditioners are all based on the idea of incomplete $LU$ factorization. They involve different storage and computational costs. In section 2, we present the truncated methods ORTHOMIN($m$) and $GMRES(m)$ methods together because they share some features. The Chebyshev Iterative Method will be presented in section 3.

The preconditioners will be given in section 4. The description of the problems and the numerical results will be presented in section 5. A discussion of these results will be given at the conclusion of the paper.

## 2. ORTHOMIN($m$), and GMRES($m$)

These methods compute an approximation to the solution as

$$x_{k+1} = x_0 + \sum_{j=1}^{k} a_j p_j$$

where $p_j$ are directions vectors and $a_j$ are the step length chosen to minimize a norm of the residual over a subspace.

ORTHOMIN($m$) computes the direction vector $p_{k+1}$ at each step so that it is $A^T A$ orthogonal to the last $p_j$, $j = k - m + 1, \ldots, k$. The choice of $a_j$ makes the residual $r_{k+1}$ orthogonal to $Ar_j$, $j = k - m + 1, \ldots, k$ so that they are conjugate (in the nonsymmetric case, conjugancy holds only on one side). This choice of $a_j$ also minimizes $||r_{k+1}||_2$ over the subspace

$$x_{k-m} + \langle p_{k-m}, \ldots, p_k \rangle$$

GMRES($m$), does not compute the approximation $x_{k+1}$ at each step, but only after a set of orthonormal vectors $\{p_k\}$, ($\{v_k\}$ in the GMRES notation), have been computed. These vectors are generated using a Gram-Schmidt procedure and form a basis of the Krylov subspace $\langle r_0, Ar_0, \ldots, A^{k+1}r_0 \rangle$. The direction vector to compute $x_k$ is calculated as a linear combination of these $\{p_k\}$. The coefficients for this combination, which could also be viewed as step lengths along the directions, are computed so that the residual at point $x_k$ is minimized over the Krylov subspace generated with the actual basis. There is not an explicit orthogonality condition on the residuals. This method is restarted every $m$ iterations. This means that a new set of orthonormal vectors have to be created, starting from $p_1 = \frac{r_0}{\|r_0\|}$ and $r_0$ being the last residual found.

Both methods are variational, that is the iterates are computed so that a norm of the residual is minimized over some subspace. For the complete methods, *i.e.* $m = N$, they both converge in at most $N$ steps. The truncated versions, ORTHOMIN($m$) and GMRES($m$), satisfy an $r$-linear error bounds available (ORTHOMIN, has actually $q$-linear bound). In addition, ORTHOMIN($m$) has guarantee of convergence when the symmetric part of $A$ ($SPA$) is positive definite and for GMRES($m$) the convergence results allows the $SPA$ to be indefinite.

## ORTHOMIN($m$)

To present the method ORTHOMIN($m$), it is convenient to first provide some insight about the Conjugate Residual method $CR$,see [4]. In the $CR$ method the approximate solution is found at every iteration as

$$x_{k+1} = x_k + a_k p_k \tag{2.1}$$

where $p_k$ are direction vectors that have to satisfy certain conjugacy conditons. Thus the process is finite, $a_k$ is the step length at stage $k$ and is chosen so that the following error function is minimized

$$E(x_{k+1}) = \|b - Ax_{k+1}\|_2 = \|b - A(x_k + a_k p_k)\|_2$$
$$= \|r_{k+1}\|_2 \tag{2.2}$$

over the affine space $x_0 + \langle p_0, \ldots, p_k \rangle$, with $p_0 = r_0$. The new residual is recursively computed as

$$r_{k+1} = r_k - a_k A p_k. \tag{2.3}$$

The direction vectors $p_k$ when $A$ is symmetric positive definite is obtained from

$$p_{k+1} = r_{k+1} + b_k p_k \tag{2.4}$$

where

$$b_k = -\frac{(Ar_{k+1}, Ap_k)}{(Ap_k, Ap_k)} \tag{2.5}$$

This choice of $b_k$ makes the direction vectors $A^T A$ orthogonal, that is

$$(Ap_k, Ap_i) = 0, \quad k \neq i \tag{2.6}$$

When $A$ is nonsymmetric the orthogonality relation (2.6) is not necessarily satisfied and in order to compute a new direction, one has to take into account all previous directions in a Gram-Schmidt process with the $A^T A$ scalar product. We have then

$$p_{k+1} = r_{k+1} + \sum_{j=0}^{k} b_j^{(k)} p_j$$

$$\text{where} \quad b_j^{(k)} = -\frac{(Ar_{k+1}, Ap_j)}{(Ap_j, Ap_j)}, \quad j \leq k \tag{2.7}$$

The approximation $x_{k+1}$ generated with this direction minimizes $E(x_{k+1})$ over the same space $x_0 + \langle p_0, p_1, \dots, p_k \rangle$ with the same step length $a_k$ as in the symmetric case. Also the property $(Ap_k, Ap_j) = 0, k \neq j$ holds. This method is known as the Generalized Conjugate Residual method, $GCR$, and it converges in as most $N$ steps, see Elman [8] and Eisenstat et al [7]. Other ways to generalize the Conjugate Residual Method, can be found in Axelsson [2], and in Jea and Young [12].

This method is very expensive in terms of storage because one has to keep all previous directions in memory. An alternative is to take a descent direction $A^T A$ orthogonal to the last $m$ directions, where $x_{k+1}$ minimizes $E(x_{k+1})$ over the affine space $x_{k-m} + \langle p_{k-m}, \dots, p_k \rangle$.

In this case only $m$ direction vectors are kept in memory. This is the ORTHOMIN($m$) method presented by Vinsome [22]. This method and the $GCR$ are equivalent to the Conjugate Residual method, if $A$ is symmetric positive definite.


Algorithm. ORTHOMIN($m$)                                                  (2.8)

  1. Start. Choose $x_0$ and compute $r_0 = b - Ax_0$.

       Set $p_0 = r_0$.

  2. Iterate. For $k = 0, \dots$ until convergence DO

$$a_k = \frac{(r_k, Ap_k)}{(Ap_k, Ap_k)}.$$

$$x_{k+1} = x_k + a_k p_k$$

$$r_{k+1} = r_k - a_k Ap_k$$

$$b_j^{(k)} = \frac{(Ar_{k+1}, Ap_j)}{(Ap_j, Ap_j)}, \quad j = k - m + 1, \dots, k$$

$$p_{k+1} = r_{k+1} + \sum_{j=j_k}^{k} b_j^{(k)} p_j$$

where $j_k = max(0, k - m + 1)$.

Convergence is attained when

$$\|r_{k+1}\|_2 < TOL \tag{2.9}$$

The storage requirements are for the vectors $x, r, Ar, \{p\}_{j_k}^{k+1}, \{Ap\}_{j_k}^{k+1}, A$ and $b$. Thus we need $(2m + 3)N$ arrays per iteration. The computational cost *(multiplications)* is $(3m + 4)N + 1mv$ where $mv$ is the cost of a matrix vector product.

We do not compute the product $Ap_{k+1}$ because these vectors are updated as follows

$$Ap_{k+1} = Ar_{k+1} + \sum_{j=j_k}^{k} b_j^{(k)} Ap_j \tag{2.10}$$

We note that when $m = 0$, $p_{k+1} = r_{k+1}$, and the resulting method is called the minimum residual method, $MR$, and in the symmetric case it resembles the gradient method.

### Convergence

Elman [8] has proven global convergence for ORTHOMIN($m$) when the symmetric part of $A$ is positive definite. He gets the following $r$-linear error bound for the residual norms

$$\|r_k\|_2 \leq \left[1 - \frac{\lambda_{min}^2(M)}{\lambda_{max}(A^T A)}\right]^k \|r_0\|_2 \tag{2.11}$$

Where $M$ is the symmetric part of $A$.

But, as pointed out by Dennis and Turner [5], Elman really obtains a $q$-linear bound

$$\|r_k\|_2 \leq \left(1 - \frac{\lambda_{min}^2(M)}{\lambda_{max}(A^T A)}\right) \|r_{k-1}\|_2. \tag{2.12}$$

If the symmetric part of $A$ is indefinite, there is no guarantee of convergence.

## GMRES$(m)$

This method based on previous work by Saad see [19,20], generates a sequence of points $x_k = x_0 + z_k$, where $z_k$ belongs to a Krylov subspace $K_k := < v, Av, \ldots, A^{k-1}v >$ where $x_0$ is an initial approximation to the solution. The updating vector $z_k$ is selected so that it minimizes over $K_k$ the error function

$$E(z) = ||b - Ax_k||_2 = ||b - A(x_0 + z_k)||_2 = ||r_0 - Az_k||_2 \qquad (2.13)$$

The method starts with Arnoldi's method. That is, using a Gram-Schmidt process it orthonormalizes a Krylov sequence $\{v_1, Av_1, \ldots, A^{k-1}v_1\}$ to generate an orthonormal matrix $V_k = [v_1, \ldots, v_k]$ whose columns span $K_k$. The vector $v_k$ is then orthogonal to all previous $v_i, i < k$. The matrix $A$ can be transformed into a Hessenberg matrix $H_k$, through the relation

$$V_k{}^T AV_k = H_k \qquad (2.14)$$

Now suppose we take $v_1 = \frac{r_0}{||r_0||}$, where $r_0$ is the initial residual $r_0 = b - Ax_0$; then Arnoldi's algorithm generates the $k$th matrix $H_k$ with elements $h_{ij}$ .

Algorithm. ARNOLDI $\qquad\qquad\qquad\qquad\qquad\qquad\qquad\qquad\qquad\qquad\qquad (2.15)$

1. Start. Choose $x_0$ and compute $r_0 = b - Ax_0$.

   Set $v_1 = \frac{r_0}{||r_0||}$.

2. Iterate. For $j = 1, 2 \ldots k$ Do:

$$h_{ij} = (Av_j, v_i), \quad i = 1, 2, \ldots, j$$

$$\hat{v}_{j+1} = Av_j - \sum_{i=1}^{k} h_{ij}v_i$$

$$h_{j+1,j} = ||\hat{v}_{j+1}||, \text{and}$$

$$v_{j+1} = \hat{v}_{j+1}/h_{j+1,j}$$

Once we have matrix $V_k$, we compute the approximation $x_k = x_0 + z_k$,

where $z_k$ can be expressed as

$$z_k = V_k y_k \qquad (2.16)$$

and the coefficient vector $y_k$ minimizes (2.13). Now, suppose that we run Arnoldi's method $k$ steps. Then we have $V_k$ and $V_{k+1}$ that satisfy the following relation

$$AV_k = V_{k+1} \, \widetilde{H}_k \qquad (2.17)$$

where the matrix $\widetilde{H}_k$ is the $k+1, k$ Hessenberg matrix generated as in (2.14), plus a row with a single element $h_{k+1,k}$ also generated in the $k$th iteration using algorithm (2.15). Then if we set $z_k = V_k y_k$ the norm to be minimized (2.13) becomes

$$E(y) = \min_{y \in \Re^k} \|r_0 - AV_k y\|_2 = \min_{y \in \Re^k} \|\beta e_1 - \widetilde{H}_k \, y\|_2 \qquad (2.18)$$

where $\beta = \|r_0\|$ and we have used the fact that $V_{k+1}$ is orthonormal. The algorithm we use to solve this minimization problem will be discussed later.

As $k$ increases, the storage needed to keep all $v_i$ becomes a fundamental problem and the number of multiplications increase at a rate $\frac{1}{2}k^2 N$.

To avoid these difficulties we restart the algorithm every $m$ steps, for $m$ fixed. That is, at the end of the $m$th iteration, we compute the approximation $x_m$ and its residual $r_m = b - Ax_m$. We reset $x_0 = x_m$ and take as the initial orthonormal vector $v_1 = \frac{r_m}{\|r_m\|}$. The algorithm for $GMRES(m)$ is given below

**Algorithm.** $GMRES(m)$ $\hspace{6cm}$ (2.19)

1. Start. Choose $x_0$ and compute $r_0 = b - Ax_0$ and $v_1 = \frac{r_0}{\|r_0\|}$.

2. Perform $m$ steps of Arnoldi's algorithm to generate $V_m$ and $\widetilde{H}_m$, starting with the actual $x_0$ and $v_1$.

3. Compute $x_m = x_0 + V_m y_m$ where $y_m$ minimizes $\|\beta e_1 - \widetilde{H}_m \, y\|, y \in \Re^m$
4. Restart. Compute $r_m = b - Ax_m$.
   $\hspace{1.5cm}$ If $\|r_m\| < TOL$, $\hspace{0.5cm}$ then $\hspace{0.5cm}$ $STOP$.
   $\hspace{1.5cm}$ Else compute $x_0 := x_m, v_1 := \frac{r_m}{\|r_m\|}$ and go to step 2.

There are several possible implementations for this algorithm; see Walker [23] and Elman et al [9]. We follow the implementation of Saad and Schultz [21] which we give below.

### $GMRES(m)$ Practical Implementation

Here we consider some of the relevant issues of the practical implementation of the method. We will outline two basic aspects; the first deals with the orthogonalization of the basis vectors $\{v_j\}$, and the second with the least squares solution of problem (2.18).

In practical implementations the Gram-Schmidt $(GS)$ orthonormalization procedure in Arnoldi's algorithm should be replaced by the modified version $(MGS)$ without explicit normalization (see Golub and Van Loan [11]). Walker [23] conducted some numerical experiments for indefinite matrices and he suggests that the more stable Householder orthogonalization should be preferred in some cases. But the Householder version is almost three times more expensive than the $MGS$ version and when using the latter to solve the problems considered in this paper, we did not have stability problems. Thus, we will use the MGS procedure.

The second issue concerns the solution of the linear least squares problem (2.18). Due to the special structure of matrix $\tilde{H}_k$, this problem can be solved factorizing $\tilde{H}_k = Q_k R_k$, with plane rotations and then backsolving the triangular system

$$R_k y_k = Q_k \beta e_1 = g_k \qquad (2.20)$$

where we have removed the last row $(k + 1)$ of zeros of $R_k$ and the last component of $g_k$.

If the factorization is updated at each step of algorithm (2.19), then the definition of functional $E(y)$ naturally provides the norm of the residual vector $r_k$ as the last component of $g_k$ without explicit calculation of $x_k$. Morever, if the algorithm has converged, *i.e* $\|r_k\| < TOL$, it stops even if $k < m$ and it then generates the solution vector, avoiding unnecessary operations. Additional savings can be obtained at step $m$ if the vector $v_{m+1}$ is not generated. Note that $v_{m+1}$ does not participate in forming $x_m$; only its norm is necessary and it is in the element $h_{m+1,m}$ of $\tilde{H}_m$. This element can be calculated [21] in the form

$$|h_{m+1,m}|^2 = \|Av_m\|^2 - \sum_{i=1}^{m} h_{i,m}^2 \qquad (2.21)$$

For small $m$ a cheaper way to compute the residual vector, when restarting the method, it is based on the fact that $r_m$ is proportional to $v_{m+1}$ and this vector is a linear combination of $Av_m$ and the previous $\{v_k\}, k < m$ (see [21] for details). The cost is $(m + 1)N$ and this approach is thus useful when this cost is not greater than the cost of the matrix-vector product in $b - Ax_m$.

A similar observation holds when calculating the element $h_{m+1,m}$ described above because it is based on the orthogonality of the vectors $\{v_k\}$, which is progressively lost as $m$ increases.

The storage requirement for Arnoldi's method is $(m+1)N$, plus one array to compute $x$. Thus the total storage is $(m + 2)N$. The average computational cost per iteration is $\left(m + 3 + \frac{1}{m}\right)N + mv$. The cost to compute $y_k$ is not included, because it is $0(m^2)$ which is negligible when $m << N$.

### Convergence

$GMRES$ is a generalization of $MINRES$ of Paige and Saunders [18] for symmetric indefinite systems. The algorithm can only break down if $h_{j+1,j} = 0$ for some $j$. But Saad and Schultz proved that this condition only holds at the solution. They showed that convergence occurs in at most $N$ steps. See also Dennis and Turner [5].

In the method $GMRES(m)$ the residual is minimized after $m$ steps and the sequence of residuals forms a nondecreasing sequence. For $m$ sufficiently large, $GMRES(m)$ converges and if $m = N$ it does so in one step. The authors also show that when the symmetric part of $A$ is positive definite, the $GMRES(m)$ method converges and they obtain the $r$-linear error bound for the residual norms

$$||r_{k+1}||_2 \leq \left[ 1 - \frac{\lambda^2 min(M)}{\lambda max(A^T A)} \right]^{m/2} ||r_0||_2 \qquad (2.22)$$

When the symmetric part of $A$ is not positive definite, they show that $GMRES(m)$ converges if $m$ is greater than a quantity that depends on the eigenvalue distribution and the condition number of a matrix $X$ defined so that $A = XDX^{-1}$. This bound is independent of the problem size (see page 867, [21]).

## 3. THE CHEBYSHEV ITERATIVE METHOD

This is an iterative method based on the use of Chebyshev polynomials in the complex plane (Manteuffel [13]). It can be applied to solve large and sparse nonsymmetric linear systems whose coefficient matrix has eigenvalues in the right half of the complex plane. In this paper we will use the adaptive version (Manteuffel [14]) that can be proven to converge when the coefficient matrix has positive definite symmetric part.

The method, starting from an initial approximation $x_0$, generates a sequence of approximations as follows

$$x_{k+1} = x_k + p_k = x_k + \sum_{i=0}^{k} \gamma_{ki} r_i \qquad (3.1)$$

Where $r_i$ is the residual at step $i$ and $p_k$ is the step direction. Defining the absolute error vector after $k$ steps as

$$\epsilon_k = x - x_k \qquad (3.2)$$

we have that

$$\epsilon_k = P_k(A)\epsilon_0 \qquad (3.3)$$

where $P_k(z)$ is a polynomial of degree $k$ such that $P_n(0) = 1$. After using norms

$$||\epsilon_k|| \leq ||P_k(A)|| \cdot ||\epsilon_0|| \qquad (3.4)$$

and the residuals also satisfy a similar relation

$$||r_k|| \leq || P_k(A)|| \cdot ||r_0|| \tag{3.5}$$

This inequality suggests that the sequence of polynomials $\{P_i(z)\}_{i=0}^k$ must be chosen monotonically tending to zero (in some norm), in order to get reductions in $||r_k||$ and $||\epsilon_k||$ at each iteration. In order to find these polynomials, the spectrum of $A$ should be known, but if we take a region containing the spectrum, and furthermore, if we are in the particular case when the region is bounded by an ellipse, with real focci and that does not contain the origin in its interior, the best polynomial $P_k(\lambda)$ has the form.

$$P_k(\lambda) = T_k((d-\lambda)/c)/T_k(d/c) \tag{3.6}$$

where $T_k$ is the $k$-th scaled and translated Chebyshev polynomial of the first kind. $P_k(\lambda)$ will be referred as the residual polynomial, $d$ is the center and $c$ is the focal distance of the ellipse.

These Chebyshev polynomials have a property of recurrence

$$\begin{aligned}
T_{k+1}(z) &= 2zT_k(z) - T_{k-1}(z) \;\;, \;\; k > 1 \\
T_0(z) &= 1 \;\;, T_1(z) = z
\end{aligned} \tag{3.7}$$

that makes the computation of the direction $p_k$ very simple, avoiding the storage of the complete set of residuals. Then (3.1) will now have the form

$$x_{k+1} = x_k + p_k = x_k + \alpha_k r_k + \beta_k p_{k-1} \tag{3.8}$$

where

$$\alpha_k = \frac{2}{c} \frac{T_k(d/c)}{T_{k+1}(d/c)} \tag{3.9}$$

$$\beta_k = \frac{T_{k-1}(d/c)}{T_{k+1}(d/c)}$$

and using the relation (3.7), we can generate $\alpha_k$ and $\beta_k$ recursively as follows:

$$\begin{aligned}
\alpha_1 &= \frac{2d}{2d^2 - c^2} \;\;, \beta_1 = d\alpha_1 - 1 \\
\alpha_k &= \left[ d - \left(\frac{c}{2}\right)^2 \alpha_{k-1} \right]^{-1} \;\;, \beta_k = d\alpha_k - 1
\end{aligned} \tag{3.10}$$

We first give the Chebyshev iteration method that assumes knowledge of the pair $(d, c)$ and afterwards we comment the adaptive procedure to compute this pair. For a given pair $(d, c)$ the basic iteration is

Algorithm. CHEBYSHEV (3.11)

1. Start. Choose $x_0$, compute $r_0 = b - Ax_0$ and $p_0 = r_0/d$

2. Iterate. For $k = 0, \ldots$ until convergence DO

$$x_{k+1} = x_k + p_k$$
$$r_{k+1} = b - Ax_k$$
$$\alpha_{k+1} = \begin{cases} 2d/(2d^2 - c^2), & \text{if} \quad k=0 \\ (d - (c/2)^2 \alpha_k)^{-1} & \text{if} \quad k > 0 \end{cases}$$
$$\beta_{k+1} = d\alpha_{k+1} - 1$$
$$p_{k+1} = \alpha_{k+1} r_{k+1} + \beta_{k+1} p_k$$

The convergence of the Chebyshev algorithm strongly depends on the pair $(d, c)$. We now briefly describe the salient aspects of Manteuffel's adaptive procedure [14] that dynamically estimates a near optimum pair that makes the convergence optimal in some sense . It starts with an initial pair $(d_0, c_0)$ (some suggestions on the initial choice can be found in [1]) and the iteration proceeds as in algorithm (3.11) until slow convergence or an increase in the residual norm is detected by proper tests. Then the adaptive phase estimates eigenvalues on an approximation of the convex hull of $A$ and a new pair is obtained by solving a $min - max$ problem. The iteration is then restarted with these new parameters.

It has been shown [13] that for large $n$

$$P_k(\lambda) \doteq S^k(\lambda) \tag{3.12}$$

where

$$S(\lambda) = \left[ (d - \lambda) + \sqrt{(d - \lambda)^2 - c^2} \right] / g \tag{3.13}$$
$$g = d + \sqrt{d^2 - c^2}$$

Applying the associated operator $S(A)$ to (3.5), the residual can be approximated as follows

$$r_k \doteq S^k(A) r_0 \tag{3.14}$$

In addition, $|S(\lambda)|$ defines an asymptotic convergence factor for each point on the complex plane $C$. In order to reduce the residual it is necessary to make the spectral radius of $S$ as small as possible, *i.e.* $d$ and $c$ should be chosen to satisfy

$$\min_{d,c} \max_{\lambda} |S(\lambda)| \tag{3.15}$$

The solution of this problem can be found in terms of the eigenvalues of $A$ that are vertices of the smallest convex polygon enclosing the spectrum of $A$. These eigenvalues form $H(A)$ the convex hull of $A$. On the other hand, the parameters $d$ and $c$ define a family of ellipses centered at $d$ and having common foci at $d+c, d-c$. When the spectrum of $A$ lies in the right half of the complex plane there is a smallest ellipse that contains it. If the closure of that ellipse does not contain the origin, the iteration will converge. As a consequence, an infinite number of smallest ellipses have this property, but the best ellipse is defined to be that with the greater rate of convergence, that is

$$-\log(\max |S(\lambda_i)|). \tag{3.16}$$

In terms of computational effort, the eigenvalue estimation is the most expensive step of the adaptive procedure and we will discuss it in some detail.

Manteuffel has shown that for $k$ sufficiently large the error $\epsilon_k$ tends to be a linear combination of the eigenvectors with the largest convergence factors, *i.e.* those whose eigenvalues lie on the ellipse furthest from the common foci. This observation also holds for the residual vector, as a consequence of (3.5). Considering dominant and subdominant eigenvectors $v_1$, and $v_2$(in the sense defined above), we have that

$$r_k = \beta_1 v_1 + \overline{\beta}_1 \overline{v}_1 + \beta_2 v_2 + \overline{\beta}_2 \overline{v}_2 + \delta \tag{3.17}$$

where $\delta$ tends to zero for large $n$, and bar denotes complex conjugate. In order to find the eigenvalues associated to $v_1$ and $v_2$, Manteuffel proposed three different methods.

Method 1 ($M1$ for short), is based on the power method which will give estimations of the eigenvalues of $A$, calculating the roots of the fourth degree orthogonal polynomial

$$P_4(z) = z^4 + c_3 z^3 + c_2 z^2 + c_1 z + c_0 \tag{3.18}$$

The coefficients of this polynomial will be found by forming the Krylov sequence associated with $r_k$ and $A$,

$$\left\{ r_k, Ar_k, A^2 r_k, A^3 r_k, A^4 r_k \right\} = \left\{ u_0, u_1, u_2, u_3, u_4 \right\} \tag{3.19}$$

and thus finding the least squares $(LS)$ solution to the system

$$Uc = u_4 \tag{3.20}$$

where $U$ is the matrix

$$U = [u_0, u_1, u_2, u_3].$$

Method 2 ($M2$) is based in the fact that the residuals naturally provide a Krylov sequence of $S(A)$,

$$u_0 = r_{k-4} \ , u_1 = r_{k-3} \ , u_2 = r_{k-2} \ , u_3 = r_{k-1} u_4 = r_k \qquad (3.21)$$

Using this vectors $u_i$, as the columns of matrix $U$ in system (3.20), one avoids the disadvantage of $M1$, that needs to compute matrix vector products $Ar$. Through the relation (3.13), the eigenvalues of $S$ can be transformed in the eigenvalues of $A$. This variant resembles the modified power method (Wilkinson [24]).

Method 3 ($M3$) uses a new operator $A'$

$$A' = 2g(dI - A) \qquad (3.22)$$

where $g$ is as in (3.13) and $d$ is the current value of the center of the ellipse.

The modified power method described in $M2$, is then used to find the dominant eigenvalues $\lambda'$ of $A'$, and then transformed in those of $A$, using the relation

$$\lambda = d - \frac{1}{2g}\lambda' \qquad (3.23)$$

For these methods, most of the computational cost lie on the $QR$ factorization of matrix $U$ for solving the $LS$ problem. This cost is approximately $25N$ multiplications. However, in method $M1$ it must be considered the cost of the four matrix-vector products. Numerically, the best method is $M3$, but in nearly symmetric or very nonsymmetric problems the method $M1$ gives better eigenvalues estimations (for a complete discussion see [1]). In this work we run the method with $M1$, $M2$ and $M3$ to see which version behaves better for our problems.

Once an approximation to the eigenvalues of $S$ which are the vertices of $H(A)$ has been obtained, the optimal pair corresponding to the best ellipse is chosen as the solution of the $min - max$ problem (3.15). This problem has a unique solution, and Manteuffel [13] describes the explicit solution of the $min - max$ problem depending upon how many eigenvalues are in the positive Hull of the spectrum.

Algorithm . ADAPTIVE CHEBYSHEV. (ACHEBY) $\qquad (3.24)$

1. Start. Choose $x_0$ and compute $r_0 = b - Ax_0$.
   Select $(d,c)$ and compute $p_0 = r_0/d$

2. Perform $k$ steps of algorithm (3.11) until convergence is attained,
   i.e.$\|r_k\| < TOL$ then $STOP$.

   Else if $\frac{\|r_{k+1}\|}{\|r_k\|} > 10^3$ or $> P_k(\lambda)$ go to step 3.

3. Adaptive Procedure. Estimate the eigenvalues of $S(A)$, solving system (3.20) to find the coefficients of the polynomial (3.18) and its roots. Then find the explicit solution of the $min - max$ problem (3.15) to get an approximation pair $(d, c)$.

4. Restart. Compute $r_0 := b - Ax_{k+1}$, $p_0 = \frac{r_0}{d}$ and go to step 2.

The storage requirement per iteration is $3N$ for the vectors $x, p, r$ ($Ax$ overwrites $r$) and $7N$ for the adaptive procedure ($x_0, r_0$ and five more vectors to estimate the eigenvalues). The computational cost is $2N + mv$ for algorithm (3.11) (we do not take into account the necessary work to find the optimal pair $(d, c)$ because it is negligible when $N$ is large, typically with $DKR$, 5% of the total cost.)

## Convergence

Manteuffel [13], has shown that the method converges if the adaptive procedure is used when the symmetric part of $A$ is positive definite.

## 4. PRECONDITIONERS

We have used three preconditioners that have been successfully applied for oil reservoir simulation. The basic idea is to approximate the "exact" $LU$ factorization of matrix $A$ by a product of triangular factors $L_a, U_a$ that are as sparse as the corresponding parts of $A$.

Then solving problem (1.1) is equivalent to solve any of the problems

$$(L_a U_a)^{-1} Ax = \overline{A}x = (L_a U_a)^{-1} b = \overline{b} \tag{4.1}$$

$$A (L_a U_a)^{-1} (L_a U_a) x = \overline{A}\overline{x} = b \tag{4.2}$$

$$L_a^{-1} A U_a^{-1} U_a x = \overline{A}\overline{x} = L_a^{-1} b = \overline{b} \tag{4.3}$$

These ways of applying the preconditioning are called left, right and split respectively. This technique has been proposed by several authors (see Dupont, Kendall and Rachford [6], Meijerink, Van der Vost [15]) and it is usually referred as an incomplete LU factorization.

Preconditionings are applied to decrease the computational effort needed to solve (1.1). If $L_a U_a$ is a good approximation to $A$, then the coefficient matrices in (4.1), (4.2) and (4.3) are closer to the identity matrix than $A$, and the methods will converge faster.

In order to illustrate different preconditioners, and because in this work we will only solve 7 banded matrices, we show in the following figure the original bands of matrix A as $\alpha_i$ and $\gamma$ as the main diagonal and the additional bands used by the preconditioners as $\beta$ and c.

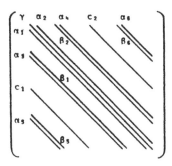

Apart from having a good preconditioner in the sense given above, the extra computational cost from preconditioning, should be as small as possible both in operations and storage. $\overline{A}$ is not actually formed, and when solving the matrix vector products that involve $\overline{A}$ in any algorithm, we solve efficiently these products using $L_a$ and $U_a$ .

In order to retain the original banded structure as much as possible, the factors $L_a, U_a$ are lower and upper triangular respectively and $D$ is diagonal, that satisfy the conditions:

1. $L_a D U_a$ and $A$ have the same element in those positions where $A$ has nonzero elements.

2. The elements of $L_a D U_a$ outside the nonzero pattern of $A$ are small or zero.

The first preconditioner that we will describe is due to Dupont, Kendall and Rachford [6] (DKR). The factors $L_a, U_a$ are formed with the corresponding strict triangular parts of matrix A. The matrix $D$ is found by forming the symbolic product $L_a D U_a$ and identifying the elements of the diagonal of this product with those of the main diagonal of a $A, \gamma$, i.e..

$$d_i^{-1} = \gamma_i - \alpha_{1i} d_j \alpha_{2j} + \alpha_{3i} d_l \alpha_{4l} + \alpha_{5i} d_p \alpha_{6p} \qquad (4.4)$$

where the indices $i, j, l, p$, denote row numbers and run depending of the position of the bands in matrix $A$.

Because in the $DKR$ factorization the product $L_a D U_a$ introduces additional bands in the position adjacent to the bands of A (at the $\beta$'s) and in two extra bands, the c's (collocated symmetrically), a better factorization can be obtained by modifying the bands adjacent to the main diagonal $\alpha_1$ and $\alpha_2$ now called $\beta_1$ and $\beta_2$ and adding bands $\beta_3, \beta_4$, $\beta_5$ and $\beta_6$, so that the product $L_a D U_a$ resembles A better. This factorization is called double descomposition $DD$. The equations that define the matrix $D$ and the extra bands, are given below

$$d_i^{-1} = \gamma_i - \beta_{1i}d_j\beta_{2j} - \beta_{3i}d_k\beta_{4k} - \beta_{5i}d_n\beta_{6n} - \alpha_{3i}d_l\alpha_{4l} - \alpha_{5i}d_p\beta_{6p}$$

$$\beta_{1i} = \alpha_{1i} - \alpha_{3i}d_l\beta_{4l} - \alpha_{5i}d_p\beta_{6p}$$

$$\beta_{2i} = \alpha_{2i} - \beta_{3i}d_k\alpha_{4k} - \beta_{5i}d_n\alpha_{6n}$$

$$\beta_{3i} = -\alpha_{3i}d_l\beta_{2l} \qquad\qquad\qquad\qquad (4.5)$$

$$\beta_{4i} = -\beta_{1i}d_j\alpha_{4j}$$

$$\beta_{5i} = -\alpha_{5i}d_p\beta_{2p}$$

$$\beta_{6i} = -\beta_{1i}d_j\alpha_{6j}$$

As in $DKR$, all the indices denote row numbers and run depending on the band positions.

The process of filling factors $L_a, U_a$ can be carried out one step more in order to cancel the error non-zero bands that still remain denoted as $c_1$ and $c_2$. This factorization is called additional bands (AB) and can be constructed by adding bands at positions $c_1$, $c_2$ and modifying bands $\alpha_1, \alpha_2$ called $\beta_1$ and $\beta_2$ and $\alpha_3$ and $\alpha_4$ now called $c_3$ and $c_4$. The equations defining it are given below

$$d_i^{-1} = \gamma_i - \beta_{1i}d_j\beta_{2j} - \beta_{3i}d_k\beta_{4k} - \beta_{5i}d_n\beta_{6n}$$
$$\qquad\qquad - c_{3i}d_lc_{4l} - c_{1i}d_mc_{2m} - \alpha_{5i}d_p\alpha_{6p}$$

$$\beta_{1i} = \alpha_{1i} - c_{3i}d_l\beta_{4l} - \alpha_{5i}d_p\beta_{6p}$$

$$\beta_{2i} = \alpha_{2i} - \beta_{3i}d_kc_{4k} - \beta_{5i}d_n\alpha_{6n}$$

$$\beta_{3i} = -c_{3i}d_l\beta_{2l} - \beta_{5i}d_nc_{2n}$$

$$\beta_{4i} = -\beta_{1i}d_jc_{4j} - c_{1i}d_m\beta_{6m}$$

$$\beta_{5i} = -\alpha_{5i}d_p\beta_{2p} \qquad\qquad\qquad\qquad (4.6)$$

$$\beta_{6i} = -\beta_{1i}d_j\alpha_{6j}$$

$$c_{1i} = -\beta_{5i}d_n\beta_{4n} - \alpha_{5i}d_pc_{4p}$$

$$c_{2i} = -c_{3i}d_l\alpha_{6l} - \beta_{3i}d_k\beta_{6k}$$

$$c_{3i} = \alpha_{3i} - \alpha_{5i}d_pc_{2p}$$

$$c_{4i} = \alpha_{4i} - c_{1i}d_m\alpha_{6m}$$

The additional computational cost per iteration is $6N * NC$ for $DKR$, $10N * NC$ for $DD$ and $12N * NC$ for $AB$ and the extra storage is $1N * NC$, $7N * NC$, $11N * NC$ respectively, where $NC$ is the order of each block element.

## 6. NUMERICAL RESULTS

### Oil Reservoir Problems

We will now describe the problems* we use for the numerical experiments. This set of problems represents some of the most common methods used in reservoir simulation and they present some of the difficulties typically encountered in these problems. They are generated from tridimensional simulation models, in a $N_x, N_y, N_z$ mesh, using a finite centered difference scheme of 7 points.

1. Pressure equation. Black oil model. Sequential Simulator. The reservoir contains barriers interfering the vertical flow that causes a heterogeneous coefficient matrix. There is also strong local contrast in the transmissibility. $N_x = 10, N_y = 10, N_z = 10$.

2. Thermal simulation of a reservoir treated with steam injection. The elements of the coefficient matrix are $6 \times 6$ blocks, where the temperature is associated with column 4 and the pressure with column 6. $N_x = 6, N_y = 6, N_z = 5$.

3. Pressure equation. Black oil model, IMPES simulator. The reservoir contains several cells with zero porous volume and strong local contrast in the transmissibility. $N_x = 35, N_y = 11, N_z = 13$.

4. Pressure equation. Black oil model, IMPES simulator. The reservoir contains several cells with zero porous volume, and flow barriers. $N_x = 16, N_y = 23, N_z = 3$.

5. Pressure equation. Black oil model. Simultaneous solution simulator. The elements of the coefficient matrix are $3 \times 3$ blocks. It is the same reservoir than problem 4.

   All these problems have seven bands, and problems 2 and 5 have block elements.

In order to give a brief description of the equations that model these problems we have followed Odeh ref.[17]. The partial differential equations for the oil, gas and water,(black oil model) wich describe the flow through the porous medium can be obtained combining the material balance and Darcy equations, where the pressure and the saturations are the variables.

$$\nabla \cdot (\lambda_o \nabla \Phi_o) - q_o = \frac{\partial}{\partial t} \left( \frac{\phi s_o}{B_o} \right), \tag{5.1}$$

$$\nabla \cdot (\lambda_g \lambda \Phi_g) + \nabla \cdot (R_{so} \lambda_o \nabla \Phi_o) + \nabla \cdot (R_{sw} \lambda_w \nabla \Phi_w) - q_g$$
$$= \frac{\partial}{\partial t} \left[ \phi \left( \frac{s_g}{B_g} + R_{so} \frac{s_o}{B_o} + R_{sw} \frac{s_w}{B_w} \right) \right] \tag{5.2}$$

$$\nabla \cdot (\lambda_w \nabla \Phi_w) - q_w = \frac{\partial}{\partial t} \left( \frac{\phi s_w}{B_w} \right) \tag{5.3}$$

The subscripts $o$, $g$ and $w$ refer to the phases oil, gas and water and $\lambda_i = \frac{k k_{ri}}{B_i \mu_i}$ are the transmissibilities of phase i, and

---

* These problems were obtained from A.H. Sherman & R. Kendall, Nolen Ass. for participation on the special session on Linear Algebra for Reservoir Simulation. SEG-SIAM-SPE Conference on Mathematical and Computational Methods in Seismic Exploration and Reservoir Modelling, Jan. 1985, Houston, Texas.

$$\frac{\partial \Phi_i}{\partial x} = \frac{\partial p}{\partial x} - g\rho_i \frac{\partial D}{\partial x} + \frac{\partial p_{c_i}}{\partial x}, \qquad i = o, g, w.$$

Here $t$=time, $s$= saturation, $q$= production rate, $p$= pressure. We also have for the saturations

$$s_o + s_g + s_w = 1$$

These equations are nonlinear because coefficients are functions of the unknowns p and s, while B and $\phi$ are functions of p. They are also coupled because the gas equation contains the oil and water saturations and $k_{ro}$ in $\lambda_o$ is a function of $s_w$ and $s_g$. These continuity equations can be manipulated to give the Pressure equation:

$$\sum_{i=1}^{3} B_i \nabla \cdot (\lambda_i \nabla \Phi_i) - q_T = \Phi C_T \frac{\partial p}{\partial t},$$

where $q_T$ is the total production rate, and $C_T$ is the total compressibility of the system.

When there is heat injection the temperature is not constant and an energy balance equation is required, and has the following form:

$$\frac{\partial}{\partial t} \sum_i \rho_i s_i U_i = \sum_i \nabla \cdot (\lambda_i H_i \nabla p) + \nabla \cdot (T_o \nabla T) - q_L - q_H, \qquad (5.4)$$

where $U$= internal energy, H= enthalpy, $q_L$ = heat loss to the formation, $q_H$ = heat sink or surce through a well, T= temperature and $T_o$ = heat conduction transmissibility.

To solve this set of equations, one could use a fully implicit method, where all equations are solved simultaneously for the unknowns for the entire grid system. The Newton–Raphson method is used to linearize. This is the case for problems 2 and 5. As a result of this method each element is a 3 x 3 submatrix, in problem 5 and in problem 2, there are 6 unknowns and each element is a 6 x 6 submatrix.

Problems 1, 3 and 4 are solved using sequential methods. The pressure equation is solved first starting with values for the coefficients referred to the beginning of the time step. In the so called sequential method, used to solve problem 1, the total flux remains constant during a time step. In problems 3 and 4 the sequential IMPES method is used. Here only the pressure equation is solved.

### Numerical Experiments

We apply the preconditioning to the system (1.1) using incomplete factorizations $DKR$, $DD$ and $AB$ as described in section 4. We limited preconditioning to the left preconditioning for all numerical experiments, because we have conducted experiments on the right preconditioning, and the numerical results were very poor.

The stopping criterion was $\frac{\|r_k\|_2}{\|r_0\|_2} < 10^{-6}$ with the initial guess $x_0 = 0$. All the experiments have been run on a UNISYS-A12 computer in single precision. The following table summarizes the work and storage required by the preconditioned methods where $7N * NC$ is the number of nonzero elements in $A$ and is the actual cost of a matrix-vector product, defined previously as $mv$.

|  | METHODS | | | PRECONDITIONERS | | |
|---|---|---|---|---|---|---|
|  | GMRES(m) | ORTHOMIN(m) | ACHEBY | DKR | DD | AB |
| WORK | $\left(m+3+\frac{1}{m}\right)7N*NC$ | $(3m+4)7N*NC$ | $2N+7N*NC$ | $6N*NC$ | $10N*NC$ | $12N*NC$ |
| STORAGE | $(m+2)N$ | $(2m+3)N$ | $10N$ | $N*NC$ | $7N*NC$ | $11N*NC$ |

*Table 1. Work and storage required by the methods and the preconditioners.*
*In problems 1, 2, 4 NC = 1, and in problems 2 and 5, NC = 6 and 3 respectively.*

We present now in table 2, the results when using Chebyshev Iterative Method, with M1, M2 and M3, the three different options that can be used in the adaptive procedure described in section 3, to find an approximation to the optimal pair.

| Problem | | 1 | | | 3 | | | 4 | | | 5 | | |
|---|---|---|---|---|---|---|---|---|---|---|---|---|---|
| Prec. | | DKR | DD | AB | DKR | DD | AB | DKR | DD | AB | DKR | DD | AB |
| M1 | OPER | 1.09 | 1.29 | 1.28 | 16.29 | 15.21 | 17.02 | 2.38 | 1.88 | 1.64 | 15.88 | 20.01 | 20.12 |
|  | ITER | 73 | 68 | 61 | 217 | 160 | 162 | 144 | 90 | 71 | 117 | 114 | 103 |
| M2 | OPER | 1.29 | 1.33 | 1.34 | *** | 16.26 | 18.49 | 2.15 | 1.84 | 1.97 | 17.30 | 20.53 | 19.73 |
|  | ITER | 86 | 70 | 64 | *** | 171 | 176 | 130 | 88 | 85 | 128 | 117 | 101 |
| M3 | OPER | 1.17 | 1.33 | 1.34 | 20. 12 | 22.06 | 15.97 | 2.15 | 2.11 | 2.36 | 17.80 | 18.77 | 20.46 |
|  | ITER | 78 | 70 | 64 | 268 | 232 | 152 | 130 | 101 | 102 | 128 | 105 | 103 |

*Table 2. Chebyshev Iterative Method, with three different options for the adaptive procedure.*
*Note: For problem 2, convergence was attained before calling the adaptive procedure.*
*\*\*\* means that no convergence was attained in 300 iterations*

From these results, we decided to use method 1, ACHEBY–M1, (the power method), for problems 1 and 3 whose preconditioned systems could be nearly symmetric. For problems 4 and 5, we will use method 3, ACHEBY–M3. We have run experiments with GMRES(2), GMRES(5), GMRES(10), ORTHOMIN(5), and the ACHEBY methods.

The storage requirement, see table 1, for ORTHOMIN is twice the requirement for GMRES; thus we feel that it is fair to compare ORTHOMIN(5) with GMRES(10), and ORTHOMIN(1) with GMRES(2). We also choose GMRES(5) to see how it compares to ORTHOMIN(5) with half storage. See table 3 for the complete set of results.

| P | PREC. | GMRES(2) OPER. | ITER. | GMRES(5) OPER. | ITER. | GMRES(10) OPER. | ITER. | ORTHOMIN(1) OPER. | ITER. | ORTHOMIN(5) OPER. | ITER. | ACHEBY OPER. | ITER. |
|---|---|---|---|---|---|---|---|---|---|---|---|---|---|
| | DKR | 1.69 | 94 | 1.49 | 71 | 1.43 | 55 | 0.98 | 49 | 1.53 | 48 | 1.09 | 73 |
| 1 | DD | 0.85 | 39 | 1.00 | 40 | 1.05 | 35 | 0.76 | 32 | 1.18 | 33 | 1.29 | 68 |
| | AB | 0.96 | 40 | 1.02 | 38 | 0.92 | 29 | 0.83 | 32 | 1.14 | 30 | 1.28 | 61 |
| | DKR | 0.98 | 11 | 0.92 | 10 | 0.88 | 9 | 1.01 | 11 | 1.04 | 10 | 1.46 | 17 |
| 2 | DD | 1.27 | 11 | 1.18 | 10 | 1.11 | 9 | 1.29 | 11 | 1.30 | 10 | 1.90 | 17 |
| | AB | 1.15 | 9 | 1.05 | 8 | 1.09 | 8 | 1.04 | 8 | 1.14 | 8 | 1.50 | 12 |
| | DKR | 26.30 | 292 | 27.43 | 261 | 26.54 | 204 | 22.42 | 224 | 27.70 | 173 | 16.29 | 217 |
| 3 | DD | 30.72 | 279 | 27.02 | 216 | 22.52 | 150 | *** | *** | 26.30 | 146 | 15.21 | 160 |
| | AB | 36.15 | 301 | 32.97 | 244 | 20.34 | 127 | 27.97 | 215 | 23.77 | 125 | 17.02 | 162 |
| | DKR | 4.82 | 243 | 2.50 | 108 | 2.78 | 97 | 1.45 | 66 | 1.90 | 54 | 2.15 | 130 |
| 4 | DD | 2.11 | 87 | 2.12 | 77 | 1.65 | 50 | 1.32 | 50 | 1.78 | 45 | 2.11 | 101 |
| | AB | 3.10 | 117 | 1.93 | 65 | 1.20 | 34 | 1.52 | 53 | 1.34 | 32 | 2.36 | 102 |
| | DKR | 17.48 | 120 | 15.87 | 102 | 16.18 | 94 | *** | *** | 12.48 | 65 | 17.80 | 128 |
| 5 | DD | 25.22 | 136 | 18.56 | 95 | 17.59 | 83 | 22.09 | 115 | 12.51 | 54 | 18.78 | 105 |
| | AB | 21.35 | 104 | 16.57 | 77 | 16.69 | 72 | 24.80 | 117 | 11.57 | 46 | 20.47 | 103 |

*Table 3. number of operations (in $10^6$) and iterations*
*for convergence. All methods, all preconditioners and all problems*
*ACHEBY-M1 was used for problems 1 and 3, and ACHEBY-M3 for problems 4 and 5*
*Note: \*\*\* means that no convergence was attained in 300 iterations*

We present these results in figs. 1 to 15, one for each problem with each preconditioner. We plot the computational effort and the error $\left(-\log\dfrac{\|r_k\|_2}{\|r_0\|_2}\right)$.

When running ACHEBY, an initial set of parameters $(1,0)$ is given, following Manteuffel [14]. The adaptive procedure is called whenever $\dfrac{\|r_{k+1}\|_2}{\|r_k\|_2} > 10^3$ or when this relative residual is greater than an estimate of $P_k(\lambda)$. In fact the adaptive routine was called very few times. We consider this information valuable, so we present these results in the following table 4.

| PROBLEM | | | | | | | | | | | | | | |
|---|---|---|---|---|---|---|---|---|---|---|---|---|---|---|
| 1–M1 | | | 2 | | | 3–M1 | | | 4–M3 | | | 5–M3 | | |
| DKR | DD | AB | DKR | DD | AB | DKR | DD | AB | DKR | DD | AB | DKR | DD | AB |
| 3 | 3 | 2 | 0 | 0 | 0 | 7 | 3 | 4 | 6 | 3 | 3 | 6 | 3 | 3 |

*Table 4. Number of calls of the adaptive procedure in the Chebyshev method*
*Note: For problem 2, convergence was attained before the adaptive procedure was used*

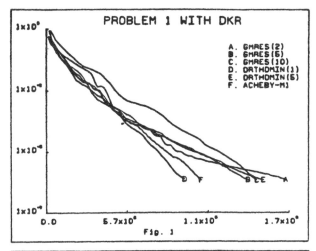

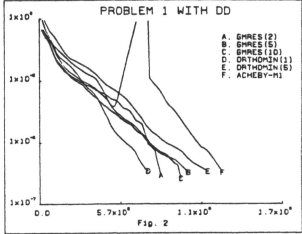

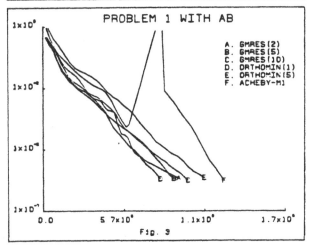

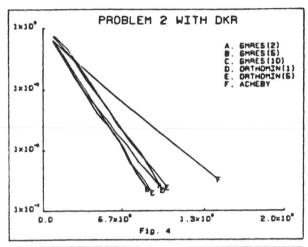

Fig. 4

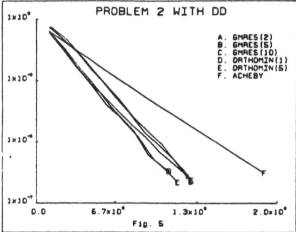

Fig. 5

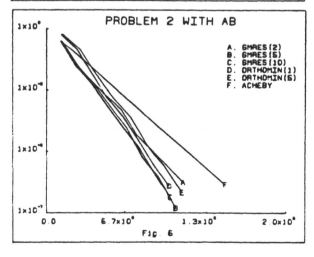

Fig. 6

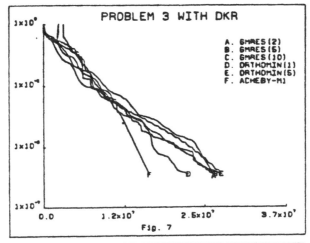

Fig. 7

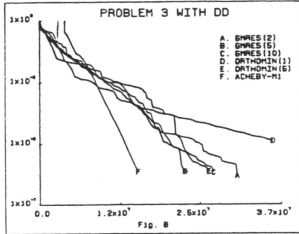

Fig. 8

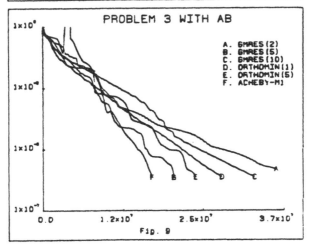

Fig. 9

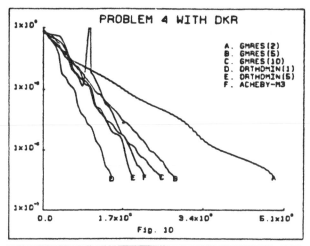

Fig. 10

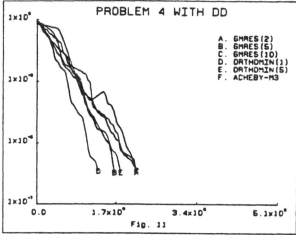

Fig. 11

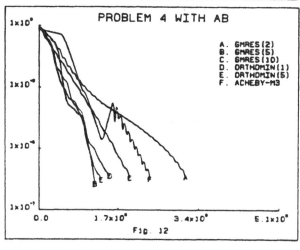

Fig. 12

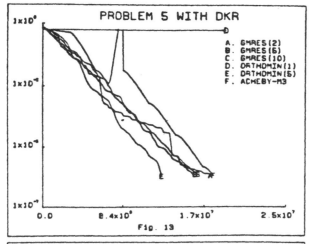

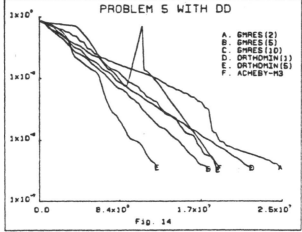

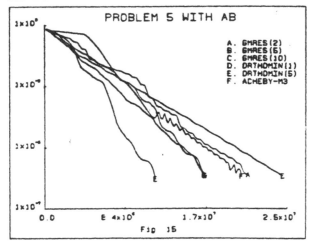

If the adaptive Chebyshev method were used to solve a problem discretized in time, the major effort to approximate the optimal pair would occur on the first time step system assuming that the matrices would not change abruptly from one step to the next. The parameters obtained in one time step could be a very good starting guess for the adaptive procedure for the next time step. In order to see how ACHEBY could behave with good knowledge of the parameters we present figures 16 to 20. In these figures, the best results obtained with GMRES($m$), ORTHOMIN($m$), and ACHEBY (over the preconditioners) are selected for a given problem (see table 5 for these set of results). We took the best results on average, for each problem. We also show in these figures the results when running ACHEBY* using a good approximation to the optimal pair $(d^*, c^*)$, (see also table 6). This approximation to the optimal pair was obtained when running it with the initial approximation $(1, 0)$.

|   |     | GMRES(2) | | ORTHO(1) | | ACHEBY | | ACHEBY* | |
|---|-----|-------|-------|-------|-------|-------|-------|-------|-------|
|   |     | OPER. | ITER. | OPER. | ITER. | OPER. | ITER. | OPER. | ITER. |
| 1 | DD  | 0.85  | 39    | 0.76  | 32    | 1.29  | 68    | 0.72  | 38    |
| 2 | DKR | 0.98  | 11    | 1.01  | 11    | 1.46  | 17    | 0.95  | 11    |
|   |     | GMRES(10) | | ORTHO(5) | | TCHEBY | | TCHEBY* | |
| 3 | AB  | 20.34 | 127   | 23.77 | 125   | 17.02 | 162   | 14.92 | 142   |
| 4 | AB  | 1.20  | 34    | 1.34  | 32    | 2.36  | 102   | 1.50  | 67    |
| 5 | DKR | 16.18 | 94    | 12.48 | 65    | 17.80 | 128   | 11.82 | 85    |

Table 5. Number of operations (in $10^6$) and iterations
for the selected best results and ACHEBY*
with near optimal parameters

|       | DKR | | DD | | AB | |
|-------|-------|-------|-------|-------|-------|-------|
|       | OPER. | ITER. | OPER. | ITER. | OPER. | ITER. |
| M1 1  | 0.96  | 64    | 0.72  | 38    | 1.02  | 49    |
| – 2   | 0.95  | 11    | 1.23  | 11    | 1.00  | 8     |
| M1 3  | 11.18 | 149   | 12.93 | 136   | 14.92 | 142   |
| M3 4  | 1.92  | 78    | 1.65  | 63    | 1.50  | 63    |
| M3 5  | 11.82 | 75    | 20.21 | 88    | 24.04 | 83    |

Table 6. Number of operations (in $10^6$) and iterations running ACHEBY*
with a good approximation to the optimal pair $(d*, c*)$.
In the adaptive routine, M1 was used for
problems 1 and 3, and M3 for problems 4 and 5.

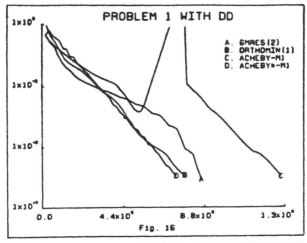

Fig. 16

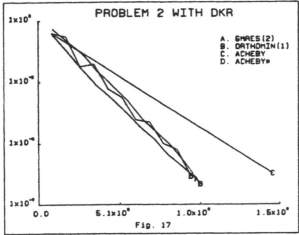

Fig. 17

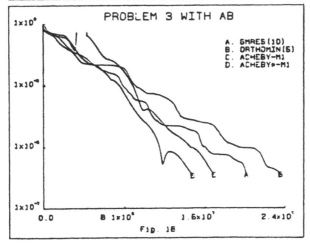

Fig. 18

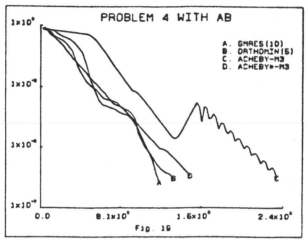

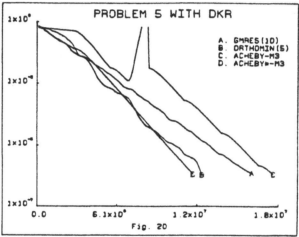

## CONCLUSIONS

As to the choice of preconditioner there is not a definite conclusion. However it seems that when using the simultaneous solutions method, as in problems 2 and 5, and perhaps due to the stability of this method, $DKR$ the cheapest and simplest, gives the best results.

When using sequential methods, an additional cost should be paid for better preconditioning. However using AB does not give much better results than DD, and the storage cost is higher.

A similar situation occurs when looking at the number of calls to the adaptive procedure when using CHEBYSHEV (table 4). Although there is need for many calls when using $DKR$, there is almost no difference for DD and AB. We conclude that AB does not seem worth using for this set of problems.

GMRES, a method not yet widely used is as good as ORTHOMIN. It is very simple to implement and is parameter free. The GMRES method is more robust than ORTHOMIN in that convergence was always attained with the former method.

For the easier problems 1 and 2, it seems worth using the low storage methods GMRES(2) and ORTHOMIN(1), but for the rest of the problems, storage should be spent to run GMRES(10) or ORTHOMIN(5).

When using CHEBYSHEV, the choice of the method to approximate eigenvalues in the adaptive procedure, is crucial, because the method could be very sensitive in some problems to the initial choice of parameters; see table 2 and also ref. [10,16] for a sensibility analysis of the method.

However, when solving time dependent problems, this method seems the most promising because as we explained earlier, in the first time step a good approximation to the optimal pair will be obtained and could be used in succeeding steps as the initial approximations. Our results show that when using ACHEBY*, starting with a very accurate pair (d*,c*), this method is significantly better than GMRES and ORTHOMIN. It is also very robust.

Acknowledgments. The authors want to thank Carlos Barrón for making the figures. The authors also wish to thank the two referees for their detailed reports and constructive remarks which assisted in clarifying the presentation.

# REFERENCES

[1] Ashby S.F., *Chebycode A FORTRAN Implementation of Manteuffel's Adaptive Chebyshev Algorithm.* Report no. UIUCDCS-R-85-1203, Department of Computer Science, University or Illinois at Urbana-Champaign, 1985.

[2] Axelsson, O. *Conjugate gradient type methods for unsymmetric an inconsistent systems of linear equations.* Lin. Alg. Appl. 29, 1980.

[3] Behie, A. and P. Vinsome. *Block iterative methods for fully implicit reservoir simulation.* SPE 10490, 1980.

[4] Chandra, R. *Conjugate gradient methods for partial differential equations,* Ph D thesis tech. rep. 129, Yale University, New Haven, CT, 1978.

[5] Dennis, J.E. and K. Turner. *Generalized Conjugate Directions,* Lin. Alg. Appl., 88/89.

[6] Duppont, T., R. Kendall, and H.H. Jr. Rachford. *An approximate factorization procedure for solving self-adjoint elliptic difference equations.* Siam J. Numer. Anal. Vol. 5, 1968.

[7] Eisenstat, S.C., H.C. Elman, and M.H. Schultz. *Variational Iterative Methods for Nonsymmetric Systems of Linear Equations.* Siam J. Numer. Anal. Vol. 20, no. 2, 1983.

[8] Elman, H.C. *Iterative Methods for Large Sparse, Nonsymmetric Systems of Linear Equations.* Ph. D. thesis technical Report 229, Yale University, New Haven, CT, 1982.

[9] Elman, H.C. Y., Saad and P.E. Saylor. *A hybrid Chebyshev Krylov Subspace Algorithm for solving Nonsymmetric Systems of Linear Equations.* Siam J. Sci. Stat. Comput. Vol. 7, no. 3, 1986.

[10] Gomez, S. and J.L. Morales, *Some results on the Chebyshev Iteration method.* Memorias del Taller de Análisis Numérico, IV Coloquio de Matemáticas del CINVESTAV, 1985.

[11] Golub, H.C. and C.F. Van Loan. *Matrix Computations,* Johns Hopkings University Press, Baltimore, 1983.

[12] Jea, K.C. and D.M. Young, *Generalized conjugate gradient acceleration of nonsymmetric iterative methods,* Lin. Alg. Appl., 34, 1980.

[13] Manteuffel, T.A., *The Chebyshev Iteration for Nonsymmetric Linear Systems,* Numer., Math., Vol. 28, 1977.

[14] Manteuffel, T.A., *Adaptive Procedure for estimating parameters for the nonsymmetric Chebyshev Iteration,* Numer., Math., Vol. 31, 1978.

[15] Meijerink, J.A. and A. Van der Vorst, *An iterative solution method for linear systems of which the coefficient matrix is a symmetric M-matrix,* Math of comp., vol. 31, no. 137, 1977.

[16] Morales, J.L. *Métodos iterativos en matrices banda.* M. Sc. in Computer Science Thesis, IIMAS, Universidad Nacional Autónoma de México. 1985.

[17] Odeh, A.S., *An overview of mathematical modeling of the behavior of hydrocarbon*

*reservoirs,* SIAM Review, Vol. 24, No. 3, 1982.

[18] Paige, C.C. and M.A. Saunders, *Solution of sparse indefinite systems of linear equations,* Siam J. Numer. Anal., 12, 1975.

[19] Saad, Y. *The Lanczos Biorthogonalization algorithm and other oblique projection methods for solving large unsymmetric systems,* Siam J. Numer. Anal., Vol. 19, no. 3, 1982.

[20] Saad, Y., *Practical use of some Krylov Subspace methods for solving indefinite and nonsymmetric linear systems,* Siam J. Sci. Stat. Comput., Vol. 5, no. 1, 1984.

[21] Saad, Y. and M.H. Schultz. *GMRES: A generalized minimal residual algorithm for solving nonsymmetric linear systems,* Siam J. Sci. Stat. Comput., Vol. 7, no. 3, 1986.

[22] Vinsome, P.K.W., *ORTHOMIN, an iterative method for solving sparse sets of simultaneous linear equations,* in Proc. Fourth Symposium on Reservoir Simulation, Society of Petroleum Engineers of AIME, 1976.

[23] Walker, H. *Implementation of the GMRES method using Householder transformations,* Siam J. Sci. Stat. Comput., 9, 1988.

[24] Wilkinson, J.H., *The algebraic eigenvalue problem.* Oxford, Clarendon Press, 1965.

# A Survey of Spline Collocation Methods for the Numerical Solution of Differential Equations

GRAEME FAIRWEATHER, Departments of Mathematics and Engineering Mechanics, University of Kentucky, Lexington, KY 40506

DANIEL MEADE, Division de Graduados e Investigacion, Instituto Tecnologico y de Estudios Superiores de Monterrey, Sucursal de Correos "J", 64849 Monterrey, N.L. MEXICO

## 1. INTRODUCTION

Over the past twenty years, spline collocation methods have evolved as valuable techniques for the solution of a broad class of problems covering ordinary and partial differential equations, functional equations, integral equations and integro-differential equations. The popularity of such methods is due in part to their conceptual simplicity, wide applicability, and ease of implementation.

Basically, a collocation method involves the determination of an approximate solution in a suitable set of functions, sometimes called trial functions, by requiring the approximate solution to satisfy the boundary conditions and the differential equation at certain points, called the collocation points. In time-dependent problems, this procedure is normally used with respect to the space variables only, in conjunction with a method of lines procedure. The collocation method is not new but dates back to the 1930's. According to Kantorovich and Akilov (1964, p. 581), the method was first proposed by Kantorovich (1934), Frazer, Jones and Skan (1937, 1938) and Frazer, Duncan and Collar (1938). The method of Kantorovich was actually a method of lines collocation procedure for the solution of a partial differential equation in two variables with collocation being applied in one variable for each fixed value of the second. The work of Frazer et al. was devoted to the solution of ordinary differential equations although mention is made by Frazer, Duncan and Collar of the applicability of collocation to the solution of partial differential equations. Collatz (1960) also discussed collocation for both ordinary and partial differential equations and provided some numerical examples. These methods have in common the choice of polynomials for trial functions. Lanczos (1938, 1956) was the first to advocate the use of collocation by polynomials at the roots of orthogonal polynomials rather than at equidistant points.

The convergence of Kantorovich's method was proved by Karpilovskaya (1953); see also Kantorovich and Akilov (1964). Shindler (1967, 1969), Vainniko (1965, 1966), and Ronto (1971) also analyzed collocation methods with polynomial trial functions. Yartsev (1967, 1968) investigated method of lines collocation for certain linear second order elliptic problems and biharmonic problems using trigonometric polynomials as trial functions. Karpilovskaya (1963a,b) analyzed a collocation method in two space variables using trigonometric polynomials of two variables, and generalized this method to a linear partial integro-differential equation with biharmonic principal part, (Karpilovskaya, 1970).

Collocation by polynomials at the roots of orthogonal polynomials is often called orthogonal collocation, particularly in the chemical engineering literature; see, for example, Finlayson (1971, 1972), Fan et al. (1971), Paterson and Cresswell (1971), Stewart and Villadsen (1969), Villadsen (1969), Villadsen and Michelsen (1978), Villadsen and Sorensen (1969), and Villadsen and Stewart (1967). Chebyshev orthogonal collocation methods are described by Fox and Parker (1968) for the solution of ordinary and partial differential equations, and integral equations. Spectral collocation methods are closely related to this form of collocation, see, for example, Gottlieb and Orszag (1977) and Karageorghis (1988).

With the upsurge of interest in spline functions in the sixties, attention turned to the use of spline functions as trial functions, and the development of the first spline collocation methods. These methods fall naturally into four categories. In the first are the methods called nodal spline collocation methods in which an approximate solution is determined by collocating at the nodes of the partition on which the splines are defined (and perhaps by satisfying additional equations). These methods are in general suboptimal in the sense that the error is of lower order than is possible by approximation in the spline space. In fact, de Boor (1966) showed that nodal cubic spline collocation for two-point boundary value problems is second order accurate and no better, whereas other projection methods with the same spline space yield fourth order accurate approximations.

The suboptimality of nodal spline collocation led researchers to seek spline collocation methods of optimal order. In their fundamental work, de Boor and Swartz (1973) showed that optimal rates of convergence can be attained by collocating at certain Gauss points in each subinterval of the partition on which the spline space is defined. This approach is called orthogonal spline collocation. A readable account of many of the developments in orthogonal spline collocation for BVODEs is given by Ascher et al. (1988).

Several researchers have succeeded in modifying nodal spline collocation so that optimal rates of convergence were achievable on uniform partitions. In these methods, either the nodal collocation solution is improved using a deferred correction-type procedure, or the optimal order approximation is determined directly by collocating a high-order perturbation of the differential equation. This class of methods is commonly known as extrapolated or modified spline collocation methods.

The fourth class consists of collocation-Galerkin methods. These are hybrid methods which combine advantages of Galerkin methods with advantages of collocation methods.

The obvious advantage of spline collocation methods over finite element Galerkin methods is that the calculation of the coefficients in the equations determining the approximate solution is very fast since no integrals need be evaluated or approximated. However, more continuity requirements are imposed on the collocation approximation which demand the use of a space of higher order spline functions. Moreover, in collocation methods more smoothness of the solution of the problem is required in order to obtain the same rate of convergence as the corresponding Galerkin method, that is, the Galerkin method using the same spline space. In this chapter, by optimal order error estimates, we shall mean estimates of the same order of accuracy as the corresponding Galerkin method, and in this case the order is referred to as the optimal rate of convergence, or the optimal order of accuracy. By optimal error estimates, we shall mean error estimates of optimal order of accuracy with minimum smoothness requirements, cf. Fairweather (1978) and Reddien (1980).

Excellent software packages based on orthogonal spline collocation have been widely available for the last decade. The package COLSYS (Ascher et al., 1979a) is a robust, general-purpose software package for solving boundary value problems for ordinary differential equations (BVODEs). In the package PDECOL (Madsen and Sincovec, 1979), which is designed to solve time dependent partial differential equations in one space variable, orthogonal spline collocation is used to discretize in space. The resulting system of ordinary differential equations is then solved using a variant of the well-known Gear package. In contrast, there are no widely available codes in existence which implement any form of finite element Galerkin method. It should also be mentioned that there are collocation-based codes for elliptic boundary value problems in ELLPACK (Rice and Boisvert, 1985), and new codes of this nature continue to be developed.

The primary purpose of this chapter is to provide a comprehensive survey of the literature on spline collocation methods for the numerical solution of differential equations. We shall review spline collocation methods for solving boundary value problems for ordinary differential equations and elliptic partial differential equations, and for initial-boundary value problems for parabolic and hyperbolic partial differential equations. The primary emphasis is on methods for BVODEs since the techniques used to solve partial differential equations are, in the main, generalizations of these. In a work of this nature it is not feasible to provide many of the mathematical or computational details. However we have made every effort to provide key references where these details may be found.

A brief outline of this chapter is as follows. Section 2 is devoted to a discussion of the four categories of spline collocation methods for BVODEs, and in Section 3, such methods for elliptic boundary value problems are described. In Section 4, the use of spline collocation in the numerical solution of initial-boundary value problems for parabolic and hyperbolic partial differential equations is discussed. These problems are discretized in the space variables using a spline collocation method and the time stepping is then performed using some standard algorithm.

There is a growing literature on spline collocation methods for the numerical solution of initial value problems for ordinary differential equations, integral equations, and integro-differential equations. A thorough treatment of these areas is beyond the scope of this work, but, in Section 5, we do provide some important references. In this section, we also mention several topics worthy of further study.

## 2. BOUNDARY VALUE PROBLEMS FOR ORDINARY DIFFERENTIAL EQUATIONS

**2.1 Introduction.** As an example, we consider the two-point boundary value problem

$$(2.1a) \qquad Lu \equiv -u'' + p(x)u' + q(x)u = f(x), \quad x \in \bar{I},$$

$$(2.1b) \quad B_0 u(0) \equiv \rho_0 u(0) + \mu_0 u'(0) = v_0, \quad B_1 u(1) \equiv \rho_1 u(1) + \mu_1 u'(1) = v_1,$$

where $\bar{I} = [0,1]$. Let $\pi = \{x_i\}_{i=0}^N$,

$$0 = x_0 < x_1 < ... < x_N = 1,$$

denote a partition of the interval $\bar{I}$ and set $h_i = x_i - x_{i-1}$, $h = \max_i h_i$, and $\bar{I}_i = [x_{i-1}, x_i]$. For a closed interval E, let $P_r(E)$ denote the set of polynomials on E of order r (that is, of degree at most r-1). Then we define

$$P(r,k;\pi) = \{v | v \in C^k(\bar{I}), \ v \in P_r(\bar{I}_i), \ i = 1,...,N\},$$

where $k \leq r-2$. Note that

$$(2.2) \qquad \dim P(r,k;\pi) = N(r-k-1) + (k+1).$$

**2.2 Nodal spline collocation.** The first spline collocation methods proposed for the solution of problems of the form (2.1) were nodal cubic spline methods. In such a method, one seeks an approximate solution $u_h$ in the space of cubic splines $P(4,2;\pi)$ which satisfies the differential equation (2.1a) at the nodes of the partition $\pi$, and the boundary conditions (2.1b), that is,

$$Lu_h(x_i) = f(x_i), \quad i = 0,..., N,$$

$$B_0 u_h(x_0) = v_0, \quad B_1 u_h(x_N) = v_1.$$

Several authors, for example, Ahlberg et al. (1967), Bickley (1968), and Albasiny and

Hoskins (1969), considered various formulations of this method which differ only in the choice of basis for the spline space. Nodal collocation with cubic splines is, however, an $O(h^2)$ method in $L^\infty$ (and no better) in general, (de Boor, 1966), and hence is suboptimal. Moreover, as an application of their analysis, Arnold and Wendland (1983) examined the nodal spline collocation method for the solution of (2.1) with Dirichlet boundary conditions and proved that the errors in the second and third derivatives of the approximate solution are of optimal order in $L^2$ but that the errors in the approximation and its first derivative are $O(h^2)$. The approach used in their analysis is related to that of Swartz and Wendroff (1974).

When higher order smoothest splines, $P(m+k,m+k-2;\pi)$, with m+k even and greater than 4, are used for the solution of second order problems, additional constraints must be imposed to define a unique solution since in the case of (2.1) there are only N+3 collocation equations and, from (2.2), the dimension of the spline space is N+m+k-1. Various conditions have been considered, by Ahlberg and Ito (1975) for smoothest splines of orders 6 and 8, and Blue (1969) for linear and nonlinear second order problems with splines of order 6. See also Sakai (1970), who derived error bounds for some simple linear problems. Lucas and Reddien (1972) and Gladwell and Mullings (1975) examined the use of spline collocation with smoothest splines for $m^{th}$ order nonlinear ordinary differential equations of the form

$$(2.3a) \qquad u^{(m)}(x) = f(x,u(x),u^{(1)}(x),...,u^{(m-1)}(x)), \quad x \in \bar{I},$$

subject to general linear boundary conditions

$$(2.3b) \qquad \sum_{j=0}^{m-1} \{\alpha_{ij}u^{(j)}(0) + \beta_{ij}u^{(j)}(1)\} = 0, \quad i = 0, 1,..., m-1.$$

In this case, k-2 additional constraints are required. For a variety of choices of boundary collocation conditions, Gladwell and Mullings proved essentially that

$$(2.4) \qquad \|D^j(u - u_h)\|_{L^\infty} \le Ch^{k-j}, \quad j = 0,...,m-1,$$

which is a suboptimal error estimate.

**2.3 Orthogonal spline collocation.** Russell and Shampine (1972) discussed in some detail theoretical and practical aspects of collocation with more general piecewise polynomial functions, namely $P(m+k,m;\pi)$, with $k \geq m$, for the solution of (2.3). Instead of specifying boundary collocation conditions, they used additional collocation points. More specifically, let

$$x_{ij} = x_{i-1/2} + \frac{1}{2}\lambda_j h_i, \quad i = 1,...,N, \quad j = 1,...,k,$$

where

$$x_{i-1/2} = \frac{1}{2}(x_i + x_{i-1}),$$

and

$$-1 \leq \lambda_1 < .... < \lambda_k \leq 1.$$

Russell and Shampine showed that if $\lambda_1 = -1$ and $\lambda_k = 1$, the $L^\infty$ error in their collocation approximation to the solution of (2.3) and its first $m-1$ derivatives is $O(h^k)$, when the solution $u \in C^{m+k}$; cf. (2.4).

In their influential paper, de Boor and Swartz (1973) demonstrated that the accuracy of the collocation approximation depends crucially on the location of the collocation points. They proved that if the solution $u \in C^{m+k+\nu}(\bar{I})$, where $\nu \leq k$ is a positive integer, and the $\{\lambda_i\}_{i=1}^k$ are chosen so that

$$\int_{-1}^{1} q(s) \prod_{i=1}^{k} (s - \lambda_i) \, ds = 0, \quad \text{for all } q \in P_\nu,$$

where $P_\nu$ denotes the set of polynomials of order $\nu$, then, for $h$ sufficiently small, the collocation approximation $u_h$ is unique and satisfies

(2.5a) $$\|D^j(u - u_h)\|_{L^\infty} \leq Ch^{k+\min(\nu,m-j)}, \quad j = 0,..., m,$$

and

(2.5b) $$|D^j(u - u_h)(x_i)| \leq Ch^{k+\nu}, \quad i = 0,...,N, \quad j = 0,..., m-1,$$

where $C$ is a positive constant independent of $h$. From (2.5a), it follows that the optimal order of accuracy in the $L^\infty$ norm, namely $O(h^{m+k})$, is achieved solely when $\nu = k$, in which case $\{\lambda_j\}_{j=1}^k$ are the zeros of the Legendre polynomial of degree $k$ on $[-1,1]$, which implies that the collocation points are the Gauss-Legendre points in each subinterval, and $u_h \in P(m+k,m-1;\pi)$. Moreover, if in this case the solution of the boundary value problem is in $C^{2k+m}(\bar{I})$, then from (2.5b) the error in the approximation to the solution and its first m-1 derivatives at the nodes of $\pi$ is $O(h^{2k})$. This is an example of superconvergence, the occurrence at certain points of a higher rate of convergence than is possible globally. If, for example, we consider the solution of (2.1) using the space of piecewise Hermite cubics, $P(4,1;\pi)$, then $O(h^4)$ accuracy is achieved for the solution and its first derivative at the nodes of $\pi$, which is a superconvergence result for the latter. In the literature, spline collocation at Gauss points is often referred to as orthogonal spline collocation. It should be noted that some of the arguments in de Boor and Swartz (1973) were improved later by these same authors (de Boor and Swartz, 1980).

The paper of de Boor and Swartz (1973) laid the foundation for the formulation and analysis of orthogonal spline collocation methods for a wide variety of problems. Russell (1974) extended their work to general nonlinear systems of first order ordinary differential equations subject to multipoint linear constraints, while Houstis (1978a) derived analogous results for systems of $m^{th}$ order ordinary differential equations. This was also done by Cerutti (1974) who even allowed the individual ordinary differential equations to have different orders. Extensions of the theory to the case of nonlinear boundary conditions were given by Reddien (1976), Weiss (1974) (who also considered first order systems), and Wittenbrink (1973). Refined error bounds for the local dependence of the error were derived by Christiansen and Russell (1978); see Herbst and Botha (1981), Oliveira (1980), and van Veldhuizen (1976) for related work. Russell and Christiansen (1978) examined adaptive mesh selection strategies for use in orthogonal spline collocation procedures. Ascher (1986) re-examined orthogonal spline collocation methods and presented the theory using the approach of Osborne (1975) and improved upon results of de Boor and Swartz (1973) and Christiansen and Russell (1978). Ascher (1986) also discussed the conditioning of the linear algebraic systems arising in the collocation method.

Hangelbroek et al. (1977) analyzed orthogonal spline collocation for the solution of boundary value problems described by integro-differential equations involving derivatives of order up to and including m with m boundary conditions. The use of orthogonal spline collocation for the solution of one such problem arising in heat transfer was described by Chawla and Chan (1979). Reddien (1979) formulated and analyzed an orthogonal spline collocation method for the solution of certain unconstrained optimal control problems. In a series of papers, de Boor and Swartz (1980, 1981a, 1981b) examined orthogonal spline collocation for the determination of an eigenvalue of an $m^{th}$ order ordinary differential equation subject to m homogeneous boundary conditions.

They show, for example, that simple eigenvalues are approximated to within $O(h^{2k})$ when the space $P(m+k,m-1;\pi)$ is employed. De Boor and Swartz (1981a) presented numerical results to illustrate the convergence results.

Collocation methods for two point boundary value problems for certain ordinary differential equations with singular coefficients were treated by de Hoog and Weiss (1978), Reddien (1973), and Russell and Shampine (1975). See also Reddien and Schumaker (1976). Singular perturbation problems were considered by Ascher (1985), Ascher and Jacobs (1989), Ascher and Weiss (1983, 1984a, 1984b), Flaherty and Mathon (1980), Flaherty and O'Malley (1984), Maier (1985), and Ringhofer (1984). See also Markowich and Ringhofer (1983), who considered a related problem. For applications of the techniques of Ascher and Weiss, see Ascher and Spudich (1986) and Spudich and Ascher (1983). It is interesting to note that Ascher (1985) and Ascher and Weiss (1983, 1984a) investigated the use of spline collocation at Lobatto points and even recommended this procedure over orthogonal spline collocation for a certain class of problems. These authors pointed out that, when comparing orthogonal collocation and the Lobatto scheme, orthogonal collocation with s collocation points should be compared to the Lobatto scheme with s+1 collocation points as the cost of implementation of these two schemes is roughly the same. Moreover, these schemes have the same rate of superconvergence at the nodes, for in the case in which the collocation points are chosen to be the Lobatto points in each subinterval, $\nu = k-2$ in (2.5b).

Reddien and Travis (1974) and Bellen and Zennaro (1984) formulated and analyzed spline collocation methods for the solution of boundary value functional differential equations. Bader (1985) presented an orthogonal spline collocation method, details of which were given for a first order nonlinear system with one retarded and one advanced functional argument. It is claimed that a generalization to systems of higher order and to a finite number of functional arguments is straightforward. Bader also described a package, FDECOL, which implements his method.

Several comparisons have been made between orthogonal spline collocation and various finite difference methods and finite element methods; see, for example, de Boor and Swartz (1977), Keast et al. (1983), Russell (1975,1977), Russell and Varah (1975), Sincovec (1977a), and Varah (1974). Keast et al. conducted a thorough computational study of efficient implementations of five finite element methods for the solution of problems of the form (2.1), and showed quite convincingly that orthogonal spline collocation is the cheapest method for a given accuracy and the easiest to implement. The B-spline package (de Boor, 1978) was used in each of the implementations together with a linear equation solver which took as much advantage as possible of the structure of the systems.

The rapid development of theoretical and computational aspects of orthogonal spline collocation methods for boundary value problems for ordinary differential equations during the seventies culminated in the development of COLSYS, a robust, general-purpose software package for the solution of mixed-order systems of ordinary differential equations subject to multipoint boundary conditions. This code, which was unveiled by Ascher et al. (1979a), implements orthogonal spline collocation with B-spline bases (de Boor, 1978), based primarily on the work of Cerutti (1974). Details of this implementation were given by Ascher et al. (1979b), and the code was described and presented by Ascher et al. (1981a,b). Briefly, in COLSYS, approximate solutions are computed on a sequence of automatically selected meshes until a user-specified set of tolerances is satisfied. A damped Newton method is used for the nonlinear iteration. The systems of linear algebraic equations which arise in the Newton method have a special block structure commonly known as almost block diagonal, (de Boor, 1978), and their coefficient matrices have the structure shown in Figure 1. In such a matrix, the first and last blocks, A and B, respectively, come from the boundary conditions while a general block $[W_{i1} \ W_{i2} \ W_{i3}]$ arises from the collocation equations on the $i^{th}$ subinterval of a partition of the domain of the problem. In COLSYS, these systems are solved using the

$$\begin{bmatrix} A & & & & & & \\ W_{11} & W_{12} & W_{13} & & & & \\ & & W_{21} & W_{22} & W_{23} & & \\ & & & \cdot & & & \\ & & & & \cdot & & \\ & & & & & \cdot & \\ & & & & W_{N1} & W_{N2} & W_{N3} \\ & & & & & & B \end{bmatrix}$$

Figure 1

package SOLVEBLOK developed by de Boor and Weiss (1980). SOLVEBLOK implements Gaussian elimination with scaled partial pivoting, and when applied to systems with matrices of the form shown in Figure 1, causes fill-in. Diaz et al. (1983a,b) developed Fortran packages, COLROW and ARCECO, specifically for the solution of linear systems with this structure. These codes implement algorithms based on a modification of the alternate row and column elimination algorithm of Varah (1976) which, in contrast to ordinary Gaussian elimination, introduces no fill-in. In this algorithm, row elimination with row pivoting (partial pivoting with row interchanges followed by elimination by rows, the standard procedure in Gaussian elimination) is alternated with column elimination with column pivoting (that is, partial pivoting by columns with column interchanges followed by elimination by columns), switching from one to the other whenever fill-in would occur otherwise. Diaz et al. (1983a) have shown that this algorithm is more efficient than ordinary Gaussian elimination as implemented in SOLVEBLOK (de Boor and Weiss, 1980), which is, admittedly, of more general applicability than either COLROW or ARCECO. It should be noted that almost block diagonal linear systems with the above structure also arise in multiple shooting techniques and certain finite difference methods for solving BVODEs.

The software package COLSYS has been compared with other packages by Denison et al. (1983), Fairweather and Vedha-Nayagam (1987), Hemker et al. (1980), and Russell (1982b), and the difficulties in making such comparisons were discussed by Russell (1982a). The efficacy of COLSYS for the solution of many problems in science and engineering has been demonstrated by several authors including Akyurtlu et al. (1986a), Ascher (1980), Ascher and Wan (1980), Bhattacharyya et al. (1986), Davis and Fairweather (1981), Davis et al. (1982), Kelkar et al. (1985), Muir et al. (1983), and Vedha-Nayagam (1987). See Carey and Finlayson (1975) and Ravaioli et al. (1985) for applications of orthogonal spline collocation with piecewise Hermite cubics.

Until recently, B-splines were the most popular choice of basis in spline collocation procedures, but it was shown by Ascher et al. (1983) that the B-spline basis is inferior, both in ease of implementation, operation counts and conditioning, to a certain monomial basis proposed by Osborne (1975), and Bader and Ascher (1987) have implemented the latter in the package COLNEW, a new version of COLSYS. Numerical examples presented by Bader and Ascher demonstrate the improvement in performance of COLNEW over COLSYS. It should be mentioned that Swartz (1988) has shown, among other things, that the B-splines can be scaled so that the conditioning problems previously investigated by Ascher et al. (1983) are no longer present. See also Paine and Russell (1986) for related work on the conditioning of spline collocation matrices.

In this new implementation of spline collocation, the structure of the matrices in the Newton method is of the form shown in Figure 2. As in Figure 1, the blocks A and B arise from the boundary conditions. The block $[H_i \ G_i]$ comes from the collocation equations on the $i^{th}$ subinterval of the partition $\pi$, and the block $[-C_i \ -D_i \ I]$ from certain

$$\begin{bmatrix}
A & & & \\
H_1 & G_1 & 0 & \\
-C_1 & -D_1 & I & \\
& & & \cdot \\
& & & & \cdot \\
& & & & & \cdot \\
& & & & H_N & G_N & 0 \\
& & & & -C_N & -D_N & I \\
& & & & & & B
\end{bmatrix}$$

Figure 2

continuity equations at the point $x_i$. In COLNEW, condensation is used to reduce the full system to one with a coefficient matrix of the form shown in Figure 3, in which the blocks $\Gamma_i$ are square, and this condensed system is then solved using SOLVEBLOK. For precise details, see Ascher et al. (1983) and Bader and Ascher (1987).

$$\begin{bmatrix}
A & & & & \\
-\Gamma_1 & I & & & \\
& \cdot & & & \\
& & \cdot & & \\
& & -\Gamma_i & I & \\
& & & \cdot & \\
& & & & \cdot \\
& & & -\Gamma_N & I \\
& & & & B
\end{bmatrix}$$

Figure 3

Majaess et al. (1989) discussed two new packages for solving certain almost block diagonal linear systems whose use in COLNEW might improve its efficiency even further. The first, ABDPACK, implements an alternate row and column elimination algorithm which avoids unnecessary fill-in when applied to the linear systems with coefficient matrices of the form shown in Figure 2. For the solution of such systems,

COLROW cannot be used without fill-in since the structures of the systems violate constraints imposed by this code. The package SOLVEBLOK can be employed but also with fill-in, the amount of which increases with the number of rows in A and the number of collocation points in each subinterval. Numerical experiments reported by Majaess et al. demonstrate the superiority of this package over the most recent version of SOLVEBLOK, namely that used in COLNEW. The second package of Majaess et al. is a modification of ABDPACK, called ABBPACK, for solving systems with coefficient matrices of the form shown in Figure 3, (which also arise in implicit Runge-Kutta methods and multiple shooting for solving BVODEs.) The use of ABBPACK in COLNEW in place of SOLVEBLOK is currently under investigation. Some major modifications to COLNEW are necessary to enable ABDPACK to be used to solve the uncondensed systems, but it appears that the potential savings justify the effort.

Software implementing orthogonal spline collocation methods for solving other types of problems have also been developed. Min-Da Ho (1983) discussed a modification to COLSYS to enable it to handle differential-algebraic systems. Zygourakis and Aris (1983) described an implementation which combines orthogonal spline collocation and monotone iteration methods. As we have already mentioned, Bader (1985) implemented his orthogonal spline collocation method for boundary value functional differential equations in the code FDECOL, which is based on COLSYS. Bader and Kunkel (1989) formulated an orthogonal spline collocation method for the solution of parameter-dependent boundary value problems and implemented this method in a code, COLCON, based on techniques used in COLNEW.

**2.4 Extrapolated/Modified collocation methods.** Extrapolated/modified collocation is a variant of the nodal cubic spline collocation which yields $O(h^4)$ approximations on uniform partitions. This method grew out of work of Fyfe (1969) in which an optimal approximation to the solution of (2.1) was determined by a deferred correction process based on the nodal collocation method. Daniel and Swartz (1975) showed how Fyfe's method could be viewed as the use of spline collocation on a perturbed differential equation satisfied by an accurate spline interpolant of the solution of the boundary value problem. They developed the theory behind this interpretation of Fyfe's method for both linear and nonlinear problems, and showed that the $L^\infty$ error is of optimal order, namely $O(h^4)$. Surprisingly, the method and results had been developed earlier by Archer (1973, 1977). For the solution of (2.1), this method, which we shall refer to as the ADS formulation of Fyfe's method, takes the form

$$-u_h''(x_i) - \frac{h^2}{12}(\theta u_h)''(x_i) + (pu_h')(x_i) + (qu_h)(x_i) = f(x_i), \quad i = 0,\ldots, N,$$

where

$$\theta u_h(x_0) = 2\Delta_x^2 u_h(x_1) - \Delta_x^2 u_h(x_2),$$

$$\theta u_h(x_i) = \Delta_x^2 u_h(x_i), \qquad i = 1,\ldots,N\text{-}1,$$

$$\theta u_h(x_N) = 2\Delta_x^2 u_h(x_{N\text{-}1}) - \Delta_x^2 u_h(x_{N\text{-}2}),$$

and $\Delta_x^2$ denotes the usual second order central difference operator. Archer (1977) showed that it is often convenient to take

$$(\theta u_h)(x_i) = u_h(x_i), \quad i = 0, N,$$

without loss of global accuracy. Hill (1973) compared Fyfe's method with the Archer-Daniel-Swartz (ADS) formulation and showed that the latter is the more efficient.

Fyfe (1970) also used his deferred correction approach to solve certain linear fourth order problems; see also Hoskins (1973). Albasiny and Hoskins (1972) derived an optimal order cubic spline method for problems of the form (2.1) but their method is more closely related to the classical Numerov technique than to the extrapolated collocation approach. Maestro and Voss (1981) used techniques similar to those of Fyfe (1969) to examine a quintic spline method for the Falkner-Skan equations. None of these authors provided rigorous error analyses.

For the solution of (2.1) with homogeneous Dirichlet boundary conditions, Archer and Diaz (1978) formulated and analyzed a family of modified collocation methods which includes both orthogonal spline collocation and extrapolated collocation as special cases.

Extending the work of Fyfe (1969), Archer (1973, 1977) and Daniel and Swartz (1975), Daniel (1977) described how deferred corrections could be used with spline collocation methods to improve the accuracy of approximations to nonlinear problems. In particular, he demonstrated how a sequence of approximations $u_h^j$, $j = 1, \ldots$, with accuracy $O(h^{4(j+1)})$ at the nodes of $\pi$ can be generated from the ADS approximation. Numerical experiments with a code implementing this scheme were described by Martin and Daniel (1977, 1981).

There is continued interest in optimal order collocation methods based on smoothest splines. For example, for general linear fourth-order problems, Irodotou-Ellina and Houstis (1988) have formulated and analyzed an $O(h^6)$ ADS-type method based on quintic splines. They also formulated a variation of this method as a deferred correction procedure; cf. Fyfe (1969). These techniques have been extended by Irodotou-Ellina (1987) to general nonlinear two-point boundary value problems.

Even degree smoothest splines on uniform partitions have also been used in collocation procedures. Collocation with quadratic splines, $P(3,1;\pi)$, was considered for the solution of second order problems by Khalifa and Eilbeck (1982), Sakai (1983), and Sakai and Usmani (1983). For a second order nonlinear problem, Kammerer et al. (1974) examined a collocation method in which the approximate solution is in the spline space $P(5,3;\pi)$, and showed that this method gives fourth order convergence on uniform grids. In these methods, the approximate solution is determined by collocating at $\{x_{j-1/2}\}_{j=1}^{N}$, the midpoints of the subintervals defined by $\pi$, (and, in the case of the method of Kammerer et al., at the endpoints $x = 0$ and $x = 1$ also) and by satisfying the boundary conditions. (See also Reddien (1977), who discussed the extension of this method to periodic boundary value problems.) Not surprisingly, such procedures do not yield optimal rates of convergence. However, for a second order nonlinear equation with linear boundary conditions, Sakai (1983) determined the $O(h^2)$ leading error term. This term can then be used in a deferred correction type setting to obtain inexpensively an improved solution, or it can be used as an error estimate. Sakai (1984) employed similar ideas in a study of nodal cubic spline collocation. Recently, Houstis, Christara and Rice (1988) formulated an extrapolated quadratic spline collocation method for the solution of (2.1) and derived an optimal order $O(h^{3-j})$ global estimate for the j-th derivative of the error, j = 0, 1, 2. Formulations based on both Fyfe's approach and the ADS approach to modified methods were examined. Superconvergence results were also derived. In particular, it is shown that superconvergence of $O(h^4)$ is achieved at the nodes of $\pi$ as well as at the mid-points, the collocation points in this case. Results of numerical experiments confirm the predicted rates of convergence.

**2.5 Collocation-Galerkin methods.** Another approach is to use a hybrid method which combines some of the efficiency and simplicity of a spline collocation method with the lower continuity of a Galerkin method in an intermediate method commonly known as a collocation-Galerkin method. The first collocation-Galerkin method was introduced by Diaz (1974, 1977) for two-point boundary value problems of the form (2.1) with homogeneous Dirichlet boundary conditions. This method consists in finding

$$u_h \in P(r+1,0;\pi) \cap \{v | v(0) = v(1) = 0\}$$

such that

$$Lu_h(x_{ij}) = f(x_{ij}), \quad i = 1,...,N; \quad j = 1,...,r-1,$$

$$A(u_h,v) = (f,v), \quad \text{for all } v \in P(2,0;\pi),$$

where

$$A(\phi,\psi) = (\phi',\psi') + (p\phi' + q\phi,\psi),$$

and the r-1 points $\{x_{ij}\}_{j=1}^{r-1}$ are the affine images of a Jacobi polynomial of degree r-1. Specifically $\{\lambda_j\}_{j=1}^{r-1}$ are the Jacobi points defined by

$$(2.6) \qquad \int_{-1}^{1} \omega(x)p(x)ds = \sum_{j=1}^{r-1} \omega_j p(\lambda_j), \quad \text{for all } p \in P_{2r-2},$$

where $\omega(x) = x(1-x)$. Diaz (1974, 1977) established optimal $L^2$ rates of convergence and superconvergence results for the approximation $u_h$. Wheeler (1977) derived optimal $L^p$-estimates for this procedure, while Dunn and Wheeler (1976) derived such estimates for any choice of (distinct) collocation points. Dunn and Wheeler also proposed a collocation-$H^{-1}$-Galerkin method in which one seeks an approximate solution $u_h \in P(r+1,-1;\pi)$ such that

$$Lu_h(x_{ij}) = f(x_{ij}), \quad i = 1,...,N; \quad j = 1,..,r-1,$$

$$A^*(u_h,v) = (f,v), \quad \text{for all } v \in P(4,1;\pi) \cap \{v | v(0) = v(1) = 0\},$$

where

$$A^*(\phi,\psi) = -(\phi,\psi'') - (\phi,(p\psi)') + (\phi,q\psi),$$

and established optimal $L^p$-estimates for any choice of the collocation points. A $C^0$-collocation-Galerkin method for linear and nonlinear singular boundary value problems was examined by Carey and Wheeler (1979) and Carey et al. (1981).

On the basis of computational complexity, the collocation-Galerkin method is intermediate between the Galerkin method and the collocation method. It has the advantage over the Galerkin procedure with the same space that the integrals involve the product of the approximate solution and a piecewise linear function; thus the integrals are simpler and, of course, there are fewer of them.

Leyk (1986) introduced a new method which is somewhat related to collocation-Galerkin methods but requires no quadratures. For the solution of (2.1) with homogeneous Dirichlet boundary conditions, this method, which Leyk named the $C^0$-collocation-like method, is defined in the following way: Find

$$u_h \in P(r+1,0;\pi) \cap \{v|v(0) = v(1) = 0\}$$

such that

(2.7)       $L u_h(x_{ij}) = f(x_{ij}), \ i = 1,...,N; \ j = 1,...,r-1,$

$$\frac{1}{w_0 \bar{h}_i} |[u_h']| + \{L u_h\}(x_i) = \{f\}(x_i), \ i = 1,...,N-1,$$

where $w_0 = 2/[r(r+1)]$, $\bar{h}_i = 0.5(h_{i+1} + h_i)$, $|[g]|(x_i) = g(x_{i^-}) - g(x_{i^+})$, the jump discontinuity in g at the node $x_i$, and $\{g\}(x_i) = \alpha_i g(x_{i^-}) + (1-\alpha_i)g(x_{i^+})$, the weighted average of g at $x_i$, with $\alpha_i = 0.5 h_i / \bar{h}_i$. In (2.7), the collocation points are the affine images of the Lobatto points on (-1,1), the zeros of the first derivative of the Legendre polynomial of degree r on (-1,1). For this method, Leyk derived optimal order error estimates in the $L^2$ and $H^1$ norms, and also superconvergence results. In the case of a self-adjoint problem, Leyk modified the method to produce one in which only symmetric systems of linear equations arise. For this scheme, optimal order $H^1$ error estimates were derived. Leyk also presented symmetric modifications of the collocation-Galerkin methods of Diaz (1977), Dunn and Wheeler (1976) and Wheeler (1977).

## 3. ELLIPTIC BOUNDARY VALUE PROBLEMS

The collocation methods which have been developed for the approximate solution of elliptic boundary value problems are, in the main, generalizations of the methods introduced in Section 2. In order to describe these, we introduce the following notation.

Let $\pi_x = \{x_i\}_{i=1}^{N_x}$ denote a partition of the interval $\bar{I}$ and set $h(\pi_x) = \max\{x_i - x_{i-1}\}$. If $\pi_x$ and $\pi_y$ are two partitions of $\bar{I}$, let $\pi$ denote the partition of the unit square $\Omega$ defined by the grid points $\pi_x \times \pi_y$. Let

$$h = \max(h(\pi_x), h(\pi_y)), \qquad \gamma_{ij} = (x_{i-1}, x_i) \times (y_{j-1}, y_j),$$

and

$$\Gamma = \{\gamma_{ij}, \ 1 \le i \le N_x, 1 \le j \le N_y\}.$$

Let $\{\lambda_k\}_{k=1}^{r-1}$, with $-1 < \lambda_1 < \dots < \lambda_{r-1} < 1$, denote the zeros of the Legendre polynomial of degree r-1 on the interval [-1,1]. For the partitions $\pi_x$ and $\pi_y$, let

$$\lambda_{ki}(\pi_x) = x_{i-1/2} + \frac{1}{2}\lambda_k(x_i - x_{i-1}), \quad 1 \le k \le r-1, \quad 1 \le i \le N_x,$$

$$\lambda_{sj}(\pi_y) = y_{j-1/2} + \frac{1}{2}\lambda_s(y_j - y_{j-1}), \quad 1 \le s \le r-1, \quad 1 \le j \le N_y,$$

and set

$$\Lambda(\pi) = \{(\lambda_{ki}(\pi_x), \lambda_{sj}(\pi_y)), \ 1 \le k,s \le r-1, \ 1 \le i \le N_x, \ 1 \le j \le N_y\}.$$

Also

$$\bar{P}(r,k;\pi) = P(r,k;\pi_x) \otimes P(r,k;\pi_y),$$

where $\otimes$ denotes the tensor product. In the following, $D_x$ and $D_y$ denote partial derivatives with respect to x and y, respectively.

The method of nodal collocation based on the tensor product of cubic splines $\bar{P}(4,2;\pi)$ was first analyzed independently by Cavendish (1972) and Ito (1972). A description of this method is given by Prenter (1975) for the boundary value problem

$$Lu \equiv -\Delta u + cu = f \text{ on } \Omega,$$

where $\Delta$ denotes the Laplacian, subject to homogeneous Dirichlet boundary conditions. In this procedure, one collocates at all but the corner nodes. It can be shown (see, for example, Prenter, 1975) that no additional information is obtained by collocating at the corner points of the square. At these nodes, one interpolates $D_x^2 D_y^2 u$. Then, on differentiating the differential equation twice with respect to x and using the fact that, because of the homogeneous Dirichlet boundary conditions, the functions u,

$D_x^2 u$, and $D_x^4 u$ are zero at the corner points, it follows that $D_x^2 D_y^2 u_h = D_x^2 f$ at these points. The collocation equations together with these corner constraints then define a unique approximate solution. Second order accuracy of this collocation scheme was proved by Cavendish (1972) and Ito (1972), and the proof was also presented by Prenter (1975). A formulation of this method for more general elliptic PDEs was considered by Ito (1972) but no error estimates were derived.

Arnold and Saranen (1984) analyzed the nodal collocation method using the tensor product of smoothest splines of arbitrary even order for equations of the form

$$(3.1) \qquad Lu \equiv \nabla \cdot (a \nabla u) + b \cdot \nabla u + cu = f \text{ in } \Omega,$$

in which the coefficients a, b, c, and the forcing function f are biperiodic functions, and derived results analogous to those of Arnold and Wendland (1983) for BVODEs. Specifically, they obtain optimal rates of convergence in $L^2$ for partial derivatives of the approximate solution which are of order at least two in one variable, while the solution itself and its gradient converge in $L^2$ at suboptimal rates.

Prenter and Russell (1976) considered linear self-adjoint elliptic partial differential equations of the form

$$(3.2) \qquad Lu \equiv -D_x(pD_x u) - D_y(qD_y u) + cu = f \text{ in } \Omega,$$

subject to homogeneous Dirichlet boundary conditions, and formulated and analyzed an orthogonal spline collocation procedure based on bicubic Hermite piecewise polynomials, that is, the space $\overline{P}(4,1;\pi)$. Their collocation approximation is defined as the element $u_h \in \overline{P}(4,1;\pi)$ which satisfies

$$Lu_h = f \text{ on } \Lambda(\pi), \quad u_h = 0 \text{ on } \partial\Omega.$$

Prenter and Russell derived optimal order estimates in both the $L^2$ and $H_0^1$ norms under the assumption of the existence and uniqueness of the collocation approximation and the assumption that the collocation approximation satisfies certain bounds. These assumptions were not required in the work of Houstis (1978b), who examined bicubic Hermite collocation for more general boundary value problems than those considered by Prenter and Russell, namely

(3.3a)        $Lu \equiv \alpha D_x^2 u + \beta D_x D_y u + \gamma D_y^2 u + \delta D_x u + \varepsilon D_y u + \zeta u = f$  in  $\Omega$,

subject to boundary conditions of the form

(3.3b)                              $Bu \equiv \mu \dfrac{\partial u}{\partial n} + \nu u = 0,$

where the coefficients $\alpha$, $\beta$, ,$\gamma$ satisfy the ellipticity condition $\beta^2 - 4\alpha\gamma < 0$. Percell and Wheeler (1980) extended the work of Prenter and Russell substantially. They established optimal order $L^2$ and $H_0^1$ error estimates for orthogonal spline collocation employing the the space $\overline{P}(r+1,1;\pi)$, with $r \geq 3$, for equations of the form (3.1) subject to homogeneous Dirichlet boundary conditions. Their analysis requires no more smoothness of the solution u than is necessary in one dimension and is free of the assumptions made by Prenter and Russell.

The collocation method with bicubic Hermites has also been formulated for general curved domains, but no analysis of the method has been provided. This method was introduced by Houstis and Rice (1977) and described in detail by Houstis et al. (1978) who compared it with three other methods on a set of 17 test problems. (See also Houstis and Rice, 1977). Another comparison involving the method was given by Houstis (1978c). Houstis et al. (1985a) described three specific collocation algorithms which use bicubic Hermite basis functions, the first for collocation on general two-dimensional domains, the second for collocation on rectangular domains with general linear boundary conditions, and the third for collocation on rectangular domains with uncoupled boundary conditions. Houstis et al. (1985b,c) presented Fortran implementations of these algorithms, which have been used to solve a wide class of elliptic problems; see, for example, Houstis et al. (1978, 1979, 1983) and Houstis and Rice (1980). These codes take into account the work of Dyksen and Rice (1984, 1986) on an appropriate ordering of the collocation equations and unknowns for rectangular domains which can be extended to general domains in a straightforward way, and the scaling of the equations.

Frind and Pinder (1979) combined the notion of isoparametric finite elements with collocation to extend the applicability of the collocation method to the solution of potential problems in general regions, and conducted computational studies of spline collocation methods and Galerkin methods. Various other numerical studies involving collocation methods have been conducted; see, for example, Chang and Finlayson (1977, 1978), Finlayson (1980), Masliyah and Kumar (1980), Dixon (1981), and Weiser et al.

(1980). Dyksen et al. (1984) presented a study of the performance of the collocation and Galerkin methods using bicubic Hermites, which was a sequel to the studies of Houstis et al. (1978) and Weiser et al. (1980). Dyksen et al. compared their implementation of collocation with the Galerkin implementation of Weiser et al. for the solution of linear self-adjoint elliptic equations on rectangular domains, and concluded that their study "strongly supports the hypothesis that (with these implementations of the methods) collocation performs better than Galerkin for computer time versus error".

Chang and Finlayson (1977, 1978) and Masliyah and Kumar (1980) considered the use of alternating direction collocation (ADC) iterative methods in their work, but provided no convergence analyses. Dyksen (1987) developed the theory for ADC using bicubic Hermites on a uniform partition for Poisson's equation on the unit square. This work was generalized by Cooper and Prenter (1989), who examined the convergence of ADC using bicubic Hermites on non-uniform partitions for a large class of linear separable elliptic partial differential equations on a rectangular domain. See also Cooper and Prenter (1988).

Zamani and Sun (1988) used bicubic Hermite collocation in the solution of a nonlinear elliptic boundary value problem modelling a compressible flow problem. The resulting nonlinear algebraic equations were solved iteratively using a successive approximation method.

For the solution of (3.3a) subject to homogeneous Dirichlet or Neumann boundary conditions, Houstis, Vavalis and Rice (1988) presented a new class of collocation methods using bicubic splines. These methods are based on generalizations of the modified collocation approaches of Fyfe (1969) and ADS. (When applied to Poisson's equation in a square, one of the new methods appears to be identical to the method proposed by Archer (1975).) Optimal order error bounds were derived for these methods and confirmed by experimental results. These results also indicate that the new methods are computationally more efficient than methods based on either collocation with bicubic Hermite piecewise polynomials (Houstis et al., 1985a,b,c) or on the Galerkin method with cubic splines. Houstis, Vavalis and Rice (1988) pointed out that, unlike general collocation based on Hermite cubics implemented by Houstis, Mitchell and Rice (1985a,b,c), the cubic spline collocation equations can be solved by various iterative methods which were studied extensively by Houstis et al. (1984) and Vavalis (1985), and their direct solution does not require pivoting. Three implementations of the methods exist in ELLPACK (Rice and Boisvert, 1985) for equation (3.3a) subject to mixed Dirichlet and Neumann boundary conditions on parts of the boundary. The modified collocation methods of Irodotou-Ellina and Houstis (1988) have also been extended by

Irodotou-Ellina (1987) to second and fourth order elliptic problems in the plane.

For the solution of elliptic problems in two space variables, two generalizations of the one-dimensional $C^0$-collocation Galerkin method have been developed. Diaz (1979a) defined a collocation-Galerkin method for Poisson's equation on the unit square, using tensor products of continuous piecewise polynomials, that is, the space $\overline{P}(r+1,0;\pi)$, and derived optimal $L^2$ and $H_0^1$ error estimates. For Dirichlet problems of the form (3.2) in which $\Omega$ is a bounded domain in the plane with a piecewise smooth boundary, Wheeler (1978) formulated and analyzed a collocation-finite element method defined on partitions of $\Omega$ which are composed of either triangles or rectangles and showed that this scheme yields optimal $L^2$ estimates.

## 4. TIME-DEPENDENT PROBLEMS

Consider the quasilinear parabolic initial boundary value problem

$$c(x,t,u)D_t u - D_x^2 u = f(x,t,u,D_x u), \ (x,t) \in (0,1) \times (0,T],$$

(4.1)
$$u(0,t) = g_0(t), \ u(1,t) = g_1(t), \ t \in (0,T],$$

$$u(x,0) = g(x), \ x \in (0,1).$$

Several authors have studied the use of nodal collocation for the spatial discretization of parabolic problems. Cavendish (1972) studied cubic spline methods with $O(h^2)$ accuracy for linear versions of (4.1). Also, a cubic spline collocation procedure for the heat equation was proposed (but not analyzed) by Papamichael and Whiteman (1973), and shown to be a member of a family of finite difference methods. Rubin and Graves (1975) developed a cubic spline collocation procedure for quasi-linear parabolic problems in one space variable, and also considered linear problems in two space variables for which they proposed a spline alternating direction method. The accuracy of these procedures was improved by Rubin and Kholsa (1976). Stys (1981) proved the convergence of an $O(h^2)$ cubic spline method for a class of nonlinear parabolic equations.

Orthogonal spline collocation methods for linear and nonlinear parabolic initial-boundary value problems in one space variable of the form (4.1) have been thoroughly analyzed by Douglas (1972) and Douglas and Dupont (1972, 1973, 1974), who considered the use of orthogonal spline collocation in space combined with various time-stepping procedures, including collocation in time. (See also Cerutti and Parter (1976) who tied together the results of de Boor and Swartz (1973) and those of Douglas and Dupont.) Hyperbolic problems have also been treated in a similar vein. Houstis (1977) considered collocation methods for second order nonlinear hyperbolic problems in one space variable of the form

$$p(x,t,u)D_t^2 u - q(x,t,u)D_x^2 u = f(x,t,u), \quad (x,t) \in (0,1) \times (0,T],$$

and derived optimal order $L^\infty$ error estimates.

Archer (1978) examined a discrete-time orthogonal collocation method for the solution of quasilinear first-order hyperbolic initial-boundary value problems of the form

$$c(x,t,u)D_t u + a(x,t,u)D_x u = f(x,t,u), \quad (x,t) \in (0,1) \times (0,T],$$

$$u(0,t) = 0, \ t \in (0,T],$$

$$u(x,0) = u_0(x), \ x \in (0,1).$$

Optimal order $L^\infty$ norm estimates for the error in the approximate solution

$$u_h: [0,T] \to P(r+1,0;\pi)$$

were derived for sufficiently smooth problems. Luskin (1979) considered a class of nonsymmetric, nonlinear hyperbolic systems with integral boundary conditions, and derived optimal order error estimates for two linearizations of a collocation process using tensor products of continuous piecewise linear functions in space and time. A collocation process in which the collocation approximation $u_h(x,t) \in P(r+1,0;\pi) \otimes P(2,0;\pi)$ was also examined. Lesaint and Raviart (1979) considered collocation methods for linear first-order hyperbolic systems and also first-order systems arising from parabolic problems in one space variable. Roughly speaking, an approximate solution is sought in a space of continuous piecewise polynomials of degree $\leq r$ defined on a partition of the $(x,t)$ domain into quadrilaterals. The collocation approximation is obtained by collocating at the product Gauss points; cf. orthogonal spline collocation in Section 3.

Orthogonal spline collocation methods have been used to solve rather complicated problems in science and engineering which involve parabolic equations in one space variable; see, for example, Chawla and Chan (1980), Chawla, Leaf and Chen (1975), Chawla et al. (1974, 1975), Nguyen et al. (1983), Pedersen and Tanoff (1982), Sincovec (1977b), and Young (1977). A modification of orthogonal spline collocation designed for the solution of convection-dominated flow problems was discussed by Allen (1983) and Allen and Pinder (1983). See also Dougherty and Pinder (1983), Mohsen and Pinder (1984, 1986), and Pinder and Shapiro (1979).

Orthogonal spline collocation methods have also been applied to the solution of nuclear engineering problems. The first work in this area was by Mason (1975), who solved the static neutron diffusion equation using piecewise cubic Hermite functions. The extension to space-time nuclear reactor dynamics was done by Meade (1982, 1983), who developed a special purpose code implementing orthogonal spline collocation with piecewise cubic Hermites for the space discretization and the θ–method for the time stepping. The efficiency of this collocation code was examined on a benchmark problem by Grossman et al. (1982) and shown to be comparable to that of a Galerkin finite element code developed by del Valle (1983). In a more thorough comparison in which both codes were compared with an implementation of a collocation-Galerkin method, del Valle et al. (1986) showed that the Galerkin code was the most efficient. In this constant-coefficient problem, the Galerkin code is aided by the fact that the calculation of the quadratures need be done only once. A time-collocation scheme was implemented for the solution of the time-dependent neutron diffusion equation by Bian (1983). The neutron transport equation was solved by the method of collocation by Grossman (1983), and by Morel (1979), who considered both symmetric and asymmetric scattering.

One of the most widely used codes for the numerical solution of time dependent problems in one space variable is PDECOL (Madsen and Sincovec, 1979). This package implements orthogonal spline collocation for the space discretization and the resulting initial value problem for a system of ordinary differential equations is solved using a variant of the well-known Gear package. In PDECOL, there arise almost block diagonal linear systems with the same structure as those occurring in COLSYS (see Figure 1) and in PDECOL these systems are solved using a band solver. Recently Keast and Muir (1987) replaced this linear equation solver with COLROW (Diaz et al. 1983a,b), and in extensive testing showed that savings of up to 50% in total execution time can be realized. PDECOL is not designed to solve initial-boundary value problems in which there are incompatibilities between the initial and boundary conditions. However the package can be modified to overcome this limitation and one such modification was applied by Akyurtlu et al. (1986b).

A different type of collocation method is one of the spatial discretizations implemented in SPRINT, (Berzins and Furzeland, 1986, and Berzins et al., 1988). In this collocation method, the approximation is a piecewise Chebyshev polynomial with $C^0$ continuity which is determined by collocating at certain Chebyshev points in each subinterval defined by a partition of the spatial domain. See Berzins and Dew (1987) for details. SPRINT is applicable to a much larger problem class than PDECOL, being designed to solve problems involving mixed systems of time-dependent algebraic, ordinary and partial differential equations.

Greenwell-Yanik and Fairweather (1986) analyzed procedures involving orthogonal spline collocation for the solution of semilinear parabolic equations of the form

$$(4.2) \qquad c(x,y,t)D_t^j u = \Delta u + b(x,y,t) \cdot \nabla u + f(x,y,t,u), \quad (x,y) \in \Omega, \ t \in [0,T],$$

with $j = 1$, subject to homogeneous Dirichlet boundary conditions and appropriate initial conditions. Optimal order error estimates were derived for a continuous time collocation method in which orthogonal spline collocation with the space $P(r+1,1;\pi)$, where $r \geq 3$, is used for the spatial discretization. Several discrete-time collocation methods were described and the analysis for the Crank-Nicolson method was presented. Greenwell-Yanik and Fairweather (1986) also formulated and analyzed a family of discrete-time collocation methods for the solution of hyperbolic equations of the form (4.2) with $j=2$.

For parabolic problems in two space variables, various alternating direction collocation methods have been proposed by Bangia et al. (1978), Botha and Celia (1982), Celia and Pinder (1984), Celia et al. (1980), Hayes (1980, 1981), and Hayes et al. (1981). These formulations are all restricted to rectangular domains in two space dimensions. The alternating direction collocation procedure was generalized to nonrectangular geometries by Celia (1983) through the definition of coordinate mappings in the context of Hermite bases. No rigorous convergence analysis has yet been produced for any of these alternating direction techniques; see, for example, Celia and Pinder (1985).

Yanik and Fairweather (1988) examined discrete time collocation methods for parabolic integro-differential equations in one-space variable of the form

$$c(x,t,u)D_t u - D_x^2 u = \int_0^t f(x,t,s,u(x,s),D_x u(x,s))ds, \ (x,t) \in (0,1) \times [0,T].$$

In the case of two space variables when the spatial region is rectangular and the coefficient c is independent of u and the function f is independent of $\nabla u$, techniques developed by Greenwell-Yanik and Fairweather (1986) can be employed to derive optimal order error estimates. The corresponding second-order hyperbolic problem was also considered by Yanik and Fairweather.

Modified collocation methods have also been developed for parabolic problems. An $O(h^4)$ method based on the work of Archer (1973) and Fyfe (1969) was examined by Cavendish and Hall (1974) for the problem

$$D_t u - D_x^2 u + cu = f(x,t).$$

It is not clear how their analysis would extend to more general problems. Archer (1977) formulated and analyzed a continuous-time modified method for the solution of problems of the form (4.1) and obtained estimates of order $h^4$.

Collocation-Galerkin methods have also been considered for the solution of time-dependent problems in one space variable. Wheeler (1977) extended the results of Diaz (1974, 1977) to certain nonlinear parabolic problems, and Kendall and Wheeler (1976) examined a Crank-Nicolson time discretization of this method. Diaz (1979b) formulated a continuous-time collocation-$H^{-1}$-Galerkin method for linear parabolic problems with time dependent coefficients in which the collocation points are given by (2.6) with $\omega(x) = x^2(1-x)^2$, and Archer and Diaz (1982) applied similar ideas to a first order hyperbolic problem. For each of these methods optimal error estimates were derived. Nakao (1981) derived superconvergence results for the collocation-$H^{-1}$-Galerkin method of Diaz (1979b) but for linear parabolic problems with coefficients which are independent of time.

## 5. CONCLUDING REMARKS

In this final section, we mention some applications of spline collocation omitted in our survey, and also indicate some topics worthy of further study.

A considerable amount of work has been done on spline collocation methods for solving initial value problems for ordinary differential equations; see Hulme (1972) for early work in this area, and Fuchs (1987) and Lie and Norsett (1989) and references cited in these papers for recent developments.

Phillips (1972) and Prenter (1973) were among the first to examine spline collocation methods for integral equations. Since then, the literature on spline collocation methods for Fredholm and Volterra-type equations has grown rapidly; see, for example, Brunner (1987a) and Brunner and van der Houwen (1986). Also, Blom and Brunner

(1986) developed a software package for solving systems of nonlinear Volterra integral equations of the second kind which implements a method introduced by these authors (Blom and Brunner, 1987). Recently, Brunner and Kauthen (1989) analyzed spline collocation methods for two-dimensional Volterra integral equations using an approach which may be considered as a generalization of that used by Brunner (1984). Several contributions to the development and analysis of collocation methods for Volterra integro-differential equations have been made by Brunner (1986, 1987, 1988). Interesting applications of collocation methods were described by Chawla et al. (1980) for the solution linear and nonlinear integral equations occurring in radiative transfer and laminar boundary layer problems, and by Brannigan and Eyre (1983) who treated singular integral equations arising in scattering theory.

A major source of integral equations is the Boundary Element Method, a widely used method for solving certain classes of problems arising in many engineering applications, such as exterior and interior boundary value problems of elasticity, fluid dynamics, electromagnetics and acoustics. In this method, the problem in question is reformulated as an integral equation or system of integral equations on the boundary of the domain. The most popular numerical technique for solving the boundary integral equations is spline collocation. While such an approach has been used in practice for more than twenty years, it was not until the early eighties that significant progress was made in the analysis of the convergence of these methods for problems in two space variables; see Arnold and Wendland (1983), Arnold and Wendland (1985), which contains an interesting mini-survey of the relevant literature, Saranen (1987, 1988) and Saranen and Wendland (1985). Atkinson (1988) provided a review of progress in the analysis of collocation methods for potential problems in three space variables. There still remains much to be done on the analysis of three dimensional problems.

Little has been done to exploit the apparent advantages of collocation-Galerkin methods. In numerical comparisons, these methods have performed as one might expect, not as well as collocation methods but better than Galerkin methods. The approach of Leyk (1986), which involves no quadratures, is an interesting alternative which seems worthy of further study for the solution of one space variable problems.

While alternating direction collocation (ADC) methods have been developed and analyzed for the iterative solution of elliptic boundary value problems (Cooper and Prenter, 1988, 1989), there are still some theoretical questions to be addressed concerning ADC methods for the solution of time dependent problems. These methods have seen some use in practice, but no rigorous error analysis of them is known to the authors.

The collocation codes COLSYS, PDECOL and those in ELLPACK have proved to be exceedingly effective in practical computations. While they have been used quite extensively, some engineers are still examining nodal spline collocation and extolling its virtues; see, for example, Doctor and Kalthia (1988). More has to be done to bring these packages, and newer software such as SPRINT (Berzins and Furzeland, 1986, Berzins et al., 1988), to the attention of researchers.

For use in the solution of large scale problems, the next generation of collocation software must take advantage of advanced architectural features of new computers, such as parallelization and vectorization. First steps in this direction have been taken by Gladwell and Hay (1989) and Houstis et al. (1987).

**Acknowledgements.** Graeme Fairweather was supported in part by the National Science Foundation Grant RII-8610671 and the Commonwealth of Kentucky through the Kentucky EPSCoR Program. Daniel Meade was supported in part by a joint National Science Foundation-CONACyT grant, and by The Third World Academy of Sciences Grant TWAS RG No. RG BC 86-10. The authors are grateful to the referees for their constructive criticism and valuable comments. The authors also wish to thank Karin Bennett, Ryan Fernandes, Alonzo Peña Piña, and Mark Robinson for assistance during the preparation of this work, and the editor of this volume, Julio Cesar Diaz, for his patience and encouragement.

## REFERENCES

Ahlberg, J. H. and T. Ito (1975), A collocation method for two-point boundary value problems, Math. Comp., 29, 761-776.

Ahlberg, J. H., E. N. Nilson and J. L. Walsh (1967), The Theory of Splines and Their Applications, Academic Press, New York.

Akyurtlu, A., J. F. Akyurtlu, C. E. Hamrin Jr. and G. Fairweather (1986a), Reformulation and numerical solution of the equations for a catalytic porous wall gas-liquid reactor, Computers Chem. Eng., 10, 361-365.

Akyurtlu, A., J. F. Akyurtlu, K. S. Denison and C. E. Hamrin Jr. (1986b), Application of the general purpose collocation software, PDECOL, to the Graetz problem, Computers Chem. Eng., 10, 213-222.

Albasiny, E. L. and W. D. Hoskins (1969), Cubic spline solutions to two-point boundary value problems, Comp. J., 12, 151-153.

Albasiny, E. L. and W. D. Hoskins (1972), Increased accuracy cubic spline solutions to two-point boundary value problems, J. Inst. Math. Appl., 9, 47-55.

Allen, M. B. (1983), How upstream collocation works, Internat. J. Numer. Methods Engrg., 19, 1753-1763.

Allen, M. B. and G. F. Pinder (1983), Collocation simulation of multiphase porous-medium flow, Soc. Pet. Eng. J., 23, 135-142.

Archer, D. (1973), Some collocation methods for differential equations, Ph.D. thesis, Rice University, Houston, TX.

Archer, D. (1975), Collocation in two dimensions - some numerical experiments, SIAM Rev., 17, 374.

Archer, D. (1977), An $O(h^4)$ cubic spline collocation method for quasilinear parabolic equations, SIAM J. Numer. Anal., 14, 620-637.

Archer, D. (1978), Collocation in $C^0$ spaces for first order hyperbolic equations. I: Optimal order global estimates for quasilinear problems, SIAM J. Numer. Anal., 15, 271-281.

Archer, D. and J. C. Diaz (1978), A family of modified collocation methods for second order two point boundary value problems, SIAM J. Numer. Anal., 15, 242-254.

Archer, D. and J. C. Diaz (1982), A collocation-Galerkin method for a first order hyperbolic equation with space and time-dependent coefficient, Math. Comp., 38, 37-53.

Arnold, D. N. and J. Saranen (1984), On the asymptotic convergence of spline collocation methods for partial differential equations, SIAM J. Numer. Anal., 21, 459-472.

Arnold, D. N. and W. L. Wendland (1983), On the asymptotic convergence of collocation methods, Math. Comp., 41, 349-381.

Arnold, D. N. and W. L. Wendland (1985), The convergence of spline collocation for strongly elliptic equations on curves, Numer. Math., 47, 317-341.

Ascher, U. (1980), Solving boundary-value problems with a spline-collocation code, J. Comput. Phys., 34, 401-413.

Ascher, U. (1985), Two families of symmetric difference schemes for singular perturbation problems, in Numerical Boundary Value ODEs, U. Ascher and R. D. Russell, eds., Birkhauser, Boston, pp. 173-191.

Ascher, U. (1986), Collocation for two-point boundary value problems revisited, SIAM J. Numer. Anal., 23, 596-609.

Ascher, U., J. Christiansen and R. D. Russell (1979a), COLSYS -- a collocation code for boundary-value problems, in Codes for Boundary-Value Problems in Ordinary Differential Equations, B. Childs et al., eds., Lecture Notes in Computer Science 76, Springer-Verlag, New York, pp.164-185.

Ascher, U., J. Christiansen and R. D. Russell (1979b), A collocation solver for mixed order systems of boundary value problems, Math. Comp., 33, 659-679.

Ascher, U., J. Christiansen and R. D. Russell (1981a), Collocation software for boundary-value ODEs, ACM Trans. Math. Software, 7, 209-222.

Ascher, U., J. Christiansen and R. D. Russell (1981b), Algorithm 569. COLSYS: Collocation software for boundary-value ODEs, ACM Trans. Math. Software, 7, 223-229.

Ascher, U. and S. Jacobs (1989) On collocation implementation for singularly perturbed two-point problems, SIAM J. Sci. Statist. Comput., 10, 533-549.

Ascher, U., R. M. M. Mattheij and R. D. Russell (1988), Numerical Solution of Boundary Value Problems for Ordinary Differential Equations, Prentice Hall, Englewood Cliffs, New Jersey.

Ascher, U., S. Pruess and R. D. Russell (1983), On spline basis selection for solving differential equations, SIAM J. Numer. Anal., 20, 121-142.

Ascher, U. and P. Spudich (1986), A hybrid collocation method for calculating complete theoretical seismograms in vertically varying media, Geophys. J. R. Astr. Soc., 86, 19-40.

Ascher, U. and F. Y. M. Wan (1980), Numerical solutions for maximum sustainable consumption growth with a multi-grade exhaustible resource, SIAM J. Sci. Statist. Comput., 1, 160-172.

Ascher, U. and R. Weiss (1983), Collocation for singular perturbation problems I: First order systems with constant coefficients, SIAM J. Numer. Anal., 20, 537-557.

Ascher, U. and R. Weiss (1984a), Collocation for singular perturbation problems II: Linear first order systems without turning points, Math. Comp., 43, 157-187.

Ascher, U. and R. Weiss (1984b), Collocation for singular perturbation problems III: Nonlinear problems without turning points, SIAM J. Sci. Statist. Comp., 5, 811-829.

Atkinson, K. E. (1988), A survey of boundary integral equation methods for the numerical solution of Laplace's equation in three dimensions, Preprint AM/88/8, Applied Mathematics Preprint, School of Mathematics, The University of New South Wales, Australia.

Bader, G. (1985), Solving boundary value problems for functional equations by collocation, in Numerical Boundary Value ODEs, U. Ascher and R. D. Russell, eds., Birkhauser, Boston, pp. 227-243.

Bader, G. and U. Ascher (1987), A new basis implementation for a mixed order boundary value ODE solver, SIAM J. Sci. Statist. Comput., 8, 483-500.

Bader, G. and P. Kunkel (1989), Continuation and collocation for parameter-dependent boundary value problems, SIAM. J. Sci. Statist. Comput., 10, 72-88.

Bangia, V. K., C. Bennett, A. Reynolds, R. Raghavan and G. Thomas (1978), Alternating direction collocation methods for simulating reservoir performance, Paper SPE 7414, 53rd SPE Fall Technical Conference and Exhibition, Houston, Texas.

Bellen, A. and M. Zennaro (1984), A collocation method for boundary value problems of differential equations with functional arguments, Computing, 32, 307-318.

Berzins, M. and P. M. Dew (1987), A note on $C^0$ Chebyshev methods for parabolic P.D.E.s, IMA J. Numer. Anal., 7, 15-37.

Berzins, M. and R. M. Furzeland (1986), A User's Manual for SPRINT - A Versatile Software Package for Solving Systems of Algebraic, Ordinary and Partial Differential Equations: Part 2- Partial Differential Equations, Report TNER.86.050, Thornton Research Centre, Shell Research Ltd., Chester, England.

Berzins, M., P. M. Dew and R. M. Furzeland (1988), Developing software for time dependent problems using the method of lines and differential algebraic integrators, Applied Numerical Mathematics, to appear.

Bhattacharyya, D., M. Jevtitch, J. T. Schrodt and G. Fairweather (1986), Prediction of membrane separation characteristics by pore distribution measurements and surface force-pore flow model, Chem. Eng. Commun., 42, 111-128.

Bian, S. H. (1983), Application of the collocation method to the solution of the time dependent neutron diffusion equation, Transport Theory and Statistical Physics, 12, 285-306.

Bickley, W. G. (1968), Piecewise cubic interpolation and two-point boundary value problems, Comp. J., 11, 206-208.

Blom, J. G. and H. Brunner (1986), Discretized collocation and iterated collocation for nonlinear Volterra integral equations of the second kind, ACM Trans. Math. Software, submitted.

Blom, J. G. and H. Brunner (1987), The numerical solution of nonlinear Volterra integral equations of the second kind by collocation and iterated collocation methods, SIAM J. Sci. Statist. Comput., 8, 806-830.

Blue, J. L. (1969), Spline function methods for nonlinear boundary value problems, Comm. ACM, 12, 327-330.

de Boor, C. (1966), The method of projections as applied to the numerical solution of two point boundary value problems using cubic splines, Ph.D. thesis, University of Michigan, Ann Arbor, MI.

de Boor, C. (1978), A Practical Guide to Splines, Applied Mathematical Sciences Vol. 27, Springer-Verlag, New York.

de Boor, C. and B. Swartz (1973), Collocation at Gaussian points, SIAM J. Numer, Anal., 10, 582-606.

de Boor, C. and B. Swartz (1977), Comments on the comparison of global methods for linear two-point boundary value problems, Math. Comp., 31, 916-921.

de Boor, C. and B. Swartz (1980), Collocation approximation to eigenvalues of an ordinary differential equation: The principle of the thing, Math. Comp., 35, 679-694.

de Boor, C. and B. Swartz (1981a), Collocation approximation to eigenvalues of an ordinary differential equation: Numerical illustrations, Math. Comp., 36, 1-19.

de Boor, C. and B. Swartz (1981b), Local piecewise polynomial projection methods for an O.D.E. which give high-order convergence at the knots, Math. Comp., 36, 21-33.

de Boor, C. and R. Weiss (1980), SOLVEBLOK: A package for solving almost block diagonal linear systems, ACM Trans. Math. Software, 6, 80-87.

Botha, J. F. and M. Celia (1982), The alternating direction collocation approximation, Proceedings of the Eighth South African Symposium of Numerical Mathematics, Durban, South Africa, July 1982, 13-26.

Brannigan, M. and D. Eyre (1983), Splines and the projection collocation method for solving integral equations in scattering theory, J. Math. Phys., 24, 177-183.

Brunner, H. (1984), Iterated collocation methods and their discretization for Volterra integral equations, SIAM J. Numer. Anal., 21, 1132-1145.

Brunner, H. (1986), High-order methods for the numerical solution of Volterra integro-differential equations, J. Comp. Appl. Math., 15, 301-309.

Brunner, H. (1987a), Implicit Runge-Kutta-Nystrom methods for general second-order Volterra integro-differential equations, Comput. Math. Applic., 14, 549-559.

Brunner, H. (1987b), Collocation methods for one-dimensional Fredholm and Volterra integral equations, in The State of the Art in Numerical Analysis, A. Iserles and M. J. D. Powell, eds., Oxford University Press, London, pp. 563-600.

Brunner, H. (1988), The numerical solution of initial-value problems for integro-differential equations, in Numerical Analysis 1987, D. F. Griffiths and G. A. Watson, eds., Pitman Research Notes in Mathematics Series 170, Longman, New York, pp. 18-38.

Brunner, H. and P. J. van der Houwen (1986), The Numerical Solution of Volterra Equations, CWI Monographs, Vol. 3, North-Holland, Amsterdam, The Netherlands.

Brunner, H. and J.-P. Kauthen (1989), The numerical solution of two-dimensional Volterra integral equations by collocation and iterated collocation, IMA J. Numer. Anal., 9, 47-59.

Carey, G. F. and B. A. Finlayson (1975), Orthogonal collocation on finite elements, Chem. Eng. Sci., 30, 587-596.

Carey, G. F., D. Humphrey and M. F. Wheeler (1981), Galerkin and collocation-Galerkin methods with superconvergence and optimal fluxes, Internat. J. Numer. Methods Engrg., 17, 939-950.

Carey, G. F. and M. F. Wheeler (1979), $C^0$-collocation-Galerkin methods, in Codes for Boundary-Value Problems in Ordinary Differential Equations, B. Childs et al., eds., Lecture Notes in Computer Science 76, Springer-Verlag, New York, pp. 250-256.

Cavendish, J. C. (1972), Collocation methods for elliptic and parabolic boundary value problems, Ph.D. thesis, University of Pittsburgh, Pittsburgh, PA.

Cavendish, J. C. and C. A. Hall (1974), $L_\infty$-convergence of collocation and Galerkin approximations to linear two-point parabolic problems, Aequationes Math., 11, 230-249.

Celia, M. A. (1983), Collocation on deformed finite elements and alternating direction collocation methods, Ph.D. thesis, Princeton University, Princeton, NJ.

Celia, M. A. and G. F. Pinder (1984), Collocation solution of the transport equation using a locally enhanced alternating direction formulation, Unification of Finite Element Methods, Ch. 13, H. Kardestuncer, ed., Elsevier Science Publishers, New York.

Celia, M. A. and G. F. Pinder (1985), An analysis of alternating-direction methods for parabolic equations, Numer. Methods Partial Differential Equations, 1, 57-70.

Celia, M. A., G. F. Pinder and L. J. Hayes (1980), Alternating direction collocation simulation of the transport equation, Proc. Third. Int. Conf. Finite Elements in Water Resources, S. Y. Wang et al., eds., Univ. Mississippi, Oxford, MS, pp. 3.36-3.48.

Cerutti, J. H. (1974), Collocation for systems of ordinary differential equations, Technical Report 230, Dept. Computer Sciences, Univ. Wisconsin, Madison.

Cerutti, J. H. and S. V. Parter (1976), Collocation methods for parabolic partial differential equations in one space dimension, Numer. Math., 26, 227-254.

Chang, P. W. and B. A. Finlayson (1977), Orthogonal collocation on finite elements for elliptic equations, Advances in Computer Methods for Partial Differential Equations - II, R. Vichnevetsky, ed., IMACS, 79-86.

Chang, P. W. and B. A. Finlayson (1978), Orthogonal collocation on finite elements for elliptic equations, Math. Comput. Simul., 20, 83-92.

Chawla, T. C. and S. H. Chan (1979), Solution of radiation-conduction problems with collocation method using B-splines as approximating functions, Int. J. Heat Mass Transfer, 22, 1657-1667.

Chawla, T. C. and S. H. Chan (1980), Spline collocation solution of combined radiation-convection in thermally developing flows with scattering, Numer. Heat Transfer, 3, 47-65.

Chawla, T. C., G. Leaf and W. Chen (1975), A collocation method using B-splines for one-dimensional heat or mass-transfer-controlled moving boundary problems, Nucl. Eng. Des., 35, 163-180.

Chawla, T. C., W. J. Minkowycz and G. Leaf (1980), Spline-collocation solution of integral equations occurring in radiative transfer and laminar boundary-layer problems, Numer. Heat Transfer, 3, 133-148.

Chawla, T. C., G. Leaf, W. L. Chen and M. A. Grolmes (1974), A collocation method using Hermite cubic splines for nonlinear transient one-dimensional heat conduction problems, Trans. Amer. Nucl. Soc., 19, 162-163.

Chawla, T. C., G. Leaf, W. L. Chen and M. A. Grolmes (1975), The application of the collocation method using Hermite cubic splines to nonlinear transient one-dimensional heat conduction problems, Trans. ASME, J. Heat Transfer, 97, 562-569.

Christiansen, J. and R. D. Russell (1978), Error analysis for spline collocation methods with application to knot selection, Math. Comp., 32, 415-419.

Collatz, L. (1960), The Numerical Treatment of Differential Equations, Springer-Verlag, Berlin.

Cooper, K. D. and P. M. Prenter (1988), Multi-boundary alternating direction collocation schemes for computing ideal flows over bodies of arbitrary shape using cartesian coordinates, Proceedings of the Los Alamos Conference on Invariant Imbedding, Marcel Dekker, New York, to appear.

Cooper, K. D. and P. M. Prenter (1989), Alternating direction collocation for separable elliptic partial differential equations, SIAM J. Numer. Anal., submitted.

Daniel, J. W. (1977) Extrapolation with spline-collocation methods for two-point boundary value problems I: Proposals and justifications, Aequationes Math., 16, 107-122.

Daniel, J. W. and B. K. Swartz (1975), Extrapolated collocation for two-point boundary-value problems using cubic splines, J. Inst. Math. Appl., 16, 161-174.

Davis, M. and G. Fairweather (1981), On the use of spline collocation methods for boundary problems arising in chemical engineering, Comput. Methods Appl. Mech. Engrg., 28, 179-189.

Davis, M. E., G. Fairweather and J. Yamanis (1982), Analysis of $SO_2$ oxidation in non-isothermal catalyst pellets using the dusty-gas model, Chem. Eng. Sci., 37, 447-452.

Denison, K.S., C. E. Hamrin Jr. and G. Fairweather (1983), Solution of boundary value problems using software packages: DD04AD and COLSYS, Chem. Eng. Commun., 22, 1-9.

Diaz, J. C. (1974), A hybrid collocation-Galerkin method for the two point boundary value problem using continuous piecewise polynomial spaces, Ph.D. thesis, Rice University, Houston, TX.

Diaz, J. C. (1977), A collocation-Galerkin method for the two point boundary value problem using continuous piecewise polynomial spaces, SIAM J. Numer. Anal., 14, 844-858.

Diaz, J. C. (1979a), A collocation-Galerkin method for Poisson's equation on rectangular regions, Math. Comp., 33, 77-84.

Diaz, J. C. (1979b), Collocation-$H^{-1}$-Galerkin method for parabolic problems with time-dependent coefficients, SIAM J. Numer. Anal., 16, 911-922.

Diaz, J. C., G. Fairweather and P. Keast (1983a), FORTRAN packages for solving certain almost block diagonal linear systems by modified alternate row and column elimination, ACM Trans. Math. Software, 9, 358-375.

Diaz, J. C., G. Fairweather and P. Keast (1983b), Algorithm 603 COLROW and ARCECO: FORTRAN packages for solving certain almost block diagonal linear systems by modified alternate row and column elimination, ACM Trans. Math. Software, 9, 376-380.

Dixon, A. G. (1981), Solution of packed-bed heat-exchanger models by orthogonal collocation using piecewise cubic Hermite functions, in Chemical Reactors, H. S. Fogler, ed., ACS Symposium Series 168, American Chemical Society, Washington, DC, pp. 287-304.

Doctor, H. D. and N. L. Kalthia (1988), Spline collocation in the flow of non-Newtonian fluids, Internat. J. Numer. Methods Engrg., 26, 413-421.

Dougherty, D. E. and G. F. Pinder (1983), A brief note on upwind collocation, Internat. J. Numer. Methods Fluids, 3, 307-313.

Douglas, J., Jr. (1972), A superconvergence result for the approximate solution of the heat equation by a collocation method, in Mathematical Foundations of the Finite Element Method with Applications to Partial Differential Equations, A. K. Aziz, ed., Academic Press, New York, pp. 475-490.

Douglas, J., Jr. and T. Dupont (1972), A finite element collocation method for the heat equation, Istituto Nazionale di Alta Matematica, Symposia Mathematica, 10, 403-410.

Douglas, J., Jr. and T. Dupont (1973), A finite element collocation method for quasilinear parabolic equations, Math. Comp., 27, 17-28.

Douglas, J., Jr. and T. Dupont (1974), Collocation Methods for Parabolic Equations in a Single Space Variable, Lecture Notes in Mathematics 385, Springer-Verlag, New York.

Dunn, R. J. and M. F. Wheeler (1976), Some collocation-Galerkin methods for two-point boundary value problems, SIAM J. Numer. Anal., 13, 720-733.

Dyksen, W. R. (1987), Tensor product generalized ADI methods for separable problems, SIAM J. Numer. Anal., 24, 59-76.

Dyksen, W. R., E. N. Houstis, R. E. Lynch and J. R. Rice (1984), The performance of the collocation and Galerkin methods with Hermite bicubics, SIAM J. Numer. Anal., 21, 695-715.

Dyksen, W. R. and J. R. Rice (1984), A new ordering scheme for Hermite bicubic collocation equations, in Elliptic Problem Solvers II, G. Birkhoff and A. Schoenstadt, eds., Academic Press, New York, 467-480.

Dyksen, W. R. and J. R. Rice (1986), The importance of scaling for the Hermite bicubic collocation equations, SIAM J. Sci. Statist. Comput., 7, 707-719.

Fairweather, G. (1978), Finite Element Galerkin Methods for Differential Equations,

Marcel Dekker, New York.

Fairweather, G. and M. Vedha-Nayagam (1987), An assessment of numerical software for solving two-point boundary-value problems arising in heat transfer, Numer. Heat Transfer, 11, 281-293.

Fan, L. T., G. K. C. Chen and L. E. Erickson (1971), Efficiency and utility of collocation methods in solving the performance equations of flow chemical reactors with axial dispersion, Chem. Eng. Sci., 26, 379-387.

Finlayson, B. A. (1971), Packed bed reactor analysis by orthogonal collocation, Chem. Eng. Sci., 26, 1081-1091.

Finlayson, B. A. (1972), The Method of Weighted Residuals and Variational Principles, Academic Press, New York.

Finlayson, B. A. (1980), Orthogonal collocation on finite elements - progress and potential, Math. Comput. Simulation, 22, 11-17.

Flaherty, J. E. and W. Mathon (1980), Collocation with polynomial and tension splines for singularly-perturbed boundary value problems, SIAM J. Sci. Statist. Comput., 1, 260-289.

Flaherty, J. E. and R. E. O'Malley (1984), Numerical methods for stiff systems of two-point boundary value problems, SIAM J. Sci. Statist. Comput., 5, 865-886.

Fox, L. and I. B. Parker (1968), Chebyshev Polynomials in Numerical Analysis, Oxford University Press, London.

Frazer, R. A., W. J. Duncan and A. R. Collar (1938), Elementary Matrices and Some Applications to Dynamics and Differential Equations, Cambridge University Press, Cambridge, England.

Frazer, R. A., W. P. Jones and S. W. Skan (1937), Approximations to functions and to the solutions of differential equations, Great Britain Air Ministry Aero. Res. Comm. Tech. Rep., 1, 517-549.

Frazer, R. A., W. P. Jones and S. W. Skan (1938), Note on approximations to functions and to solutions of differential equations, Phil. Mag., 25, 740-746.

Frind, E. O. and G. F. Pinder (1979), A collocation finite element method for potential problems in irregular domains, Internat. J. Numer. Methods Engrg., 14, 681-701.

Fuchs, P. M. (1987), On the stability of spline-collocation methods of multivalue type, BIT, 27, 374-388.

Fyfe, D. J. (1969), The use of cubic splines in the solution of two-point boundary value problems, Comput. J., 12, 188-192.

Fyfe, D. J. (1970), The use of cubic splines in the solution of certain fourth order boundary value problems, Comput. J., 13, 204-205.

Gladwell, I. and R. I. Hay (1989), Vector- and parallelisation of ODE BVP codes, Parallel Computing, to appear.

Gladwell, I. and D. J. Mullings (1975), On the effect of boundary conditions in collocation by polynomial splines for the solution of boundary value problems in ordinary differential equations, J. Inst. Math. Appl., 16, 93-107.

Gottlieb, D. and S. A. Orszag (1977), Numerical Analysis of Spectral Methods: Theory and Applications, Regional Conference Series in Applied Mathematics, Vol. 26, SIAM, Philadelphia, PA.

Greenwell-Yanik, C. E. and G. Fairweather (1986), Analyses of spline collocation methods for parabolic and hyperbolic problems in two space variables, SIAM J. Numer. Anal., 23, 282-296.

Grossman, L. M. (1983), A collocation method for the neutron transport equation, Transport Theory and Statistical Physics, 12, 307-321.

Grossman, L. M., J. P. Hennart and D. Meade (1982), Finite element collocation methods for space-time reactor dynamics, Trans. Amer. Nucl. Soc., 41, 311-312.

Hangelbroek, R. J., H. G. Kaper and G. K. Leaf (1977), Collocation methods for integro-differential equations, SIAM J. Numer. Anal., 14, 377-390.

Hayes, L. J. (1980), An alternating-direction collocation method for finite element approximations on rectangles, Comput. Math. Appl., 6, 45-50.

Hayes, L. J. (1981), A comparison of alternating-direction collocation methods for the transport equation, New Concepts in Finite Element Analysis, T. J. R. Hughes et al., eds, AMD-Vol. 44, American Society of Mechanical Engineers, New York, pp. 169-177.

Hayes, L. J., G. Pinder and M. Celia (1981), Alternating-direction collocation for rectangular regions, Comput. Meth. Appl. Mech. Engrg., 27, 265-277.

Hemker, P. W., H. Schippers, and P. M. De Zeeuw (1980), Comparing some aspects of two codes for two-point boundary-value problems, Report NW 98/80, Mathematisch Centrum, Amsterdam.

Herbst, B. M. and J. F. Botha (1981), Computable error estimates for the collocation method applied to two-point boundary value problems, IMA J. Numer. Anal., 1, 489-497.

Hill, T. R. (1973), The solution of two-point boundary-value problems by extrapolated collocation with cubic splines, M.A. thesis, University of Texas, Austin, TX.

de Hoog, F. R. and R. Weiss (1978), Collocation methods for singular boundary value problems, SIAM J. Numer. Anal., 15, 198-217.

Hoskins, W. D. (1973), Cubic spline solutions to fourth-order boundary value problems, Comm. ACM, 16, 382-385.

Houstis, E. N. (1977), Application of method of collocation on lines for solving nonlinear hyperbolic problems, Math. Comp., 31, 443-456.

Houstis, E. N. (1978a), A collocation method for systems of nonlinear ordinary differential equations, J. Math. Anal. Appl., 62, 24-37.

Houstis, E. N. (1978b), Collocation methods for linear elliptic problems, BIT, 18, 301-310.

Houstis, E. N. (1978c), The complexity of numerical methods for elliptic partial differential equations, J. Comp. Appl. Math., 4, 191-197.

Houstis, E. N., C. C. Christara and J. R. Rice (1988), Quadratic-spline collocation methods for two-point boundary value problems, Internat. J. Numer. Methods Engrg., 26, 935-952.

Houstis, E. N., R. E. Lynch, J. R. Rice and T. S. Papatheodorou (1978), Evaluation of numerical methods for elliptic partial differential equations, J. Comput. Phys. 27, 323-350.

Houstis, E. N., W. F. Mitchell and T. S. Papatheodorou (1979), A $C^1$-collocation method for mildly nonlinear elliptic equations, in Advances in Computer Methods for Partial Differential Equations III, R. Vichnevetsky, ed., IMACS, pp. 18-27.

Houstis, E. N., W. F. Mitchell and T. S. Papatheodorou (1983), Performance evaluation of algorithms for mildly nonlinear elliptic problems, Internat. J. Numer. Methods Engrg., 19, 665-709.

Houstis, E. N., W. F. Mitchell and J. R. Rice (1985a), Collocation software for second-order elliptic partial differential equations, ACM Trans. Math. Software, 11, 379-412.

Houstis, E. N., W. F. Mitchell and J. R. Rice (1985b), Algorithm 637 GENCOL: Collocation on general domains with bicubic Hermite polynomials, ACM Trans. Math. Software, 11, 413-415.

Houstis, E. N., W. F. Mitchell and J. R. Rice (1985c), Algorithm 638 INTCOL and HERMCOL: Collocation on rectangular domains with bicubic Hermite polynomials, ACM Trans. Math. Software, 11, 416-418.

Houstis, E. N. and J. R. Rice (1977), Software for linear elliptic problems on general two dimensional domains, Advances in Computer Methods for Partial Differential Equations-II, R. Vichnevetsky, ed., IMACS, pp. 7-12.

Houstis, E. N. and J. R. Rice (1980), An experimental design for the computational evaluation of partial differential equation solvers, in Production and Assessment of Numerical Software, M. Delves and M. Hennell, eds., Academic Press, London, pp. 57-66.

Houstis, E. N., J. R. Rice and E. A. Vavalis (1984), Spline-collocation methods for elliptic partial differential equations, Advances in Computer Methods for Partial Differential Equations-V, R. Vichnevetsky and R. S. Stepleman, eds., IMACS, pp. 191-194.

Houstis, E. N., J. R. Rice and E. A. Vavalis (1987), Parallelization of a new class of cubic spline collocation methods, Advances in Computer Methods for Partial Differential Equations-VI, R. Vichnevetsky and R. S. Stepleman, eds., IMACS, pp. 167-174.

Houstis, E. N., E. A. Vavalis and J. R. Rice (1988), Convergence of $O(h^4)$ cubic spline collocation methods for elliptic partial differential equations, SIAM J. Numer.

Anal., 25, 54-74.

Hulme, B. L. (1972), One-step piecewise polynomial Galerkin methods for initial value problems, Math. Comp., 26, 415-426.

Irodotou-Ellina, M. (1987), Spline collocation methods for high order elliptic boundary value problems, Ph.D. thesis, University of Thessalonika, Greece.

Irodotou-Ellina, M. and E. N. Houstis (1988), An $O(h^6)$ quintic spline collocation method for fourth order two-point boundary value problems, BIT, 28, 288-301.

Ito, T. (1972), A collocation method for boundary value problems using spline functions, Ph.D. thesis, Brown Univ., Providence, RI.

Kammerer, W. J., G. W. Reddien and R. S. Varga (1974), Quadratic interpolatory splines, Numer. Math., 22, 241-259.

Kantorovich, L. V. (1934), On a new method of approximate solution of partial differential equations, Dokl. Akad. Nauk. SSSR, 4, 532-536.

Kantorovich L. V. and G. P. Akilov (1964), Functional Analysis in Normed Spaces, Pergamon Press, Oxford.

Karageorghis, A. (1988), Chebyshev spectral methods for solving two-point boundary value problems arising in heat transfer, Comput. Methods. Appl. Mech. Engrg., 70, 103-121.

Karpilovskaya, E. B. (1953), On the convergence of an interpolation method for ordinary differential equations, Uspekhi Mat. Nauk, 8, 111-118.

Karpilovskaya, E. B. (1963a), Convergence of the collocation method, Soviet Math. Dokl., 4, 1070-1073.

Karpilovskaya, E. B. (1963b), Convergence of a collocation method for certain boundary-value problems of mathematical physics, Sibirsk. Mat. Z., 4, 632-640.

Karpilovskaya, E. B. (1970), A method of collocation for integro-differential equations with biharmonic principal part, U.S.S.R. Comp. Math. Phys., 10, No. 6, 240-246.

Keast, P., G. Fairweather and J. C. Diaz (1983), A computational study of finite element methods for second order linear two-point boundary value problems, Math. Comp., 40, 499-518.

Keast, P. and P. H. Muir (1987), EPDCOL: A more efficient PDECOL code, Technical Report 1987CS-6, Division of Computing Science, Dalhousie University, Halifax, Nova Scotia.

Kelkar, C. P., C. E. Hamrin, Jr. and G. Fairweather (1985), Letter to the Editor, AIChE J., 31, 348-349.

Kendall, R. P. and M. F. Wheeler (1976), A Crank-Nicolson-$H^{-1}$-Galerkin procedure for parabolic problems in a single space variable, SIAM J. Numer. Anal., 13, 861-876.

Khalifa, A. K. A. and J. C. Eilbeck (1982), Collocation with quadratic and cubic splines, IMA J. Numer. Anal., 2, 111-121.

Lanczos, C. (1938), Trigonometric interpolation of empirical and analytical functions, J. Math. Phys., 17, 123-195.

Lanczos, C. (1956), Applied Analysis, Prentice Hall, Englewood Cliffs, New Jersey.

Lesaint, P. and P. A. Raviart (1979), Finite element collocation methods for first order systems, Math. Comp., 33, 891-918.

Leyk, Z. (1986), A $C^0$-collocation-like method for two-point boundary value problems, Numer. Math., 49, 39-53.

Lie, I. and S. P. Norsett (1989), Superconvergence for multistep collocation, Math. Comp., 52, 65-79.

Lucas, T. R. and G. W. Reddien (1972), Some collocation methods for nonlinear boundary value problems, SIAM J. Numer. Anal., 9, 341-356.

Luskin, M. (1979), An approximation procedure for nonsymmetric, nonlinear hyperbolic systems with integral boundary conditions, SIAM J. Numer. Anal., 16, 145-164.

Madsen, N. K. and R. F. Sincovec (1979), Algorithm 540. PDECOL, General collocation software for partial differential equations, ACM Trans. Math. Software, 5, 326-351.

Maestro, R. A. and D. A. Voss (1981), A quintic spline collocation procedure for solving the Falkner-Skan boundary-layer equation, Comput. Methods Appl. Mech. Engrg., 25, 129-148.

Maier, M. R. (1985), Numerical solution of singular perturbed boundary value problems using a collocation method with tension splines, in Numerical Boundary Value ODEs, U. Ascher and R. D. Russell, eds., Birkhauser, Boston, pp. 207-225.

Majaess, F., P. Keast and G. Fairweather (1989), Packages for solving almost block diagonal linear systems arising in spline collocation at Gaussian points with monomial basis functions, in Proceedings of the Workshop on Scientific Systems, Shrivenham, England, July 1988, J. C. Mason and M. G. Cox, eds., Clarendon Press, Oxford, England, to appear.

Markowich, P. A. and C. A. Ringhofer (1983), Collocation methods for boundary value problems on "long" intervals, Math. Comp., 40, 123-150.

Martin, A. J. and J. W. Daniel (1977), Extrapolation with spline-collocation methods for two-point boundary value problems II: $C^2$-cubics with detailed results, Report CNA-125, Center for Numerical Analysis, University of Texas, Austin, TX.

Martin, A. J. and J. W. Daniel (1981), Extrapolation with spline-collocation methods for two-point boundary value problems II: $C^2$-cubics, Aequationes Math., 22, 39-41.

Masliyah, J. H. and D. Kumar (1980), Application of orthogonal collocation on finite elements to a flow problem, Math. Comput. Simulation, 12, 49-54.

Mason, J. H. (1975), Collocation Methods for the Solution of the Static Neutron Diffusion Equation, Doctoral thesis, MIT.

Meade, D. (1982), Collocation Methods for Space-Time Nuclear Reactor Dynamics, Ph.D. thesis, Univ. of California, Berkeley.

Meade, D. (1983), Solution of the neutron group-diffusion equations by orthogonal collocation with cubic Hermite interpolants, Transport Theory and Statistical Physics, 12, 271-284.

Min-Da Ho (1983), A collocation solver for systems of boundary-value differential/algebraic equations, Computers Chem. Eng., 7, 735-737.

Mohsen, M. F. N. and G. F. Pinder (1984), Orthogonal collocation with 'adaptive' finite elements, Internat. J. Numer. Methods Engrg., 20, 1901-1910.

Mohsen, M. F. N. and G. F. Pinder (1986), Collocation with 'adaptive' finite elements in Buckley-Leverett problem, Internat. J. Numer. Methods Engrg., 23, 121-131.

Morel, J. E. (1979), A collocation method for the solution of the neutron transport equation with both symmetric and asymmetric scattering, Ph.D. thesis, Univ. of New Mexico, Albuquerque, NM.

Muir, P. H., G. Fairweather and M. Vedha-Nayagam (1983), The effect of Prandtl number on heat transfer from an isothermal rotating disk with blowing at the wall, Int. Comm. Heat Mass Transfer, 10, 287-297.

Nakao, M. (1981), Some superconvergence estimates for a collocation-$H^{-1}$-Galerkin method for parabolic problems, Mem. Fac. Sci. Kyushu Univ. Ser. A, 35, 291-306.

Nguyen, V. V., G. F. Pinder, W. G. Gray and J. F. Botha (1983), Numerical simulation of uranium in-situ mining, Chem. Eng. Sci., 38, 1855-1862.

Oliveira, F. A. (1980), Collocation and residual correction, Numer. Math., 36, 27-31.

Osborne, M. R. (1975), Collocation, difference equations, and stitched function representations, Topics in Numerical Analysis II, J. J. H. Miller, ed., Academic Press, New York, pp. 121-132.

Paine, J. and R. D. Russell (1986), Conditioning of collocation matrices and discrete Green's functions, SIAM J. Numer. Anal., 23, 376-392.

Papamichael, N. and J. R. Whiteman (1973), A cubic spline technique for the one dimensional heat conduction problem, J. Inst. Math. Appl., 11, 111-113.

Paterson, W. R. and D. L. Cresswell (1971), A simple method for the calculation of effectiveness factors, Chem. Eng. Sci., 26, 605-616.

Pedersen, H. and M. Tanoff (1982), Spline collocation method for solving parabolic pde's with initial discontinuities: application to mixing with chemical reaction, Computers Chem. Eng., 6, 197-207.

Percell, P. and M. F. Wheeler (1980), A $C^1$ finite element collocation method for elliptic equations, SIAM J. Numer. Anal., 17, 605-622.

Phillips, J. L. (1972), The use of collocation as a projection method for solving linear operator equations, SIAM J. Numer. Anal., 9, 14-28.

Pinder, G. E. and A. Shapiro (1979), A new collocation method for the solution of the convection-dominated transport equation, Water Resources Research, 15, 1177-1182.

Prenter, P. M. (1973), A collocation method for the numerical solution of integral equations, SIAM J. Numer. Anal., 10, 570-581.

Prenter, P. M. (1975), Splines and Variational Methods, Wiley-Interscience, New York.

Prenter, P. M. and R. D. Russell (1976), Orthogonal collocation for elliptic partial differential equations, SIAM J. Numer. Anal., 13, 923-939.

Ravaioli, U., P. Lugli, M. A. Osman and D. K. Ferry (1985), Advantages of collocation methods over finite differences in one-dimensional Monte Carlo simulations of submicron devices, IEEE Transactions on Electron Devices, ED-32, 2097-2101.

Reddien, G. W. (1973), Projection methods and singular two point boundary value problems, Numer. Math., 21, 193-205.

Reddien, G. W. (1976), Approximation methods for two-point boundary value problems with nonlinear boundary conditions, SIAM J. Numer. Anal., 13, 405-411.

Reddien, G. W. (1977), Approximation methods and alternative problems, J. Math. Anal. Appl., 60, 139-149.

Reddien, G. W. (1979), Collocation at Gauss points as a discretization in optimal control, SIAM J. Control Optim., 17, 298-306.

Reddien, G. W. (1980), Projection methods for two-point boundary value problems, SIAM Rev., 22, 156-171.

Reddien, G. W. and L. L. Schumaker (1976), On a collocation method for singular two point boundary value problems, Numer. Math., 25, 427-432.

Reddien, G. W. and C. C. Travis (1974), Approximation methods for boundary value problems of differential equations with functional arguments, J. Math. Anal. Appl., 46, 62-74.

Rice, J. R. and R. F. Boisvert (1985), Solving Elliptic Problems using ELLPACK, Springer-Verlag, New York.

Ringhofer, C. (1984), On collocation schemes for quasilinear singularly perturbed boundary value problems, SIAM J. Numer. Anal., 21, 864-882.

Ronto, N. I. (1971), Application of the method of collocation to solve boundary value problems, Ukrain. Mat. Z., 23, 415-421.

Rubin, S. G. and R. A. Graves (1975), Viscous flow solutions with a cubic spline approximation, Comput. & Fluids, 3, 1-36.

Rubin, S. G. and P. K. Khosla (1976), Higher-order numerical solutions using cubic splines, AIAA Journal, 14, 851-858.

Russell, R. D. (1974), Collocation for systems of boundary value problems, Numer. Math., 23, 119-133.

Russell, R. D. (1975), Efficiencies of B-spline methods for solving differential equations, Proc. Fifth Manitoba Conference on Numerical Math., 599-617.

Russell, R. D. (1977), A comparison of collocation and finite differences for two-point boundary value problems, SIAM J. Numer. Anal., 14, 19-39.

Russell, R. D. (1982a), Difficulties in evaluating differential equation software, in Numerical Analysis, J. P. Hennart, ed., Lecture Notes in Mathematics 909, Springer-Verlag, New York, pp. 175-184.

Russell, R. D. (1982b), Global codes and their comparison, in Numerical Integration of Differential Equations and Large Linear Systems, J. Hinze, ed., Lecture Notes in Mathematics 968, Springer-Verlag, New York, pp. 256-268.

Russell, R. D. and J. Christiansen (1978), Adaptive mesh selection strategies for solving boundary value problems, SIAM J. Numer. Anal., 15, 59-80.

Russell, R. D. and L. F. Shampine (1972), A collocation method for boundary value problems, Numer. Math., 19, 1-28.

Russell, R. D. and L. F. Shampine (1975), Numerical methods for singular boundary value problems, SIAM J. Numer. Anal., 12, 13-36.

Russell, R. D. and J. M. Varah (1975), A comparison of global methods for linear two-point boundary value problems, Math. Comp., 29, 1007-1019.

Sakai, M. (1970), Spline interpolation and two-point boundary value problems, Mem. Fac. Kyushu Univ. Ser. A, 24, 17-34.

Sakai, M. (1983), Quadratic spline approximation for boundary value problem, Rep. Fac. Sci. Kagoshima Univ. Math. Phys. Chem., 16, 1-14.

Sakai, M. (1984), A posteriori improvement of cubic spline approximate solution of two-point boundary value problem, Publ. Res. Inst. Math. Sci., 20, 137-149.

Sakai, M. and R. Usmani (1983), Quadratic spline solution and two-point boundary value problems, Publ. Res. Inst. Math. Sci., 19, 7-13.

Saranen, J. (1987), On the convergence of the spline collocation with discontinuous data, Math. Methods Appl. Sci., 9, 59-75.

Saranen, J. (1988), The convergence of even degree spline collocation solution for potential problems in smooth domains of the plane, Numer. Math., 53, 499-512.

Saranen, J. and W. L. Wendland (1985), On the asymptotic convergence of collocation methods with spline functions of even degree, Math. Comp., 45, 91-108.

Shindler, A. A. (1967), Some theorems of the general theory of approximate methods of analysis and their application to the collocation, moments and Galerkin methods,

Siberian Math. J., 8, 302-314.

Shindler, A. A. (1969), Rate of convergence of the enriched collocation method for ordinary differential equations, Siberian Math. J., 10, 160-163.

Sincovec, R. F. (1977a), On the relative efficiency of higher order collocation methods for solving two-point boundary value problems, SIAM J. Numer. Anal., 14, 112-123.

Sincovec, R. F. (1977b), Generalized collocation methods for time-dependent, nonlinear boundary-value problems, Soc. Pet. Eng. J., 17, 345-352.

Spudich, P. and U. Ascher (1983), Calculation of complete theoretical seismograms in vertically varying media using collocation methods, Geophys. J. R. Astr. Soc., 75, 101-124.

Stewart, W. E. and J. Villadsen (1969), Graphical calculation of multiple steady states and effectiveness factors for porous catalysts, AIChE J., 15, 28-34.

Stys, T. (1981), The method of collocation by cubic splines for nonlinear parabolic equations, Bull. Acad. Polon. Sci. Ser. Sci. Math., 29, 91-98.

Swartz, B. (1988), Conditioning collocation, SIAM J. Numer. Anal., 25, 124-147.

Swartz, B. and B. Wendroff (1974), The relation between the Galerkin and collocation methods using smooth splines, SIAM J. Numer. Anal., 11, 994-996.

Vainniko, G. M. (1965), On the stability and convergence of the collocation method, Differential Equations, 1, 186-194.

Vainniko, G. M. (1966), The convergence of the collocation method for non-linear differential equations, USSR Comput. Math. and Math. Phys., 6, 47-58.

del Valle, E. (1983), Application of the Galerkin finite element method to reactor dynamic problems, Transport Theory and Statistical Physics, 12, 251-269.

del Valle, E., J. C. Diaz and D. Meade (1986), Comparison of variational methods for the solution of the dynamic group-diffusion problem: one dimensional case, in Variational Methods in Geosciences, Y. K. Sasaki, ed., Elsevier, Amsterdam, The Netherlands, pp. 243-248.

Varah, J. M. (1974), A comparison of some numerical methods for two-point boundary value problems, Math. Comp., 28, 743-755.

Varah, J. M. (1976), Alternate row and column elimination for solving certain linear systems, SIAM J. Numer. Anal., 13, 71-75.

Vavalis, E. A. (1985), High order spline collocation methods for elliptic partial differential equations, Ph.D. thesis, Univ. of Thessaloniki, Thessaloniki, Greece.

Vedha-Nayagam, M., P. Jain and G. Fairweather (1987), The effect of surface mass transfer on buoyancy-induced flow in a variable-porosity medium adjacent to a horizontal heated plate, Int. Comm. Heat Mass Transfer, 14, 495-506.

van Veldhuizen, M. (1976), A refinement process for collocation approximations, Numer. Math., 26, 397-407.

Villadsen, J. V. (1969), Selected Approximation Methods for Chemical Engineering Problems, Instituttet for Kemiteknik, Numerisk Institut, Danmarks Teknise Hojskole.

Villadsen, J. V. and M. L. Michelsen (1978), Solution of Differential Equation Models by Polynomial Approximation, Prentice Hall, Englewood Cliffs, New Jersey.

Villadsen, J. V. and J. P. Sorensen (1969), Solution of parabolic partial differential equations by a double collocation method, Chem. Eng. Sci., 24, 1337-1349.

Villadsen, J. V. and W. E. Stewart (1967), Solution of boundary-value problems by orthogonal collocation, Chem. Eng. Sci., 22, 1483-1501.

Weiser, A., S. C. Eisenstat and M. H. Schultz (1980), On solving elliptic equations to moderate accuracy, SIAM J. Numer. Anal., 17, 908-929.

Weiss, R. (1974), The application of implicit Runge-Kutta and collocation methods to boundary-value problems, Math. Comp., 28, 449-464.

Wheeler, M. F. (1977), A $C^0$-collocation-finite element method for two-point boundary value problems and one space dimensional parabolic problems, SIAM J. Numer. Anal., 14, 71-90.

Wheeler, M. F. (1978), An elliptic collocation-finite element method with interior penalties, SIAM J. Numer. Anal., 15, 152-161.

Wittenbrink, K. A. (1973), High order projection methods of moment- and collocation-type for nonlinear boundary value problems, Computing, 11, 255-274.

Yanik, E. G. and G. Fairweather (1988), Finite element methods for parabolic and hyperbolic integro-differential equations, Nonlinear Anal., 12, 785-809.

Yartsev, Yu. P. (1967), Convergence of the collocation method on lines, Differential Equations, 3, 838-842.

Yartsev, Yu. P. (1968), The method of line collocation, Differential Equations, 4, 481-485.

Young, L. C. (1977), A preliminary comparison of finite element methods for reservoir simulation, Advances in Computer Methods for Partial Differential Equations-II, IMACS (AICA), 307-320.

Zamani, N. G. and W. Sun (1988), Collocation finite element solution of a compressible flow, Math. Comput. Simulation, 30, 243-251.

Zygourakis, K. and R. Aris (1983), Monotone iteration methods with adaptive collocation for solving coupled systems of nonlinear boundary value problems, Computers Chem. Eng., 7, 183-193.

# Index

Printed and bound by CPI Group (UK) Ltd, Croydon, CR0 4YY

21/10/2024

01777097-0013